Perinatal Stem Cells

Zhong Chao Han
Tsuneo A. Takahashi
Zhibo Han • Zongjin Li
Editors

Perinatal Stem Cells

Biology, Manufacturing and Translational Medicine

Springer

Editors
Zhong Chao Han
Beijing Engineering Laboratory
of Perinatal Stem Cells
Jiangxi Engineering Research Center for
Stem Cell, Health & Biotech Co.
Beijing
China

Zhibo Han
State Key Lab of Experimental
Hematology
Institute of Hematology and Blood
Diseases Hospital
Chinese Academy of Medical Sciences
Tianjin
China

Tsuneo A. Takahashi
Institute for Frontier Medical Sciences
Kyoto University
Kyoto
Japan

Zongjin Li
School of Medicine
Nankai University
Tianjin
China

ISBN 978-981-13-2702-5 ISBN 978-981-13-2703-2 (eBook)
https://doi.org/10.1007/978-981-13-2703-2

Library of Congress Control Number: 2019930130

This Springer imprint is published by the registered company Springer Nature Singapore Pte Ltd.
The registered company address is: 152 Beach Road, #21-01/04 Gateway East, Singapore 189721, Singapore

Contents

1 Introduction of Perinatal Tissue-Derived Stem Cells

Zongjin Li and Zhong Chao Han

Abstract

Stem cells represent a population of cells with the potential applications including regenerative medicine and tissue engineering owing to their proliferation and differentiation ability. A wide variety of stem or progenitor cells, including adult bone marrow stem cells, endothelial progenitor cells, mesenchymal stem cells (MSCs), resident cardiac stem cells, and embryonic stem cells (ESCs), have been shown to have positive effects in preclinical studies and therefore hold promise for treating and curing debilitating and deadly diseases. Several of these types of stem cells have been tested in early-stage clinical trials, such as MSCs, human embryonic stem cells (hESCs), and induced pluripotent stem cells (iPSCs). However, ethical and technological issues have limited the applications of hESCs or iPSCs in this field. Perinatal tissue-derived stem cells have been isolated from the umbilical cord, Wharton's jelly, placenta, amniotic membrane, and amniotic fluid, which are normally discarded as medical waste. This chapter will provide a general introduction of perinatal tissue-derived stem cells and their potential clinical application.

1.1 Introduction

Stem cells are termed as cells with the ability to self-renew and differentiate into certain cells in vivo and in vitro. The primary roles of stem cells in a living organism are to maintain and repair the tissue injury in the organs or tissues they are resided in [1]. Stem cell therapy has shown great promise for regenerative repair of injured or diseased tissues [1–3]. According to the sources and development stages, stem cells can be classified into four major categories:

- Adult stem cells are all stem cells derived from living humans.
- Embryonic stem cells are stem cells derived from embryos.
- Induced pluripotent stem cells are a type of pluripotent stem cell that can be generated directly from adult cells.
- Perinatal stem cells are those derived from perinatal tissues.

Based on the development stages, perinatal tissue-derived stem cells are between stages

Z. Li
School of Medicine and Key Laboratory of Bioactive Materials, Ministry of Education, College of Life Sciences, Nankai University, Tianjin, China
e-mail: zongjinli@nankai.edu.cn

Z. C. Han (✉)
Beijing Engineering Laboratory of Perinatal Stem Cells, Jiangxi Engineering Research Center for Stem Cell, Health & Biotech Co., Beijing, China
e-mail: hanzhongchao@health-biotech.com

Z. C. Han et al. (eds.), *Perinatal Stem Cells*, https://doi.org/10.1007/978-981-13-2703-2_1

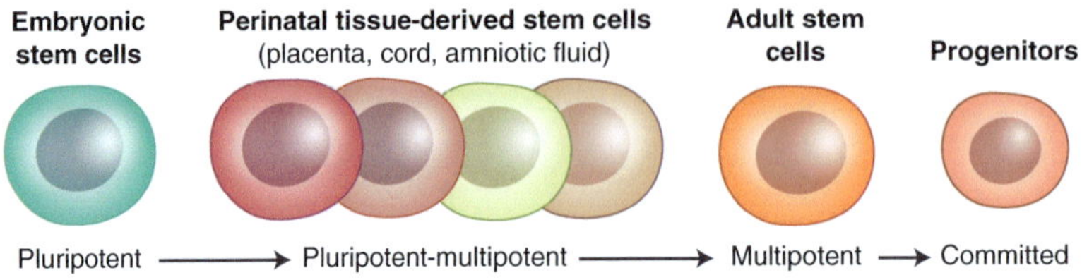

Fig. 1.1 Current model for the developmental position and potency of the perinatal stem cells from several perinatal sources emerging during gestation. Reproduced with permission from [16]

of embryonic and adult stem cells (Fig. 1.1). Perinatal tissue-derived stem cells include cells isolated from umbilical cord blood, amniotic fluid, and placenta, as well as cells from postgestational maternal peripheral blood [4–6].

Stem cell-based regenerative medicine has appeared as a potential therapeutic alternative for the treatment of degeneration diseases or injury, such as coronary artery disease, peripheral arterial disease, Parkinson's disease, type I diabetes mellitus, Huntington's disease, muscle damage, and many others [3, 7]. A wide variety of stem or progenitor cells, including adult bone marrow stem cells, endothelial progenitor cells, mesenchymal stem cells (MSCs), resident cardiac stem cells, and hESCs, have been shown to have positive effects in preclinical studies and therefore hold promise for *treating* and *curing* debilitating and deadly *diseases* [8]. Moreover, several of these types of stem cells have been tested in early-stage clinical trials, such as MSCs, hESCs, and iPSCs [9]. The most pressing difficulties for hESCs are ethical issues and teratoma formation [10]. iPSCs, without ethical concerns, has quickly become the leading alternative to hESCs. However, the use of iPSCs has several drawbacks, such as genomes integration with viral methods and low efficiency of reprogramming. Moreover, even viral-free methods can induce genetic and epigenetic abnormalities during prolonged in vitro cell culture [11, 12].

MSCs are fibroblast-like multipotent cells that can be isolated from various tissue sources, including the bone marrow, adipose tissue, and perinatal related tissues [13]. MSCs are an attractive candidate for stem cell therapy owing to their accessibility, pluripotency, and expansive potential. Furthermore, MSCs are free of both ethical concerns and teratoma formation. Bone marrow-derived MSCs (BM-MSCs) are among the most frequently used cell type in regenerative studies. However, the disadvantages in the proliferative capacity of BM-MSCs and the way to isolate BM-MSCs are accompanied by a risk of infection and painful for patients, which means finding alternatives is essential in clinical application [14]. Moreover, specific cell populations, such as hematopoietic stem cells, endothelial cells, and epithelial cells, are highly needed for designed therapeutic purposes [15]. Fortunately, several kinds of cell population have been isolated successfully from perinatal tissues. For example, MSCs and endothelial progenitor cells can be isolated from Wharton's jelly of the umbilical cord and cord blood.

1.2 Perinatal Related Tissues

The perinatal period means "around the time of birth," time period from the 20th week of gestation through the neonatal period (the first 28 days of life) [16]. The perinatal related tissues refer to these tissues being discarded at the time of birth, such as the placenta, umbilical cord, cord blood, amniotic membrane, and amniotic fluid. Those tissues contain early mesenchymal stem cells, hematopoietic stem cells, and many others, which are generally discarded after delivery and have a greater differentiation potency than adult stem cells, such as those from the bone marrow. Because of the limitations of the adult stem cell sources outlined above, perinatal sources of stem cells may yet prove to represent the most important and potentially useful source of stem cells. Perinatal related stem cells have been isolated from cord blood to tissue, including umbilical cord, placenta, and amniotic fluid [17].

1.2.1 Umbilical Cord

The umbilical cord is a connection between the developing placenta and the embryo [4]. The mean length of the umbilical cord is estimated at 55–66 cm, as determined by large-scale studies. The umbilical cord is made up of mucous connective tissue called Wharton's jelly, instead of normal connective tissue and skin. The umbilical cord in transverse section reveals two umbilical arteries and a single umbilical vein embedded in Wharton's jelly (Fig. 1.2). Mesenchymal stem cells derived from Wharton's jelly (WJ-MSCs) are considered to be convenient and readily alternative source of MSCs. Umbilical cord blood, indeed placental blood, can be extracted once from placenta, which is rich in endothelial progenitor and hematopoietic stem cells that can be used to treat leukemia and cardiovascular diseases. Hematopoietic stem cells from card blood carry an advantage over a bone marrow transplantation because the donor does not need to be an exact match for the recipient [19].

1.2.2 The Structure of the Placenta

The mature human placenta is a discoid organ 20–25 cm in diameter, 3 cm thick, and weighing 400–600 g. It is composed of chorionic plate (fetal surface) covered by a layer of amnion and chorion, from which stem villi arise from umbilical vessels surrounded by connective tissue and fibrin [20]. The basal plate (maternal surface), insertion site of the anchoring villi into maternal endometrium, is composed of placental villi, including capillaries, fetal macrophages (Hofbauer cells), and fibroblasts surrounded by trophoblast and syncytiotrophoblast layers (Fig. 1.3).

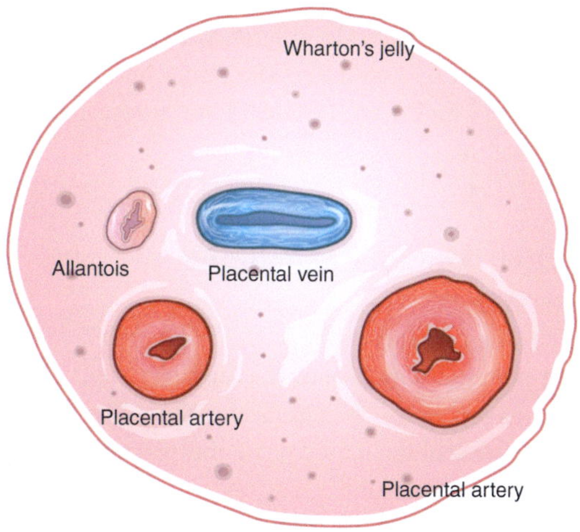

Fig. 1.2 A cross-sectional image of a postpartum umbilical cord showing the major structures [18]

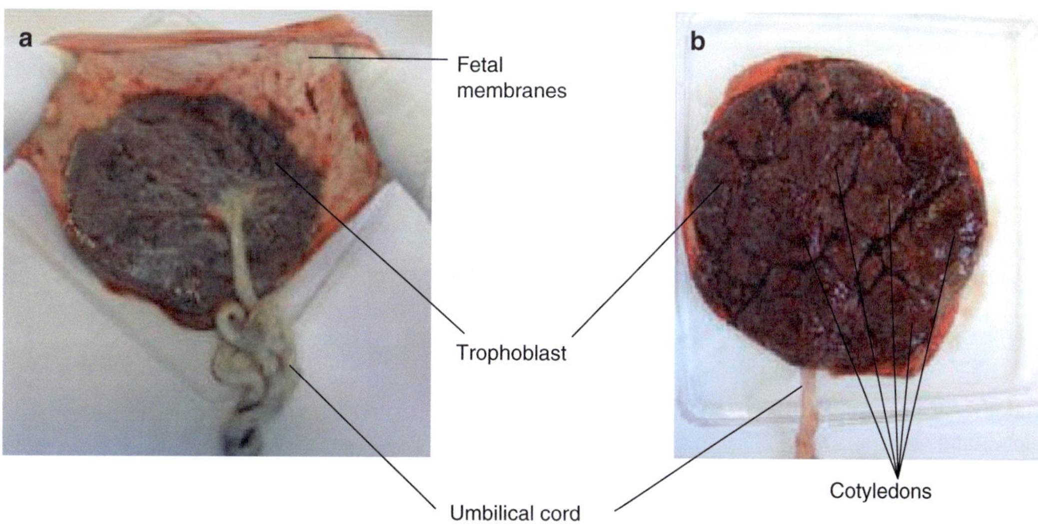

Fig. 1.3 Human term placenta. (**a**) Fetal surface. It is possible to distinguish the trophoblast, the fetal membranes which continue from the edge of the placenta, and the umbilical cord. (**b**) Maternal surface. The placenta is subdivided into irregular lobed structures, termed cotyledons. Reproduced with permission from [20]

1.2.3 Amniotic Fluid

The amniotic fluid is the protective liquid contained by the amniotic sac of a gravid amniotes. This fluid serves as a cushion for the growing fetus but also serves to facilitate the exchange of nutrients, water, and biochemical products between the mother and fetus. The volume of amniotic fluid increases with the growth of fetus. From the 10th to the 20th week, it increases from 25 to 400 mL approximately. There is about 1 L of amniotic fluid at birth. Recent studies show that amniotic fluid contains a considerable quantity of stem cells. These amniotic stem cells are pluripotent and able to differentiate into various tissues, which may be useful for future human application [21].

1.2.4 Amniotic Membrane

The amniotic membrane is the innermost layer of the placenta which lines the amniotic cavity and is a two cell layers membrane [22]. The membrane inserts on the edge of the placental disc, runs on the chorionic plate, and is a membranous sac of amniotic fluid. Amnion epithelial cells and amnion MSCs have been isolated from amniotic membrane for regenerative applications.

1.3 Perinatal Stem Cells

Since the perinatal related tissues are discarded at the time of birth, stem cells from these tissues represent a noninvasive, simple, and safe means for harvesting powerful stem cell types. Perinatal stem cells have been isolated from cord blood to tissue, including the umbilical cord, placenta, amniotic membrane, and amniotic fluid, for pre- and/or clinical applications. The types of perinatal stem cells include hematopoietic stem/endothelial progenitor cells from umbilical cord blood and MSCs from Wharton's jelly of the umbilical cord, amniotic fluid, and placenta. Moreover, multipotent cells derived from the human term placenta and the endothelial progenitor cells collected from the umbilical cord vein also have been isolated. MSCs from perinatal tissue sources, such as Wharton's jelly, placental blood and placental tissue, amnion and amniotic fluid, and umbilical cord blood, express multipotent stem cell markers such as Oct-4, Sox-2, and Nanog, which demonstrate that perinatal tissues represent the more primitive sources of stem cells for translational applications. Moreover, MSCs from perinatal related tissues have greater and faster expansion potential than BM-MSCs in vitro. Perinatal related tissues offer an abundant resource for developing stem cell banks.

1.3.1 Classification of Perinatal Stem Cells

Based on their origin, perinatal stem cells can be divided into three groups: stem cells from the amniotic fluid, placenta, and umbilical cord. Placenta has three sources of stem cells, amnion, villi, and blood, whereas umbilical cord stem cells come from two sources, cord blood and Wharton's jelly (Table 1.1).

1.3.2 Umbilical Cord Blood

Umbilical cord blood has been isolated prior to/immediately following birth. The majority of umbilical cord blood are hematopoietic stem

Table 1.1 The classification of perinatal tissue-derived stem cells based on origin

Perinatal tissues-derived stem cells	Placenta	Placental hematopoietic stem cell
		Placental mesenchymal stem cell
		Chorionic stem cell
		Trophoblast stem cell
	Amnion	Amniotic membrane stem cell, epithelial, mesenchymal
	Amniotic fluid	Epithelial stem cell
		Mesenchymal stem cell
	Umbilical cord	Wharton's jelly stem cell
		Umbilical vein endothelial cell

cells, 100,000 stem cells per mL. Alternative to the bone marrow, umbilical cord blood transplantation is curative for malignant and nonmalignant diseases like Fanconi's anemia, aplastic anemia, leukemia, and metabolic and other congenital disorders [23]. Both public and private cord blood banks, by collecting of cord blood from the umbilical cord and placenta immediately after birth, and then processing and storing the cord blood unit containing stem cells, have developed in response to the future potential application in treating many life-threatening diseases of the blood and others. Other than cord blood stem cells, cells from Wharton's jelly have been suggested as displaying the stemness phenotype which is the mesenchymal stromal cells or MSCs, and this kind of MSCs has been widely used in clinical application. Some private companies offer banking service of Wharton's jelly-derived MSCs.

1.3.3 Amniotic Fluid

Amniotic fluid fills the sac that surrounds and protects a developing fetus in the uterus. Researchers have identified stem cells in samples of amniotic fluid drawn from pregnant women during a procedure called amniocentesis, a test conducted to test for abnormalities. Moreover, collection and characterization of amniotic fluid from scheduled caesarean section (C-section) deliveries are possible and have been conserved the stem cells extracted from amniotic fluid in private stem cells banks [21]. Epithelioid cells were the major cell type in amniotic fluid, which reveal a highly proliferative population of adherent cells capable of producing therapeutic doses of MSCs [21].

1.3.4 Amniotic Membrane

Epithelial cells and MSCs have been isolated from amniotic membrane for regenerative applications [22]. However, the resurgence of interest in the use of amniotic membrane is for wound treatment of eye and burns. The use of amniotic membranes on the ocular surface is now a well-established therapy that can speed healing, particularly for severe inflammatory conditions.

1.3.5 Placental Stem Cells

The placenta is an organ that provides oxygen and nutrients for a growing fetus. These nutrients include blood rich in stem cells, which are passed to the child through the umbilical cord [24, 25]. The terminal placenta is, from a physiological viewpoint, a worn-out organ with no further use, and stem cells can be isolated from both amnion and placental villi (Fig. 1.4). Placental blood, indeed cord blood, has been used for stem cell transplants to treat many life-threatening diseases and is being researched in the emerging field of regenerative medicine. Therefore, stem cells from placental blood could be purified and cryopreserved to enrich the stem cell population isolated from the umbilical cord. However, there

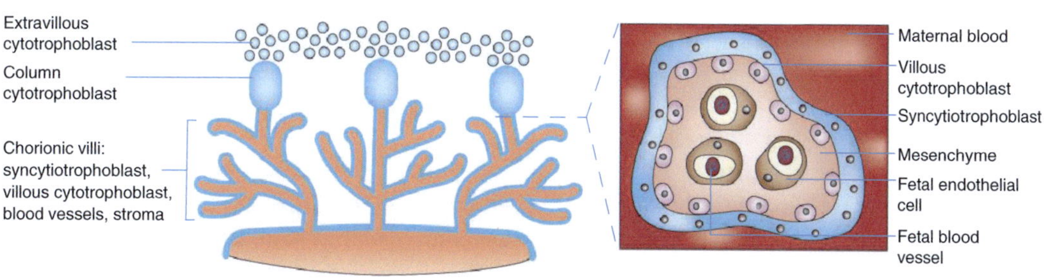

Fig. 1.4 Structure of the human placenta. The inset image shows a cross section through a chorionic villus: trophoblast-derived structures (blue) and mesoderm-derived tissues (orange). The inset images illustrate the number and type of cell layers between the maternal and fetal blood. Reproduced with permission from [25]

may not be enough stem cells in the umbilical cord to successfully treat their disorder when a patient receives a cord blood transplant.

Parents around the world are paying large sums of money to preserve cord blood from their newborns believing that they are providing an additional level of insurance for their children in case of a life-threatening disease. There is no doubt that the stem cells from cord blood can be used to treat hematopoietic system disorders. However, little attention is paid to the dosage of stem cells (number of cells per kilogram of body weight). Given the limited number of cord blood stem cells present in one umbilical cord and limited expansion ability, not only adults but also children older than 5–7 years cannot be treated even if they have their cord blood preserved [26].

The placenta is a large, highly vascularized hematopoietic tissue that functions during the embryonic and fetal development. Low oxygen concentration condition in the placenta may favor placental and fetal stem cells, as many adult stem cells are found in low oxygen niches, with stemness maintained through hypoxia-dependent pathways. Placenta contains several times more stem cells than the umbilical cord blood, which will provide the greatest chance for transplant success and make stem cell transplants available to more people [4–6]. The human placenta could represent an accessible supplemental source of cells for therapeutic strategies in addition to umbilical cord blood cells [27].

Anatomically, the placenta can be divided into several regions (Fig. 1.4). Other than hematopoietic stem cells, placenta-derived mesenchymal stem cells reside in chorionic villi and the fetal membranes of the term placenta and express pluripotency stem cell markers such as Oct-4, Nanog, Sox-2, SSEA-4, TRA-1-60, and Tra-1-8 [28, 29].

1.3.6 Special Population of Perinatal Stem Cells

A number of studies have shown that MSCs derived from different tissues share many similarities but also exhibit some differences. Using identical culture conditions, major differences were observable in the frequencies, proliferation and differentiation potentials, as well as biological functions [5]. The unique subpopulation of MSCs possessing specific differentiation potency may contribute to designed therapeutic strategies [14]. VCAM-1+ placenta chorionic villi-derived mesenchymal stem cells display potent pro-angiogenic activity [6] and might be preferred in clinical application for therapeutic angiogenesis [30].

According to placental circulation, about 150 mL of maternal blood fills into intervillous space and bath the villi (Fig. 1.4), and maternal cells also can be isolated for future applications.

1.4 Prospective of Perinatal Stem Cells

Stem cell banking of perinatal stem cells, especially cord blood, has become a popular business, and private cord blood banks are mushrooming and flourishing around the world. Cord blood transplantation has been used to treat hematopoietic system disorders, and the first was in 1988 for Fanconi's anemia treatment. Due to the close ontogenic relationship to embryonic stem cells, stem cells derived from different placental regions as well as from umbilical cord blood and amniotic fluid have immunoprivileged characteristics, possess a broader plasticity, and proliferate faster than adult MSCs. Moreover, the human placenta is normally discarded after birth, and cells can be isolated avoiding any ethical concerns [31]. Furthermore, perinatal stem cells represent stem cell types which combine some properties of pluripotent embryonic stem cells with other properties of multipotent mesenchymal stem cells. Stem cell banking either public or private allows us to collect stem cells from the umbilical cord, amniotic fluid, and placenta when the baby is born, so we can take advantage of future possibilities in stem cell therapy.

References

1. Feng G, Cui J, Zheng Y, Han Z, Xu Y, Li Z. Identification, characterization and biological significance of very small embryonic-like stem cells (VSELs) in regenerative medicine. Histol Histopathol. 2012;27:827–33.
2. Liu N, Qi X, Han Z, Liang L, Kong D, Han Z, et al. Bone marrow is a reservoir for cardiac resident stem cells. Sci Rep. 2016;6:28739.
3. Du W, Tao H, Zhao S, He ZX, Li Z. Translational applications of molecular imaging in cardiovascular disease and stem cell therapy. Biochimie. 2015;116:43–51.
4. Lu LL, Liu YJ, Yang SG, Zhao QJ, Wang X, Gong W, et al. Isolation and characterization of human umbilical cord mesenchymal stem cells with hematopoiesis-supportive function and other potentials. Haematologica. 2006;91:1017–26.
5. Yang ZX, Han ZB, Ji YR, Wang YW, Liang L, Chi Y, et al. CD106 identifies a subpopulation of mesenchymal stem cells with unique immunomodulatory properties. PLoS One. 2013;8:e59354.
6. Du W, Li X, Chi Y, Ma F, Li Z, Yang S, et al. VCAM-1(+) placenta chorionic villi-derived mesenchymal stem cells display potent pro-angiogenic activity. Stem Cell Res Ther. 2016;7:49.
7. Li Z, Han Z, Wu JC. Transplantation of human embryonic stem cell-derived endothelial cells for vascular diseases. J Cell Biochem. 2009;106:194–9.
8. Li Z, Wilson KD, Smith B, Kraft DL, Jia F, Huang M, et al. Functional and transcriptional characterization of human embryonic stem cell-derived endothelial cells for treatment of myocardial infarction. PLoS One. 2009;4:e8443.
9. Xu X-L, Yi F, Pan H-z, Duan S-l, Ding Z-c, Yuan G-h, et al. Progress and prospects in stem cell therapy. Acta Pharmacol Sin. 2013;34:741–6.
10. Li Z, Suzuki Y, Huang M, Cao F, Xie X, Connolly AJ, et al. Comparison of reporter gene and iron particle labeling for tracking fate of human embryonic stem cells and differentiated endothelial cells in living subjects. Stem Cells. 2008;26:864–73.
11. Li Z, Hu S, Ghosh Z, Han Z, Wu JC. Functional characterization and expression profiling of human induced pluripotent stem cell- and embryonic stem cell-derived endothelial cells. Stem Cells Dev. 2011;20:1701–10.
12. Shi Y, Inoue H, Wu JC, Yamanaka S. Induced pluripotent stem cell technology: a decade of progress. Nat Rev Drug Discov. 2017;16:115–30.
13. Wei X, Yang X, Han Z-P, Qu F-F, Shao L, Shi Y-F. Mesenchymal stem cells: a new trend for cell therapy. Acta Pharmacol Sin. 2013;34:747–54.
14. Tao H, Han Z, Han ZC, Li Z. Proangiogenic features of mesenchymal stem cells and their therapeutic applications. Stem Cells Int. 2016;2016:1314709.
15. Tong L, Zhao H, He Z, Li Z. Current perspectives on molecular imaging for tracking stem cell therapy. In: Erondu OF, editor. Medical imaging in clinical practice. London: InTech; 2013.
16. Pappa KI, Anagnou NP. Novel sources of fetal stem cells: where do they fit on the developmental continuum? Regen Med. 2009;4:423–33.
17. Yen BL, Huang HI, Chien CC, Jui HY, Ko BS, Yao M, et al. Isolation of multipotent cells from human term placenta. Stem Cells. 2005;23:3–9.
18. Hill M. UNSW embryology. Placenta histology. [Internet] 2010. [updated 26 May 2010; cited 7 Mar 2011]. https://embryology.med.unsw.edu.au/embryology/index.php/Placenta_Development.
19. Zhai QL, Qiu LG, Li Q, Meng HX, Han JL, Herzig RH, et al. Short-term ex vivo expansion sustains the homing-related properties of umbilical cord blood hematopoietic stem and progenitor cells. Haematologica. 2004;89:265–73.
20. Evangelista M, Soncini M, Parolini O. Placenta-derived stem cells: new hope for cell therapy? Cytotechnology. 2008;58:33–42.
21. Pierce J, Jacobson P, Benedetti E, Peterson E, Phibbs J, Preslar A, et al. Collection and characterization of amniotic fluid from scheduled C-section deliveries. Cell Tissue Bank. 2016;17:413–25.
22. Niknejad H, Peirovi H, Jorjani M, Ahmadiani A, Ghanavi J, Seifalian AM. Properties of the amniotic membrane for potential use in tissue engineering. Eur Cells Mater. 2008;7:88–99.
23. Kurtzberg J, Laughlin M, Graham ML, Smith C, Olson JF, Halperin EC, et al. Placental blood as a source of hematopoietic stem cells for transplantation into unrelated recipients. N Engl J Med. 1996;335:157–66.
24. Gekas C, Dieterlen-Lievre F, Orkin SH, Mikkola HK. The placenta is a niche for hematopoietic stem cells. Dev Cell. 2005;8:365–75.
25. Rossant J, Cross JC. Placental development: lessons from mouse mutants. Nat Rev Genet. 2001;2:538–48.
26. Gekas C, Rhodes KE, Van Handel B, Chhabra A, Ueno M, Mikkola HK. Hematopoietic stem cell development in the placenta. Int J Dev Biol. 2010;54:1089–98.
27. Dzierzak E, Robin C. Placenta as a source of hematopoietic stem cells. Trends Mol Med. 2010;16:361–7.
28. Makhoul G, Chiu RC, Cecere R. Placental mesenchymal stem cells: a unique source for cellular cardiomyoplasty. Ann Thorac Surg. 2013;95:1827–33.
29. Antoniadou E, David AL. Placental stem cells. Best Pract Res Clin Obstet Gynaecol. 2016;31:13–29.
30. Du WJ, Chi Y, Yang ZX, Li ZJ, Cui JJ, Song BQ, et al. Heterogeneity of proangiogenic features in mesenchymal stem cells derived from bone marrow, adipose tissue, umbilical cord, and placenta. Stem Cell Res Ther. 2016;7:163.
31. Malek A, Bersinger NA. Human placental stem cells: biomedical potential and clinical relevance. J Stem Cells. 2011;6:75–92.

2 The Regenerative and Reparative Potential of Amniotic Membrane Stem Cells

Mirja Krause, Jonathan Lozano, and Rebecca Lim

Abstract

This book chapter will provide an in-depth description of the regenerative and reparative potential of key perinatal stem and stem-like cells from the amniotic membrane. Drawing from the paradigm of paracrine-mediated signaling, this chapter will highlight current dogmas that claim the primary mechanisms of repair. These include immunomodulation via the expression of anti-inflammatory, as well as de novo pro-resolution and stem cell niche factors. Reports to date on the ability of exogenously delivered stem cells and stem-like cells to maintain and regenerate niche stem cells during repair will be described. Additionally, newly recognized roles of stem cell-derived extracellular vesicles in tissue repair and regeneration will be discussed. Finally, to highlight their clinical potential, recent research with a focus on extracellular vesicles derived from perinatal stem cells will be summarized. Current challenges and the status quo of the development of standard techniques to manufacture clinical-grade vesicles will be provided.

M. Krause · R. Lim (✉)
The Ritchie Centre, Hudson Institute of Medical Research, Clayton, VIC, Australia

Department of Obstetrics and Gynaecology, Monash University, Clayton, VIC, Australia
e-mail: Rebecca.Lim@monash.edu

J. Lozano
The Ritchie Centre, Hudson Institute of Medical Research, Clayton, VIC, Australia

Laboratory C-238, Department of Genomic Medicine and Environmental Toxicology, Biomedical Research Institute, National Autonomous University of Mexico, Mexico City, Mexico

2.1 Introduction

The potential of stem cells in regenerative medicine has placed them in the spotlight and has led to increased interest in the field. Stem cells have been thought to replace and/or repair damaged or dysfunctional cells. They are present in virtually all tissues and organs throughout the human body and can be generally divided into three age-specific groups: embryonic, fetal, and adult [1]. In general, stem cells are characterized by their ability to maintain clonality, proliferative capacity, and plasticity [2]. For an overview, see Fig. 2.1.

Adult or endogenous stem cells are undifferentiated cells found in differentiated tissues or organs (e.g., bone marrow, liver, brain, and blood) in specialized niches with their primary role being tissue maintenance and repair [3]. These stem cells are multipotent, that is, they have the capability to differentiate into various cell types. This makes them very appealing for regenerative medicine. However, due to their limited numbers within the tissues of origin and their poor proliferative capacity ex vivo, it is challenging to obtain them in larger quantities.

Z. C. Han et al. (eds.), *Perinatal Stem Cells*, https://doi.org/10.1007/978-981-13-2703-2_2

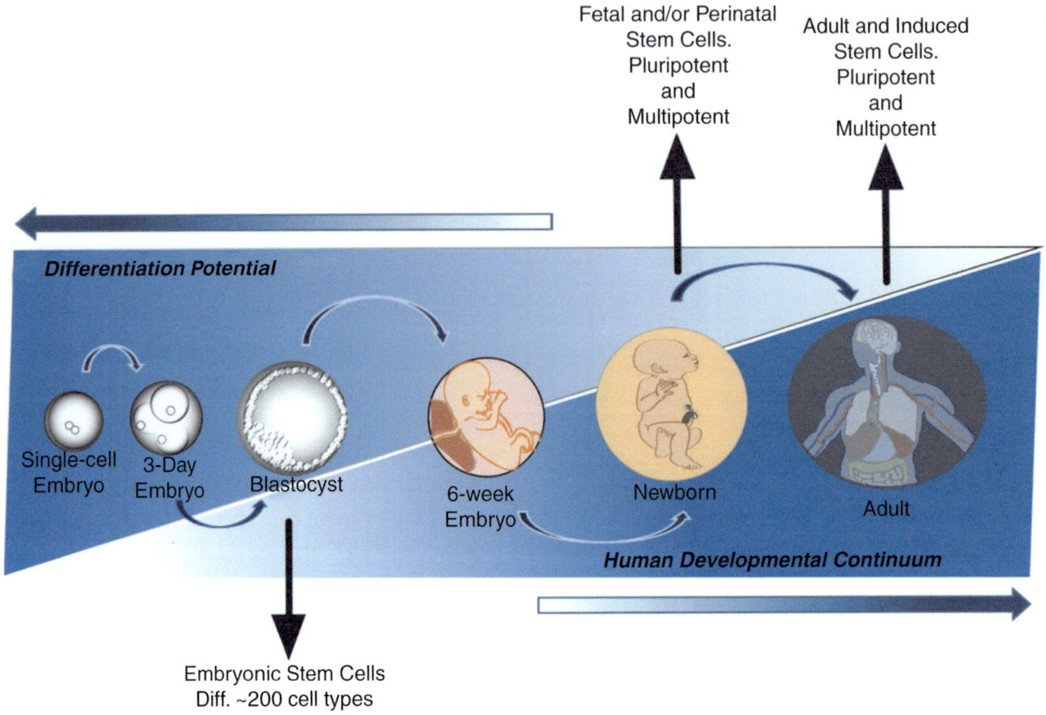

Fig. 2.1 Human stem cells. An overview of the stages during human development where various broad classes of stem cells are present. Embryonic stem cells (left) are pluripotent and can differentiate into cells that represent all three germ layer. Fetal/perinatal stem cells (center) are pluripotent or multipotent and can be obtained from fetal tissue or gestational tissue, respectively. Adult stem cells (right) are multipotent but manipulated into induced pluripotent stem cells

Mesenchymal stem cells (MSCs) and hematopoietic stem cells (HSCs) are two well-studied adult stem cell types. MSCs are multipotent—they are able to differentiate into a variety of lineages such as osteoblasts, myocytes, chondrocytes, and adipocytes, while HSCs can differentiate into all blood cell types. Presently, comprehensive research and many clinical trials are employing MSCs as therapeutics in regenerative medicine. The NIH database for clinical trials lists over 5000 trials involving MSCs (www.clinicaltrials.gov). Treatment of diseases or medical conditions such as type 1 diabetes, kidney disease, organ transplantations, cancer, graft-versus host disease, and many more is being investigated.

In contrast to MSCs and HSCs, embryonic stem cells (ESCs) originate from the inner cell mass of a blastocyst. They are pluripotent and can be expanded indefinitely ex vivo without the loss of their differentiation potential [1]. However, their allogeneic nature can lead to immune rejection when employing ESCs to treat diseases. Their inclination to form tumors in vivo remains a concern [4]. Additionally, ethical and political controversies hamper the progress of embryonic stem cell research due to the fact that ESCs can only be collected from embryos.

A newer addition to the family of stem cells are the induced pluripotent stem cells (iPSCs) which involve genetic reprogramming to acquire embryonic stem cell-like properties, and in doing so, they become pluripotent. The reprogramming of mature somatic cells is achieved through the induction of pluripotency factors using viruses and proteins or the addition of small molecules [5]. The Yamanaka factors used to induce pluripotency, namely, c-myc, oct3/4, Klf4, and Sox2, are proto-oncogenes which carry a risk of cancer when these cells are used therapeutically [6]. However, iPSCs can be derived from diseased patients, hence making them great candidates to be used as tools in drug development and the in vitro modeling of diseases. This topic has been recently reviewed by Avior et al. [7]. iPSCs have been used to screen drugs to treat a wide range of neurological diseases such as Alzheimer's, bipolar disorder, Down syndrome, Huntington disease, Parkinson's, and schizophrenia [7]. The potential of iPSCs to inform origins of disease and assist in drug discovery has been recognized by several initiatives worldwide aiming to generate banks of iPSCs from diseased patients in order to learn more about the developmental origins of disease and cellular pathways underlying the disease. Some of these international initiatives include the CIRM (California Institute for Regenerative Medicine, USA), the StemBANCC (Stem cells for Biological Assays of Novel drugs and prediCtive toxiCology, EU), and the EBiSC (European Bank for induced pluripotent Stem Cells). Over 50% of these approaches have led to the suggestion of patient-specific therapies, lending themselves to the next generation of personalized medicine.

Australian-based stem cell company, Cynata, has built their *Cymerus Platform Technology* on iPSCs (www.cynata.com). The production of Cynata's iPSCs utilizes a non-integrating episomal reprogramming method [8] which does not require viruses, but instead employs plasmids to transfer the Yamanaka factors. Cynata uses iPSCs as a starting material to produce MSCs and other cell therapy products at a commercial scale. Cellular Dynamics International (CDI, www.cellulardynamics.com) a US-based company was one of the first ones to use iPSCs for the creation of several different MSC cell lines for clinical applications (e.g., neurons, cardiomyocytes, astrocytes, hepatoblasts). The Riken Institute in Japan has driven innovation and development in the field of iPSCs. In 2014, the Riken Institute was the first to start a clinical trial using autologous iPSCs to treat age-related macular degeneration [9]. This was largely enabled by recent legislations passed by the Japanese government to accelerate the development process of regenerative medicine products. Concurrently, Japan funded and developed a network of universities and research centers whose remit was to provide clinical-grade iPSCs (Kyoto University) and develop iPSC-based therapies (Keio University, CiRA, Riken Institute, and Osaka University) [10].

Finally, there are the fetal or perinatal stem cells, which can be derived from postembryonic perinatal tissues as well as embryo-derived extrafetal products [2]. These include the placenta, placental membranes, amniotic fluid, umbilical cord, and cord blood. For an overview on the cells and if they have been used in tissue banking and clinical trials, see Table 2.1. Their properties can vary significantly depending on their tissue of origin. It is worth noting that there are several features that set them apart from adult and embryonic stem cells. For example, perinatal cells do not spontaneously differentiate in culture like ESCs, they are more resistant to hypoxia [11], their phenotype remains more stable in culture after serial passaging, and they tolerate cryopreservation better than adult stem cells [12]. Furthermore, perinatal stem cells show immunosuppressive properties in vitro and in vivo by altering innate and adaptive immune responses [2, 5]. They can be transplanted without displaying in vivo tumorigenicity [4, 13, 14] or immune rejection [15].

Another important advantage of perinatal stem cell lies in their accessibility, since the cells are harvested from the placenta and umbilical cord after childbirth. With an estimated 130 million annual births worldwide (www.worldometers.info), perinatal tissues are the most plentiful reservoir of stem cells with the potential to be applied across a variety of clinical conditions. Commonly the isolation methods are simple and the cell yield is high. Furthermore, these tissues are generally considered medical waste, and thus, their use is unencumbered by ethical concerns. An overview of studies focused on potential therapeutic application of fetal stem cells has been published describing a range of target organ systems including the brain, spinal cord, heart, lung, liver, and kidney to pancreas, bones, muscles, and eyes [16]. While the majority of these studies referred to the application of umbilical cord HSCs, several studies also made use of amniotic stem cells such as hAMSCs, hAFSCs, and

Table 2.1 Perinatal stem cells

Tissue of origin	Perinatal stem cell type	Potency	Tissue or cell banking (Y or N)	Clinical trials (Y or N)
Placenta				
Chorion	Human chorionic MSCs (hCMSCs)	Multipotent	Y	N
Chorion	Chorionic trophoblastic cells (hCTCs)		N	N
Amnion	Human amniotic MSCs (hAMSCs)	Multipotent	Y	N
Amnion	Human amniotic epithelial cells (hAECs)	Pluripotent	Y	Y
Amniotic fluid				
	Amniotic fluid MSCs (AF-MSCs)	Multipotent	Y	N
	Amniotic fluid stem cells (AFSCs)	Pluripotent	Y	N
	Heterogeneous cell population	Multi- or pluripotent	N	N
Umbilical cord				
Wharton's jelly	Umbilical cord MSCs (UCMSCs), also called Wharton's jelly MSCs (WJMSCs)	Multipotent	Y	N
Umbilical cord blood	Umbilical cord blood MSCs (UCB-MSCs)	Multipotent	Y	Y
Umbilical cord blood	Heterogeneous cell population	Pluripotent	N	N

hAECs. The properties of perinatal stem cells, currently ongoing research, and the great potential these cells hold for regenerative medicine will be discussed in detail below.

There are specific considerations to be made in order to tap into the therapeutic potential of stem cells, particularly at a commercial scale. These include the need for efficient and easy isolation methods, as well as accurate characterization protocols while employing animal-free culture products. Cell expansion must be scalable in order to meet sizable commercial demands. It is also essential to have procedures for cryopreservation of the stem cells, using nontoxic agents that impede neither plasticity nor their proliferation capacity. All production protocols must be compliant with good manufacturing practices. One of the critical challenges of using stem cells in therapeutics is the need for large-scale cell manufacturing while maintaining their plasticity. Traditional 2D static tissue culture systems cannot meet these demands, and hence, scalable culture systems are desirable. Physiological and hydrodynamic factors can influence proliferation rates and the differentiation of the stem cells [17, 18]. These can be controlled and maintained in bioreactor cultivation systems such as spinner flasks, hollow-fiber bioreactors, or wave bioreactors among others. These systems as well as their advantages and disadvantages have been reviewed by Liu et al. [18]. The systems used for stem cell cultivation and their advantages and disadvantages are summarized in Table 2.2.

Despite the abovementioned challenges, several cell manufacturing and cell banking platforms have been successfully established in the last three decades. For example, the cell banking platform Genea Biocells (www.geneabiocells.com) which focuses on pluripotent human embryonic stem cells was the first to create clinical-grade hESC cell lines [41]. In terms of adult stem cells, American CryoStem (www.americancryostem.com) was established to collect, process, and cryopreserve autologous adipose tissue-derived stem cells (ADSCs). Athersys's MultiStem platform (www.athersys.com) offers "off-the-shelf" bone marrow mesenchymal stem cells (BM-MSCs). The first private stem cell bank storing amniotic stem cells was opened in 2009 by Biocell Center (www.biocellcenter.com) [42]. Most of these cell banking platforms are accredited by nonprofit organizations such as the American Association of Blood Banks (AABB) and the Foundation for the Accreditation of Cell Therapy (FACT), which are responsible for developing and implementing standards for stem

Table 2.2 Overview of different stem cell cultivation systems and their advantages and disadvantages

Type of cell culture	Bioreactor type	Advantages	Disadvantages	Application	References
Mixed	Spinner flasks + scaffolds	Increase efficiency of scaffold cell seeding and survival, compared to static culture	High shear stress	Culture of MSCs	[19]
	Rotating wall bioreactor	Low shear stress; good mass transfer; controlled oxygenation	Limited in size, hard to scale up	Culture of MSCs, HSCs	[20]
	Wave bioreactor + scaffolds	Low stress, maintains cells in suspension. The disposable bag system has great advantages of safety.	Expensive	Culture of HSCs	[19, 21, 22]
	Microcarrier-based bioreactors	Increase attachment surface area, easy to scale up, serve as cell delivery systems	Difficult to harvest, rough external surface could damage cells	Culture of MSCs	[23, 24]
Static	Flasks, dishes, and bags	Controlled cell population	Only suitable for low dose and small patient population, labor intensive	Culture of adherent stem cell types	[24, 25]
	Multilayered vessels	Large surface area, simplest system to scale up	Difficult to monitor and control. Difficult to achieve uniform distribution and hard to harvest. Cell heterogeneity	Culture of MSCs	[26–29]
Perfusion	Bioreactors with 3D scaffolds	Allows sufficient mass transport and oxygen supply, regulate cell spatial organization, cell cycle progression, cell proliferation and differentiation	Challenging harvest	Culture of MSCs, ESCs	[30–32]
	Microencapsulation-based bioreactors	Protects cells from hydrodynamic forces. Preserves 3D environment. Prevents excessive aggregation. Good for large scale	Cell release and separation is challenging	Culture of MSCs, ESCs	[18, 33–35]
	Hollow-fiber membrane bioreactor	High surface-to-volume ratio. The ability to operate with continuous nutrient supply grants homogeneicity. Better differentiation outcome than static 2D culture	Difficult to scale up. Lower cell density compared to T-Flasks	Culture of MSCs, ESCs	[18, 36, 37]
	Fibrous bed bioreactor	Provide 3D environment. High surface-to-volume ratio	Difficult to harvest	Culture of ESCs	[18, 30]
	Column bioreactors	High surface-to-volume ratio. Better for osteogenic differentiation than spinner flask and rotating-wall bioreactors	Pressure variation leads to heterogeneity	Culture of MSCs	[18, 19, 38, 39]
	Microfluidic bioreactors	Adapts fast and easily to meet cell requirements. Excellent for differentiation. Rapidly optimize culture conditions for tissue engineering applications. Scalable	Difficult to harvest	Culture of MSCs	[19, 40]

cell banking. The AABB's focus lies in cellular therapies that specifically concern HSCs, and they accredit all major blood banks in the USA, including umbilical cord blood (UCB).

The following sections of the review will highlight the potential of perinatal stem cells in translational medicine in particular through examples of clinical trials and commercialization. Furthermore, it will discuss the challenges that remain in the production of clinical-grade stem cells. Standard procedures and accurate quality control measures for perinatal stem cell isolation, manufacture, and cryopreservation are urgently needed and are currently being developed. Finally, it will conclude with an outlook into the future, what is next in the field of the application of stem cells in translational medicine.

2.2 Key Perinatal Stem Cell Types

2.2.1 Umbilical Cord/Cord Blood Stem Cells

The umbilical cord tissue is commonly referred to as Wharton's jelly, and it is a proteoglycan-rich connective tissue that surrounds the umbilical vessels (two arteries and one vein) [43]. The nomenclature of the tissue and its derived stem cells is not standardized. Therefore, in some publications the umbilical cord matrix/mesenchymal stem cells (UCMSCs) are considered to be the same as Wharton's jelly mesenchymal stem cells (WJMSCs). Despite the ability to divide the stroma of the umbilical cord tissue, the subamnion, Wharton's jelly, and perivascular zone, the term "Wharton's jelly cells" is often extended to all umbilical stromal cells [44]. Thus, there still is a need for standardization, and one needs to be aware that the terms currently cannot be used interchangeably. For the purposes for this book chapter, we will use the terms as cited in their original reference.

WJMSCs show markers of all three germ layers and can differentiate down mesenchymal lineages including adipogenic, chondrogenic, and osteogenic cell types [1, 45]. Since WJMSCs do not show somatic mutations nor express HLA-DR antigens, they are considered more primitive than adult MSCs [46]. WJMSCs are immune-privileged, demonstrate high proliferative potential, are able to suppress immune cells, and display regenerative properties. Their anti-inflammatory and anti-fibrotic properties have been exploited in the regeneration of liver tissue [47]. WJMSCs express low levels of MHC class I (HLA-ABC) enabling them to avoid natural killer cells [48]. Furthermore, they do not express MHC class II (HLA-DR) nor antigens involved in T- and B-cell activation such as CD80, CD86, CD40, and CD40L. Hence, WJMSCs also do not induce an allogenic T-cell response [48]. WJMSCs secrete high levels of IL-10 and TGF-β, typical anti-inflammatory molecules. A detailed overview on surface markers was recently summarized in a review [48]. The review also discusses clinical application areas of the cells in diseases and tissue regeneration such as cardiovascular diseases, liver disease, cancer therapy, peripheral nerve repair, and cartilage regeneration. However, additional investigations are needed, and improved protocols have to be developed in order to apply WJMSCs in cell-based therapy and tissue engineering.

The umbilical cord blood contains umbilical cord blood hematopoietic stem cells (UCB-HSCs), and mesenchymal stem cells (UCB-MSCs) [49]. UCB-HSCs have already been used for several years in clinical settings for hematopoietic cell transplantation [50, 51]. Their proliferative potential is about eight times higher than that of BM-MSCs due to longer telomeres [52], and they retain their differentiation potential after freeze-thawing [53]. A recent review on the advances in methods and procedures for cord blood (CB) harvest and cost-effective strategies [54] describes the biology of CB and emphasized its immune privilege. Specifically, anti-inflammatory properties were highlighted as the main reason for its indirect therapeutic benefits for regenerative medicine applications [54]. The collection and processing procedure of CB are described in minute detail pointing out that proper methodology has a direct effect on the reproducible clinical use of CB. It is worth noting that a major limitation lies in the inadequate public awareness of the opportunities to donate CB and tissue, combined with a scarcity of public stem cell banks. However, UCB banking (both public and private) and its transplantation have grown exponentially since the establishment of the first UCB bank in 1991, and public awareness is growing [3].

Allogeneic HSC transplantation lends itself to the treatment of diseases such as leukemia, lymphoma, and different genetic disorders (anemia, erythrocyte enzyme deficiencies, glycogen storage disorders, adrenoleukodystrophies) [55]. A major advantage of using HSCs from the UC compared to those from the bone marrow is that the UCB is easily collected at birth and can be made available when a transplant is needed. Moreover, a less rigorous antigen tissue matching is required for UCB-HSCs compared to BM-HSCs [56]. However, the likelihood of privately stored UCB being used is estimated at only 0.01% [2].

The success of UCB engraftment is largely dose-dependent. This is an important consideration for clinical applications given that there is a limited amount of UCB that can be collected (about 70 mL per UC) [2]. It is equally important to note that genetic diseases cannot be treated with an autologous UCB-transplant, as the mutations persist in the UCB [2]. It is for reasons such as these, in addition to the low probability of accessing one's own UCB, that organizations such as the American Society of Blood and Marrow Transplantation and the American Academy of Pediatrics recommend the donation of UCB to a public UCB bank instead of private storage [2].

To date, over 300 trials are listed in the NIH clinical trial database of which a majority focuses on UCB transplants and only a handful pertain to the application of stem cells from the UC tissue (www.clinicaltrials.gov). Additionally, the potential for UCB-derived stem cells to treat different non-hematopoietic diseases has been investigated for several years [57], with some progressing to clinical trials (www.clinicaltrials.gov)—specifically for the experimental treatment of autism (trial ID: NCT02176317), cerebral palsy (trial ID: NCT01193660; NCT01072370), and vascular disease (trial ID: NCT02287831) (www.clinicaltrials.gov).

2.2.2 Placental/Placental Membrane Stem Cells

The placenta connects the developing fetus to the uterine wall, providing oxygen as well as a wide range of hormones, cytokines, transcription factors, and growth factors [58]. The placenta also facilitates waste elimination, nutrient uptake, and gas exchange. The fetal side of the placenta is bordered by the chorionic plate and umbilical cord. The chorionic and amniotic membranes are also located on this side of the placenta. The maternal side of the placenta is attached to the uterus and bordered by the decidual plate. There are two distinct cellular regions in the placenta: the chorion and the amnion; the chorion is made of chorionic trophoblastic cells and chorionic mesenchymal cells, while the amnion consists of amniotic mesenchymal cells and amniotic epithelial cells [58]. Figure 2.2 provides a detailed overview of the placental anatomy. Four different stem and stem-like cell types have been isolated from the placenta: human chorionic mesenchymal stromal cells (hCMSCs), chorionic trophoblastic cells (hCTCs), human amniotic mesenchymal stromal cells (hAMSCs), and human amniotic epithelial cells (hAECs). For an overview on perinatal stem cells, see Fig. 2.3. hCMSCs and hCTCs are poorly studied compared to hAMSCs and hAECs. In the case of hCMSCs, this might be due to their reduced proliferative capacity. hAMSCs and hAECs are described in greater detail below. Specialized protocols have been developed for their isolation as well as for their cultivation and in vitro differentiation which has been summarized in a recent review [58].

Over a decade ago, the placenta was reported to harbor HSCs [59]. However, many questions about these HSCs remain. For example, it is unknown if these cells migrate to the tissue or if they are produced in situ [60]. Also the fate of placental HSCs is unclear: whether they supply the need of hematopoietic cells needed during the development of a vascular network between the fetus and the mother, or if instead they serve as a stock of HSCs that can migrate to intraembryonic sites during later stages of gestation [60].

Placenta-derived cell populations have been described in an earlier review [15] where the different placental compartments are located (see also Figs. 2.2 and 2.3), the cell populations of the placental tissue and the placental immunology. Despite this review being a decade old, it recognizes the potential of placental cells in clinical applications, such as hepatic regeneration, cardiac repair, and the treatment of neurological disorders, and briefly discusses aspects of cell and tissue banking.

Chorionic as well as amniotic stem cells, in particular hCMSCs and hAMSCs, are multipotent and can differentiate into several different cell types such as osteocytes, skeletal myocytes, adipocytes, and chondrocytes. They carry typical stem cell markers such as CD166, CD105, CD90, CD73, CD44, CD29, CD10, and HLA-ABC [58, 61, 62]. In contrast the hAECs are pluripotent and carry stem cell markers, such as OCT-4, NANOG, SOX-2, and TRA-1-60, and can be cultured in serum-free medium [14, 58, 63].

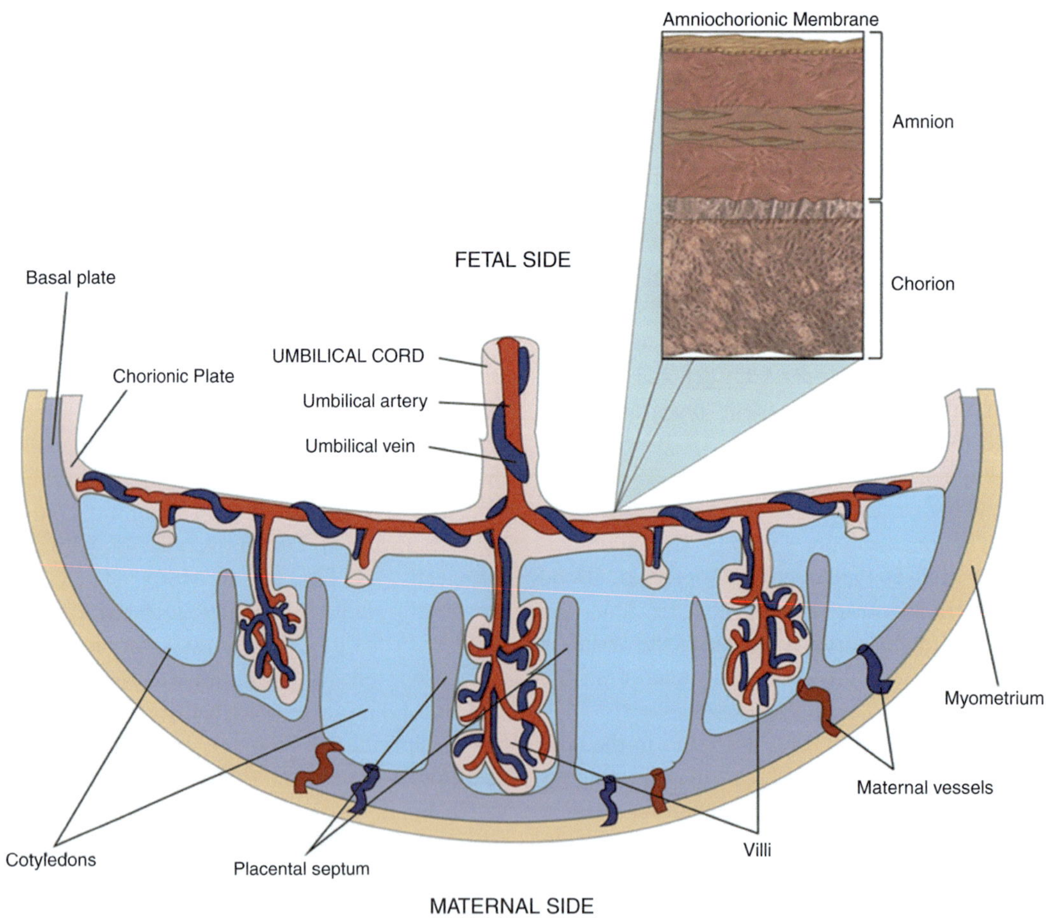

Fig. 2.2 Placental anatomy. The fetal side of the placenta is bordered by the chorionic plate and umbilical cord. The amniochorionic membranes (consisting of the amnion and the chorion) are also located on this side of the placenta. The maternal side of the placenta is attached to the uterus and consists of three layers: the basal plate, the myometrium in the middle, and the outermost border, the decidual plate, which is connected to the uterine wall

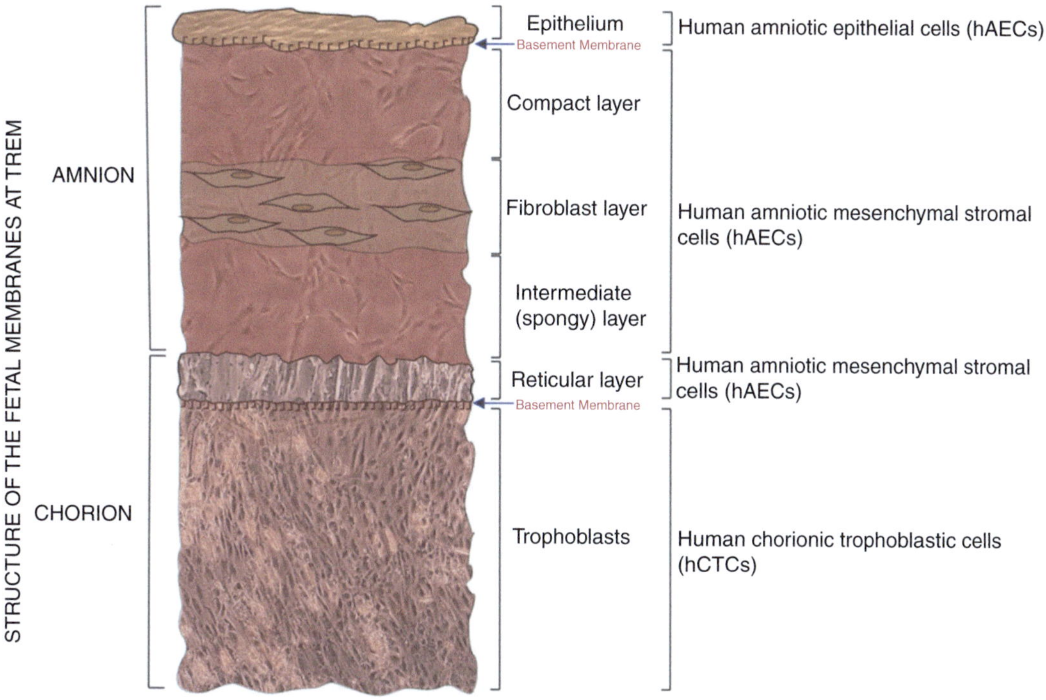

Fig. 2.3 Structure of the fetal membranes at term. The cellular layers of the amniochorionic membranes are shown in detail. The different types of perinatal stem cells that can be found in the tissues are listed for each cellular layer (right)

2.2.3 Amniotic Fluid Stem Cells

The developing embryo is enclosed by fetal membranes, which comprise of the innermost amniotic membrane (human amniotic membrane—HAM) and the overlaying chorionic membrane. Together, these fetal membranes contain the amniotic fluid, which is mildly acidic (pH 7.0). The amniotic fluid contains electrolytes, proteins, carbohydrates, and lipids. Most importantly, it also contains a heterogeneous cell population that displays markers from all three germ layers [64] and can be multi- or pluripotent [65]. According to their morphology, growth, and biochemical features, amniotic fluid-derived cells can be classified as epithelioid (E-type), fibroblastic (F-type), and amniotic fluid-specific (AF-type) [66]. The stem cells that can be derived from the amniotic fluid are the amniotic fluid mesenchymal stromal cells (AF-MSCs) and amniotic fluid-derived stem cells (AFSCs). They are both multipotent and carry markers such as CD73, CD90 and CD105 which are also common to BM-MSCs. In contrast to ESCs and iPSCs, AFSCs do not have the tendency to form teratomas in vivo [67, 68]. Despite early recognition of the therapeutic potential of these stem cells [69], there remains an absence of standardized methods for isolating and characterizing specific cell populations [70]. This is of particular importance given the many unknowns regarding the heterogeneity of cell populations present in the amniotic fluid. Some of the cell populations do not replicate under commonly employed culture conditions in vitro [70, 71].

2.2.4 Amniotic Membrane-Derived Stem/Stem-Like Cells

The human amnion or amniotic membrane is a strong, elastic, and transparent membrane containing different types of collagen [72]. It acts as a protective shield against infections, toxins, and traumas. It encompasses three different cellular layers: epithelium, basement membrane, and the mesenchyme [72]. Historically, the HAM has been used in clinical applications for over a century. The first report describing the medical use of fetal membranes was as biological dressings for skin wounds [73]. The possibility to de-epithelialize the HAM has availed it for use in ophthalmology, dermatology, and gynecology [74]. HAM is inexpensive and easily obtained from full-term placentas which are otherwise commonly discarded as medical waste. Its antimicrobial, anti-angiogenic, and anti-tumorigenic properties in addition to its mechanical characteristics make it an attractive biomaterial.

These days, HAMS are used as allografts, autografts, and scaffolds in tissue engineering [74]. Two types of stem cells have been isolated from the HAM, namely, hAMSCs and hAECs. hAMSCs are isolated from the mesenchymal germ layer, while hAECs are isolated from the ectodermal germ layer. Several features make HAM-derived stem cells attractive for regenerative medicine, including that the HAM is usually discarded after childbirth and thus can be easily harvested to isolate stem cells [72]. Most importantly though, the HAM-derived stem cells are immune-privileged and reflect the antimicrobial, anti-tumorigenic, immunomodulatory, and anti-inflammatory properties of its tissue of origin [75–77].

hAECs can be easily isolated from the amnion via trypsinization [78]. The cells express neither MSC nor HSC markers [63] but instead express early stem cell markers such as NANOG, OCT-4, SOX-2, and SSEA4 [14, 63, 79]. Similar to ESCs, hAECs are pluripotent and have been shown to differentiate into cells from all three germ layer lineages (e.g., astrocytes, osteocytes, cardiomyocytes, hepatic cells, lung cells) [79, 80]. hAECs also do not express class IA and class II human leukocyte antigens (HLAs) which makes them immune-privileged [79, 81]. They do, however, express HLA-G, a nonclassical HLA class I antigen [82], which is believed to protect the developing fetus from recognition by the maternal immune system [83]. Upon xenogenic and allogenic transplantation, no acute immune rejection has been observed [84–88]. Furthermore, hAECs were also reported to possess immunomodulatory features such as suppression of T-cell response [82, 89], T- and B-cell proliferation [90], as well as reduced B-cell and macrophage infiltration [82, 91].

Amnion-derived stem cells have been shown to possess wound healing properties and promote angiogenesis [92, 93]. They secrete soluble factors such as prostaglandin PGE_2, TGF-β, Fas-L, AFP, TRAIL, and MIF, which are important mediators in an inflammatory response [90, 94]. Their regenerative and immunomodulatory properties have been studied in the context of liver fibrosis [95, 96] and lung disease [97]. The world's first clinical trial using hAECs was recently completed in Australia (trial ID: ACTRN12614000174684; www.anzctr.org.au). The cells' established safety record in wound healing facilitate continual clinical translations of research studies. Many features of the immunomodulatory and regenerative properties are still unknown. There is, however, accumulating evidence suggesting that the secretion of certain factors guides the host's immune response toward repair and additional mechanisms of repair exist. Whether amnion-derived stem cells affect endogenous progenitor cells expediting regenerative processes and repair mechanisms as has been observed for MSCs [98] remains to be determined.

2.3 Future Outlook: Perinatal Stem Cells

According to the International Registry in Organ Donation and Transplantation's database, several thousands of people are benefiting from transplants every year worldwide (www.irodat.org). It is estimated that millions of people could benefit over their lifetime from regenerative medicine (stem cell therapy and tissue engineering). Perinatal stem cells have emerged as being distinct from other stem cell types. Many of them are pluripotent, do not show tumorigenicity, and are immune-privileged. These attributes make the cells ideal candidates for allo- and even xeno-transplantation in regenerative therapies. Studies have shown that perinatal stem cells potentially can be applied in therapies targeting diseases from nearly every organ system of the body. As discussed above, several commercial companies are setting up platforms to further develop the large-scale production of perinatal stem cells. Recently, Pluristem Therapeutics Inc. (Israel; www.pluristem.com) offered placenta-derived stem cells for the treatment of critical limb ischemia and incomplete engraftment of transplanted hematopoietic cells. The first product has already completed phase I and II clinical testing. Celgene Cellular Therapeutics (USA; www.cellgene.com) is currently testing placenta-derived adherent cells for Crohn's disease and peripheral artery disease with diabetic foot ulcer in clinical trials.

Perinatal stem cell research was previously focused on their cellular origins, phenotypes, and other stem cell characteristics. This focus has now shifted toward preclinical and clinical applications, storage and banking protocols as well as facilities, manufacturing techniques, and the development of intellectual property enabling commercialization. Increasingly, research circumstantiates the enormous potential of perinatal stem cells and supports their transition from the laboratory into the clinics.

2.3.1 Extracellular Vesicles from Perinatal Stem Cells

Certainly, the application of perinatal stem cells in cell therapies and regenerative medicine is very promising. However, many aspects of the in vivo effects are not yet completely understood. It has become clear that the transplanted cells engraft and differentiate locally. However, the rate of cell engraftment is often low, thereby raising questions around the necessity for cell engraftment. For example, a study employing hAECs to treat chronic lung injury [99, 100] was apparently efficacious [97] despite apparent lack of cell engraftment, thereby indicating that other mechanisms of action must exist. The transplanted cells secrete factors that lead to paracrine effects which influence the immune response of the recipient and can, for example, affect endogenous stem cells to elicit repair [101].

The potential for extracellular vesicles (EVs) as potential carriers of these paracrine signals has garnered much attention in the stem cell research arena. In the past 8 years, there have been over 5000 reports suggesting the paracrine effects of exogenously delivered stem cells are exerted through the release of EVs. Various cell types secrete a wide range of EVs that differ in size, cargo, and function into the intracellular space. They are classified according to their size and origin. Microvesicles bud directly from the plasma membrane and are up to 1000 nm, apoptotic bodies originate from dying cells and are 800–5000 nm, and exosomes are the smallest vesicle type with 30–120 nm and are predominantly released from the endosomal compartment [102, 103]. EVs carry miRNA, mRNA, and proteins from their cells of origin as cargo and interact with target cells subsequently modifying their phenotype and function [104]. These effects are reportedly recapitulated by EVs, and administration of EVs alone display similar, if not the same, regenerative effects as their cell of origin. Furthermore, EVs can be stored easily without losing function, as was shown in detail for EVs from granulocytes [105].

The development of EV-based therapeutics has been recently reviewed by Ohno et al. [106]. The review divides the application of EVs in a clinical setting into two categories: (1) biological medicines and (2) drug delivery systems. Biological medicines utilize native EVs with EVs derived from different types of stem cells being of particular interest. The review summarizes clinical trials of EV-based therapies: listing ten trials in phase I and two in phase II focused on different kinds of cancer, meningitis, and diabetes. Although no serious side effects were reported in any of the phase I trials, many uncertainties remain especially concerning long-term safety and xenotransplantation. Long-term experience and standardized isolation methods are lacking. It is currently impossible to exactly determine and physically separate different EV classes and subclasses.

To date the role of EVs from stem cells has been reported in ESCs, MSCs [107], endothelial progenitor cells (EPCs) [108], neural progenitor cells (NPCs) [109], and cardiac progenitor cells (CPCs) [110, 111]. Given that migration of the stem cells themselves to the site of injury is not essential for efficacy [112–115], EVs from different types of stem cells were investigated and found to display reparative, regenerative, and protective therapeutic functions reflective of the cells themselves (recently reviewed in Han et al. [111]). Stem cell EVs carry ESC-associated transcription factors such as NANOG, OCT-4, HoxB4, and also MSC markers: CD105, prominin-1/CD133, and kit [116]. Additionally, molecules of important cellular pathways have been associated with stem cell-derived EVs including WNT, β-catenin, and hedgehog [116]. The therapeutic potential of stem cell EVs has been investigated in various tissue injury models, for example, acute kidney injury, CCl4-induced liver injury, and acute myocardial ischemia (recently summarized in a review [117]). All studies showed a potential efficacy of the EVs such as protection against injury, accelerated repair, promotion of recovery, and improvement of symptoms. Rani et al. summarized clinical studies where EVs were used to treat not just tissue injuries but various medical conditions (e.g., myocardial infarction, renal injury, chronic kidney disease, hepatoma growth, bladder tumor growth, breast cancer) [118]. The EVs employed in these studies were derived from various stem cell types, being mainly BM-MSCs, but also from perinatal stem cells such as hWJMSCs and hUCBMSCs. Due to their regenerative immunomodulatory properties, also EV therapies in regenerative medicine look promising. Notably, their safety profile is superior compared to that of stem cells since immune rejection does not appear to be an issue [119].

Several studies show the immunosuppressive and reparative properties of MSC-derived EVs. A study by Schaefer et al. [120] showed that EVs from BM-MSCs prevent pulmonary arterial hypertension in rats. Two other studies showed similar effects on acute lung injury induced by the *E. coli* endotoxin in mice [121, 122]. BM-MSC-derived EVs were also used in a study in 2014 [123] to treat graft-versus-host disease (GVHD) in a single patient, where EVs were reported to have similar effects compared to a BM-MSC transplant. A reduced pro-inflammatory cytokine response could be observed in the patient, and the clinical GVHD symptoms were significantly improved. A study by Khan et al. [124] applied EVs from ESCs to treat myocardial infarction in mice. In vascular disorders the transplantation of the ESCs themselves can lead to the occurrence of teratomas [125], which was not observed when ESC-derived EVs were administered instead of the cells.

In the last 5 years, also the therapeutic potential of EVs derived from perinatal stem cells has been investigated. Zhao et al. [126] reported on the efficacy of exosomes from hUCMSCs (hUCMSC-EX) in a rat model of acute myocardial infarction. The administration of hUCMSC-EX lowered cardiac fibrosis and improved cardiac systolic function. In another study, hUCMSC-EX were applied to in vitro and in vivo models of liver fibrosis [127]. hUCMSC-EX alleviated the inflammation in carbon tetrachloride-induced liver fibrosis with apparent protection of hepatocytes and inhibition of epithelial-to-mesenchymal transition. In the most recent study, Xiao et al. [128]were able to reverse the increased inflammation present in a rat model of burn injury or macrophages exposed to LPS using hUCMSC-EX. Moreover, the authors found that exosomes overexpressing miR-181c, which was deregulated in early stage burn rats [129], further reduced inflammation. Kilpinen et al. [130] showed that hUCBMSC-derived EVs protect against ischemic acute kidney injury. Preconditioning of the cells with the inflammation cytokine, IFNγ, however, significantly changes the composition of the proteins found in and on the vesicles and results in a loss of the protective feature. This suggests that the production and secretion of EVs is strongly influenced by the cell culture conditions. This needs to be considered when studying the clinical potential of EVs. The abovementioned in vivo studies are summarized in Table 2.3.

Table 2.3 Examples of studies showing potential therapeutic applications of perinatal stem cell-derived EVs

Cell type/tissue from which EVs originate	Disease/medical condition studied	Reference
UBC blood	Ischemic acute kidney injury	[130]
hAECs	Uteroplacental communication during pregnancy	[131]
Amniotic fluid	pregnancy	[132]
Amniotic fluid	Ovarian follicular atresia	[133]
hUCMSCs	Transplantation safety	[134]
hUCMSC	Burn-induced inflammation	[128]
hUCMSC	Acute myocardial ischemic injury	[126]
hUCMSC	Liver fibrosis	[127]

MSC-derived EVs show great potential for clinical applications. Some companies such as Esperite (Esperite's subunit: The Cell Factory (www.cell-factory.com)) are the first to have patented and commercialized EV-based products. Their products are applied in the field of neurology (epilepsy), orthopedics (e.g., rheumatoid arthritis), and gastroenterology (e.g., Crohn's disease). However, more research and clinical trials to assess the safety and long-term effects are still needed. Furthermore, standards have to be developed and implemented for the safe and efficacious implementation of EV-based therapeutics.

In 2015, members of the *International Society for Extracellular Vesicles* (ISEV) and the *European Cooperation in Science and Technology (COST) program European Network on Microvesicles and Exosomes in Health and Disease* (ME-HaD) published the first guidelines on safety and regulatory requirements for the pharmaceutical production and clinical application of EVs [135]. The paper summarizes recent developments in the field while covering aspects of safety and regulatory requirements, as well as production and quality control processes. As a basis, it lists seven registered clinical trials investigating EV-based therapies. The guidelines emphasized the importance of close collaboration between researchers and clinicians, adequate and appropriate infrastructure and technology, a quality management system, and compliance with GxP standards that take into account both donor and recipient safety when manufacturing clinical-grade EVs. Furthermore, it compares the application of EVs to that of whole cells. It defines the pharmaceutical category of EVs (biological medicine) which is relevant to understand which regulatory frameworks play a role for the manufacturing and application of EV-based therapeutics. Notably, issues of safety have been raised previously [106, 134]. For example, there is not enough data to determine if EVs, since they carry cellular components from their donor cells, might also carry tumorigenic potential [106]. No long-term studies are yet available to confirm the long term safety of EV-based treatments. Furthermore, viruses and other pathogens have been shown to hijack EVs to hide from the host's immune system [106]. It is not known if diseases can spread between organisms via the transmission of EVs. Further studies and the development of safe procedures are still needed to clarify these safety parameters.

Even though several clinical trials incorporating EVs are ongoing, none of them yet includes perinatal stem cell-derived EVs. Current research on the application of these EVs as therapeutics shows great potential. Due to their special characteristics, research should soon move from the laboratory toward the clinical evaluation and clinical trials, respectively. In combination with the development of standards for clinical-grade production of EVs, in general this will successfully advance the translation of EVs (stem cell-derived as well as perinatal stem cell-derived) into the clinic.

2.4 Conclusion

Perinatal stem cells have special reparative and regenerative properties making them attractive to cell therapy. Concerning stem cells from the amniotic membrane, the long-standing safety record of the amnion used in ophthalmological and wound healing paves the way for an easier translation of cell-based therapies into the clinic.

The fact that perinatal stem cells are immune-privileged only adds to their enormous potential. Stem cells appear to exert some of their effects by initiating a change in the host's own repair mechanisms as well as its immune system responses. That exosomes may package key signaling factors secreted by stem cells responsible for these effects opens up a new avenue of therapeutics. With these in mind, we can expect to see an increase in the application of perinatal stem cells and their secretome in the future.

References

1. Marcus AJ, Woodbury D. Fetal stem cells from extra-embryonic tissues: do not discard. J Cell Mol Med. 2008;12(3):730–42.
2. Fauza DO, Bani M, editors. Fetal stem cells in regenerative medicine. Stem cell biology and regenerative medicine. New York: Springer; 2016.
3. McGuckin CP, Forraz N. Umbilical cord blood stem cells--an ethical source for regenerative medicine. Med Law. 2008;27:147–65.
4. Ilancheran S, Moodley Y, Manuelpillai U. Human fetal membranes: a source of stem cells for tissue regeneration and repair? Placenta. 2009;30(1):2–10.
5. Atala A, Murphy SV, editors. Perinatal stem cells. New York: Springer; 2014.
6. Takahashi K, Yamanaka S. Induction of pluripotent stem cells from mouse embryonic and adult fibroblast cultures by defined factors. Cell. 2006;126(4):663–76.
7. Avior Y, Sagi I, Benvenisty N. Pluripotent stem cells in disease modelling and drug discovery. Nat Rev Mol Cell Biol. 2016;17(3):170–82.
8. Schlaeger TM, et al. A comparison of non-integrating reprogramming methods. Nat Biotechnol. 2015;33(1):58–63.
9. Cyranoski D. Next-generation stem cells cleared for human trial. Nature News. 2014. http://www.nature.com/news/next-generation-stem-cells-cleared-for-human-trial-1.15897.
10. Azuma K, Yamanaka S. Recent policies that support clinical application of induced pluripotent stem cell-based regenerative therapies. Regen Ther. 2016;4(2016):36–47.
11. Turner CGB, Fauza DO. Fetal tissue engineering. Clin Perinatol. 2009;36(2):473–88.
12. Spinelli V, Guillot PV, De Coppi P. Induced pluripotent stem (iPS) cells from human fetal stem cells (hFSCs). Organogenesis. 2014;9(2):101–10.
13. De Coppi P, et al. Isolation of amniotic stem cell lines with potential for therapy. Nat Biotechnol. 2007;25(1):100–6.
14. Miki T, et al. Stem cell characteristics of amniotic epithelial cells. Stem Cells. 2005;23(10):1549–59.
15. Parolini O, et al. Concise review: isolation and characterization of cells from human term placenta: outcome of the first international workshop on placenta derived stem cells. Stem Cells. 2008;26(2):300–11.
16. Deeney SM, Crombleholme TM. Historical perspectives. In: Fetal stem cells in regenerative medicine. Stem cell biology and regenerative medicine. New York: Springer; 2016. p. 3–32.
17. King JA, Miller WM. Bioreactor development for stem cell expansion and controlled differentiation. Curr Opin Chem Biol. 2007;11(4):394–8.
18. Liu N, et al. Stem cell engineering in bioreactors for large-scale bioprocessing. Eng Life Sci. 2014;14(1):4–15.
19. Godara P, McFarland CD, Nordon RE. Design of bioreactors for mesenchymal stem cell tissue engineering. J Chem Technol Biotechnol. 2008;83(4):408–20.
20. Chen X, et al. Bioreactor expansion of human adult bone marrow-derived mesenchymal stem cells. Stem Cells. 2006;24(9):2052–9.
21. Timmins NE, et al. Clinical scale ex vivo manufacture of neutrophils from hematopoietic progenitor cells. Biotechnol Bioeng. 2009;104(4):832–40.
22. Hami LS, et al. GMP production and testing of Xcellerated T Cells™ for the treatment of patients with CLL. Cytotherapy. 2004;6(6):554–62.
23. Yang Y, Rossi FM, Putnins EE. Ex vivo expansion of rat bone marrow mesenchymal stromal cells on microcarrier beads in spin culture. Biomaterials. 2007;28(20):3110–20.
24. Liu N, et al. Three-dimensional neural differentiation of embryonic stem cells with ACM induction in microfibrous matrices in bioreactors. Biotechnol Prog. 2013;29(4):1013–22.
25. Kirouac DC, Zandstra PW. The systematic production of cells for cell therapies. Cell Stem Cell. 2008;3(4):369–81.
26. Panchalingam KM, et al. Bioprocessing strategies for the large-scale production of human mesenchymal stem cells: a review. Stem Cell Res Ther. 2015;6(1):225.
27. Bartmann C, et al. Two steps to functional mesenchymal stromal cells for clinical application. Transfusion. 2007;47(8):1426–35.
28. Schallmoser K, et al. Rapid large-scale expansion of functional mesenchymal stem cells from unmanipulated bone marrow without animal serum. Tissue Eng Part C Methods. 2008;14(3):185–96.
29. Mareschi K, et al. Multipotent mesenchymal stromal stem cell expansion by plating whole bone marrow at a low cellular density: a more advantageous method for clinical use. Stem Cells Int. 2011. https://doi.org/10.1155/2012/920581
30. Ouyang A, Ng R, Yang ST. Long-term culturing of undifferentiated embryonic stem cells in conditioned media and three-dimensional fibrous matrices

without extracellular matrix coating. Stem Cells. 2007;25(2):447–54.

31. Li Y, et al. Human cord cell hematopoiesis in three-dimensional nonwoven fibrous matrices: in vitro simulation of the marrow microenvironment. J Hematother Stem Cell Res. 2001;10(3):355–68.
32. Kim J, Ma T. Perfusion regulation of hMSC micro-environment and osteogenic differentiation in 3D scaffold. Biotechnol Bioeng. 2012;109(1):252–61.
33. Serra M, et al. Microencapsulation technology: a powerful tool for integrating expansion and cryo-preservation of human embryonic stem cells. PLoS One. 2011;6(8):e23212.
34. Wilson JL, McDevitt TC. Stem cell microencapsulation for phenotypic control, bioprocessing, and transplantation. Biotechnol Bioeng. 2013;110(3):667–82.
35. Siti-Ismail N, et al. The benefit of human embryonic stem cell encapsulation for prolonged feeder-free maintenance. Biomaterials. 2008;29(29):3946–52.
36. Gerlach JC, et al. Adipogenesis of human adipose-derived stem cells within three-dimensional hollow fiber-based bioreactors. Tissue Eng Part C Methods. 2011;18(1):54–61.
37. Roberts I, et al. Scale-up of human embryonic stem cell culture using a hollow fibre bioreactor. Biotechnol Lett. 2012;34(12):2307–15.
38. Bancroft GN, et al. Fluid flow increases mineralized matrix deposition in 3D perfusion culture of marrow stromal osteoblasts in a dose-dependent manner. Proc Natl Acad Sci. 2002;99(20):12600–5.
39. Bancroft GN, Sikavitsas VI, Mikos AG. Technical note: design of a flow perfusion bioreactor system for bone tissue-engineering applications. Tissue Eng. 2003;9(3):549–54.
40. Toh Y-C, et al. A novel 3D mammalian cell perfusion-culture system in microfluidic channels. Lab Chip. 2007;7(3):302–9.
41. Crook JM, et al. The generation of six clinical-grade human embryonic stem cell lines. Cell Stem Cell. 2007;1(5):490–4.
42. Prasongchean W, Ferretti P. Amniotic fluid stem cells for the repair of prenatal and perinatal defects. In: Perinatal stem cells. New York: Springer; 2014. p. 115–23.
43. Taghizadeh RR, Cetrulo KJ, Cetrulo CL. Wharton's Jelly stem cells: future clinical applications. Placenta. 2011;32:S311–5.
44. Anzalone R, et al. Wharton's jelly mesenchymal stem cells for the treatment of type 1 diabetes. In: Perinatal stem cells. New York: Springer; 2014. p. 313–23.
45. Fong C-Y, et al. Human Wharton's jelly stem cells have unique transcriptome profiles compared to human embryonic stem cells and other mesenchymal stem cells. Stem Cell Rev Rep. 2010;7(1):1–16.
46. Oliver-Vila I, et al. Evaluation of a cell-banking strategy for the production of clinical grade mesenchymal stromal cells from Wharton's jelly. Cytotherapy. 2016;18(1):25–35.
47. Sheers I, et al. Cell therapy for the treatment of metabolic liver disease: an update on the umbilical cord derived stem cells candidates. Open Tissue Eng Regener Med J. 2011;4:48–53.
48. Bastawrous M, et al. Wharton's jelly stem cells. In: Fetal stem cells in regenerative medicine. Stem cell biology and regenerative medicine. New York: Springer; 2016. p. 257–76.
49. Kim D-W, et al. Wharton's jelly-derived mesenchymal stem cells: phenotypic characterization and optimizing their therapeutic potential for clinical applications. Int J Mol Sci. 2013;14(6):11692–712.
50. Mayani H, Lansdorp PM. Biology of human umbilical cord blood-derived hematopoietic stem & progenitor cells. Stem Cells. 1998;16(3):153–65.
51. Roura S, Gálvez-Montón C, Bayes-Genis A. Bioluminescence imaging: a shining future for cardiac regeneration. J Cell Mol Med. 2013;17(6):693–703.
52. Schuller CE, Jankowski K, MacKenzie KL. Telomere length of cord blood-derived CD34+ progenitors predicts erythroid proliferative potential. Leukemia. 2007;21:983–91.
53. Broxmeyer HE, et al. Hematopoietic stem/progenitor cells, generation of induced pluripotent stem cells, and isolation of endothelial progenitors from 21- to 23.5-year cryopreserved cord blood. Blood. 2011;117(18):4773–7.
54. Harris DT. Umbilical cord blood stem cell populations. In: Fetal stem cells in regenerative medicine. Stem cell biology and regenerative medicine. New York: Springer; 2016. p. 241–55.
55. Horwitz ME, Chao N. Umbilical cord hematopoietic stem cell transplantation. In: Hematopoietic stem cell transplantation. Totowa: Humana Press; 2008. p. 267–88.
56. Sullivan MJ. Banking on cord blood stem cells. Nat Rev Cancer. 2008;8(7):555–63.
57. Harris DT, et al. The potential of cord blood stem cells for use in regenerative medicine. Expert Opin Biol Ther. 2007;7(9):1311–22.
58. Shammary Al M, Moretti FM. Placental stem cells and culture methods. In: Fetal stem cells in regenerative medicine. Stem cell biology and regenerative medicine. New York: Springer; 2016. p. 277–92.
59. Serikov V, et al. Human term placenta as a source of hematopoietic cells. Exp Biol Med. 2009;234(7):813–23.
60. Barcena A, et al. The human term placenta as a source of transplantable hematopoietic stem cells. In: Perinatal stem cells. New York: Springer; 2014. p. 171–81.
61. Portmann Lanz CB, et al. Placental mesenchymal stem cells as potential autologous graft for pre- and perinatal neuroregeneration. Am J Obstet Gynecol. 2006;194(3):664–73.
62. Zhang X, et al. Mesenchymal progenitor cells derived from chorionic villi of human placenta for

cartilage tissue engineering. Biochem Biophys Res Commun. 2006;340(3):944–52.
63. Murphy S, et al. Amnion epithelial cell isolation and characterization for clinical use, Chapter 1: Unit 1E.6. Hoboken: Wiley; 2007. https://doi.org/10.1002/9780470151808.sc01e06s13
64. Di Bernardo J, Kunisaki SM. Amniotic fluid stem cell populations. In: Fetal stem cells in regenerative medicine. Stem cell biology and regenerative medicine. New York: Springer; 2016. p. 167–79.
65. Prusa AR, et al. Oct-4-expressing cells in human amniotic fluid: a new source for stem cell research? Hum Reprod. 2003;18(7):1489–93.
66. Hoehn H, et al. Cultivated cells from diagnostic amniocentesis in second trimester pregnancies. I. Clonal morphology and growth potential. Pediatr Res. 1974;8(8):746–54.
67. Kim EY, Lee K-B, Kim MK. The potential of mesenchymal stem cells derived from amniotic membrane and amniotic fluid for neuronal regenerative therapy. BMB Rep. 2014;47(3):135–40.
68. Petsche Connell J, et al. Amniotic fluid-derived stem cells for cardiovascular tissue engineering applications. Tissue Eng Part B Rev. 2013;19(4):368–79.
69. Kaviani A, et al. The amniotic fluid as a source of cells for fetal tissue engineering. J Pediatr Surg. 2001;36(11):1662–5.
70. Brazzo JA III. Historical perspectives. In: Fetal stem cells in regenerative medicine. Stem cell biology and regenerative medicine. New York: Springer; 2016. p. 133–47.
71. Prusa AR, Hengstschläger M. Amniotic fluid cells and human stem cell research: a new connection. Med Sci Monit. 2002;8(11):253–7.
72. Sanjurjo-Rodríguez C, et al. Mesenchymal stem cells from human amniotic membrane. In: Perinatal stem cells. New York: Springer; 2014. p. 191–8.
73. Davis J. Skin transplantation with a review of 550 cases at the Johns Hopkins Hospital. Johns Hopkins Hospital Rep. 1910;15:307–96.
74. Díaz-Prado S, et al. Human amniotic membrane: a potential tissue and cell source for cell therapy and regenerative medicine. In: Emerging trends in cell and gene therapy. Totowa: Humana Press; 2013. p. 55–78.
75. Insausti C, et al. Amniotic membrane-derived stem cells: immunomodulatory properties and potential clinical application. Stem Cells Cloning. 2014;7(2):53–63.
76. Kim GJ. Treatment of liver disease using placental stem cells: feasibility of placental stem cells in liver diseases: potential implication of new cell therapy-based strategies for hepatic diseases. In: Perinatal stem cells. New York: Springer; 2014. p. 159–70.
77. Kubo M, et al. Immunogenicity of human amniotic membrane in experimental xenotransplantation. Invest Ophthalmol Vis Sci. 2001;42(7):1539–46.
78. Murphy SV, et al. Isolation, cryopreservation and culture of human amnion epithelial cells for clinical applications. J Vis Exp. 2014;94:e52085.
79. Ilancheran S, et al. Stem cells derived from human fetal membranes display multilineage differentiation potential. Biol Reprod. 2007;77(3):577–88.
80. Izumi M, et al. Quantitative comparison of stem cell marker-positive cells in fetal and term human amnion. J Reprod Immunol. 2009;81(1):39–43.
81. Banas RA, et al. Immunogenicity and immunomodulatory effects of amnion-derived multipotent progenitor cells. Hum Immunol. 2008;69(6):321–8.
82. Li H, et al. Immunosuppressive factors secreted by human amniotic epithelial cells. Invest Ophthalmol Vis Sci. 2005;46(3):900–7.
83. Kovats S, et al. A class I antigen, HLA-G, expressed in human trophoblasts. Science. 1990;248(4952):220.
84. Akle CA, et al. Immunogenicity of human amniotic epithelial cells after transplantation into volunteers. Lancet. 1981;318(8254):1003–5.
85. Avila M, et al. Reconstruction of ocular surface with heterologous limbal epithelium and amniotic membrane in a rabbit model. Cornea. 2001;20(4):414.
86. Bailo M, et al. Engraftment potential of human amnion and chorion cells derived from term placenta. Transplantation. 2004;78(10):1439.
87. Sankar V, Muthusamy R. Role of human amniotic epithelial cell transplantation in spinal cord injury repair research. Neuroscience. 2003;118(1):11–7.
88. Yuge I, et al. Transplanted human amniotic epithelial cells express connexin 26 and Na-K-adenosine thriphosphatase in the inner ear. Transplantation. 2004;77(9):1452.
89. Lim R, et al. Amniotic membrane stem cell populations. In: Fetal stem cells in regenerative medicine. Stem cell biology and regenerative medicine. New York: Springer; 2016. p. 149–66.
90. McDonald CA, et al. The Potential of human amnion epithelial cells as an immunomodulatory and neuroregenerative treatment for multiple sclerosis. In: Perinatal stem cells. New York: Springer; 2014. p. 231–42.
91. Kamiya K, et al. Topical application of culture supernatant from human amniotic epithelial cells suppresses inflammatory reactions in cornea. Exp Eye Res. 2005;80(5):671–9.
92. Kim S-W, Zhang H-Z, Guo L, et al. Amniotic mesenchymal stem cells enhance wound healing in diabetic NOD/SCID mice through high angiogenic and engraftment capabilities. PLoS One. 2012;7(7):e41105.
93. Kim S-W, Zhang H-Z, Kim CE, et al. Amniotic mesenchymal stem cells have robust angiogenic properties and are effective in treating hind limb ischaemia. Cardiovasc Res. 2012;93(3):328–534.
94. Liu YH, et al. Amniotic epithelial cells from the human placenta potently suppress a mouse model of multiple sclerosis. PLoS One. 2012;7(4):e35758.

95. Manuelpillai U, et al. Transplantation of human amnion epithelial cells reduces hepatic fibrosis in immunocompetent CCl_4-treated mice. Cell Transplant. 2010;19(9):1157–68.
96. Zhang D, Jiang M, Miao D. Transplanted human amniotic membrane-derived mesenchymal stem cells ameliorate carbon tetrachloride-induced liver cirrhosis in mouse. PLoS One. 2011;6(2):e16789.
97. Murphy S, et al. Human amnion epithelial cells prevent bleomycin-induced lung injury and preserve lung function. Cell Transplant. 2011;20(6):909–23.
98. Tropea KA, et al. Bronchioalveolar stem cells increase after mesenchymal stromal cell treatment in a mouse model of bronchopulmonary dysplasia. Am J Physiol Lung Cell Mol Physiol. 2012;302(9):L829–37.
99. Hodges RJ, Jenkin G, et al. Human amnion epithelial cells reduce ventilation-induced preterm lung injury in fetal sheep. Am J Obstet Gynecol. 2012;206(5):448.e8–448.e15.
100. Hodges RJ, Lim R, et al. Amnion epithelial cells as a candidate therapy for acute and chronic lung injury. Stem Cells Int. 2012; https://doi.org/10.1155/2012/709763.
101. Bollini S, et al. The regenerative role of the fetal and adult stem cell secretome. J Clin Med. 2013;2(4):302–27.
102. Raposo G, Stoorvogel W. Extracellular vesicles: exosomes, microvesicles, and friends. J Cell Biol. 2013;200(4):373–83.
103. Théry C. Exosomes: secreted vesicles and intercellular communications. F1000 Biol Rep. 2011;3:15.
104. Colombo M, Raposo G, Théry C. Biogenesis, secretion, and intercellular interactions of exosomes and other extracellular vesicles. Annu Rev Cell Dev Biol. 2014;30(1):255–89.
105. Lőrincz ÁM, et al. Effect of storage on physical and functional properties of extracellular vesicles derived from neutrophilic granulocytes. J Extracell Vesicles. 2014;3 https://doi.org/10.3402/jev.v3.25465.
106. Ohno S-I, Drummen G, Kuroda M. Focus on extracellular vesicles: development of extracellular vesicle-based therapeutic systems. Int J Mol Sci. 2016;17(2):172.
107. Bruno S, et al. Mesenchymal stem cell-derived microvesicles protect against acute tubular injury. J Am Soc Nephrol. 2009;20(5):1053–67.
108. Deregibus MC, et al. Endothelial progenitor cell–derived microvesicles activate an angiogenic program in endothelial cells by a horizontal transfer of mRNA. Blood. 2007;110(7):2440–8.
109. Marzesco A-M, et al. Release of extracellular membrane particles carrying the stem cell marker prominin-1 (CD133) from neural progenitors and other epithelial cells. J Cell Sci. 2005;118(13):2849–58.
110. Barile L, et al. Extracellular vesicles from human cardiac progenitor cells inhibit cardiomyocyte apoptosis and improve cardiac function after myocardial infarction. Cardiovasc Res. 2014;103(4):530–41.
111. Han C, et al. Exosomes and their therapeutic potentials of stem cells. Stem Cells Int. 2015;2016(7-8):1–11.
112. da Silva Meirelles L, Caplan AI, Nardi NB. In search of the in vivo identity of mesenchymal stem cells. Stem Cells. 2008;26(9):2287–99.
113. Dai W, et al. Allogeneic mesenchymal stem cell transplantation in postinfarcted rat myocardium. Circulation. 2005;112(2):214–23.
114. Iso Y, et al. Multipotent human stromal cells improve cardiac function after myocardial infarction in mice without long-term engraftment. Biochem Biophys Res Commun. 2007;354(3):700–6.
115. Noiseux N, et al. Mesenchymal stem cells overexpressing Akt dramatically repair infarcted myocardium and improve cardiac function despite infrequent cellular fusion or differentiation. Mol Ther. 2006;14(6):840–50.
116. Nawaz M, et al. Extracellular vesicles: evolving factors in stem cell biology. Stem Cells Int. 2015;2016(1):1–17.
117. Zhang B, et al. Focus on extracellular vesicles: therapeutic potential of stem cell-derived extracellular vesicles. Int J Mol Sci. 2016;17(2):174.
118. Rani S, et al. Mesenchymal stem cell-derived extracellular vesicles: toward cell-free therapeutic applications. Mol Ther. 2015;23(5):812–23.
119. Chen TS, et al. Enabling a robust scalable manufacturing process for therapeutic exosomes through oncogenic immortalization of human ESC-derived MSCs. J Transl Med. 2011;9(1):47.
120. Schäfer A, et al. Disconnecting the Yin and Yang relation of epidermal growth factor receptor (EGFR)-mediated delivery: a fully synthetic, EGFR-targeted gene transfer system avoiding receptor activation. Hum Gene Ther. 2011;22(12):1463–73.
121. Chen J-Y, et al. Therapeutic effects of mesenchymal stem cell-derived microvesicles on pulmonary arterial hypertension in rats. Acta Pharmacol Sin. 2014;35(9):1121–8.
122. Zhu YG, et al. Human mesenchymal stem cell microvesicles for treatment of escherichia coli endotoxin-induced acute lung injury in mice. Stem Cells. 2014;32(1):116–25.
123. Kordelas L, et al. MSC-derived exosomes: a novel tool to treat therapy-refractory graft-versus-host disease. Leukemia. 2014;28(4):970–3.
124. Khan M, et al. Embryonic stem cell–derived exosomes promote endogenous repair mechanisms and enhance cardiac function following myocardial infarction novelty and significance. Circ Res. 2015;117(1):52–64.
125. Caspi O, et al. Transplantation of human embryonic stem cell-derived cardiomyocytes improves myocardial performance in infarcted rat hearts. J Am Coll Cardiol. 2007;50(19):1884–93.
126. Zhao Y, et al. Exosomes derived from human umbilical cord mesenchymal stem cells relieve

acute myocardial ischemic injury. Stem Cells Int. 2015;2015(1):1–12.

127. Li T, et al. Exosomes derived from human umbilical cord mesenchymal stem cells alleviate liver fibrosis. Stem Cells Dev. 2013;22(6):845–54.

128. Li X, et al. Exosome derived from human umbilical cord mesenchymal stem cell mediates MiR-181c attenuating burn-induced excessive inflammation. EBioMedicine. 2016;8:72–82.

129. Haijun Z, et al. Detection of the MicroRNA expression profile in skeletal muscles of burn trauma at the early stage in rats. Ulus Travma Acil Cerrahi Derg. 2015;21(4):241–7.

130. Kilpinen L, et al. Extracellular membrane vesicles from umbilical cord blood-derived MSC protect against ischemic acute kidney injury, a feature that is lost after inflammatory conditioning. J Extracell Vesicles. 2013;2:269.

131. Sheller S, et al. Amnion-epithelial-cell-derived exosomes demonstrate physiologic state of cell under oxidative stress. PLoS One. 2016;11(6):e0157614.

132. Munro A, et al. Risk of persistent and recurrent cervical neoplasia following incidentally detected adenocarcinoma-in-situ. Am J Obstet Gynecol. 2016;216(3):272–7.

133. Xiao G-Y, et al. Exosomal miR-10a derived from amniotic fluid stem cells preserves ovarian follicles after chemotherapy. Sci Rep. 2016;6:23120.

134. Sun L, et al. Safety evaluation of exosomes derived from human umbilical cord mesenchymal stromal cell. Cytotherapy. 2016;18(3):413–22.

135. Lener T, et al. Applying extracellular vesicles based therapeutics in clinical trials – an ISEV position paper. J Extracell Vesicles. 2015;4 https://doi.org/10.3402/jev.v4.30087.

The Stemness of Perinatal Stem Cells

3

Yan Zhang, Zongjin Li, and Na Liu

Abstract

Perinatal tissues provide a list of new sources for stem cell derivation, such as amniotic fluid, fetal membranes (amnion and chorion), umbilical cord, and placental tissue. The perinatal stem cells represent an intermediate cell type which has recently been described to combine qualities of both adult stem cell and ESCs and possess a broad multipotent plasticity. This chapter will focus on the stemness of several types of perinatal stem cells and provide a review of the comparison of the differentiation potential between the perinatal stem cells, adult stem cells, and embryonic stem cells.

3.1 Introduction

Stem cells have the ability of self-renewal and multiple differentiation potential. They are the progenitor cells of all kinds of functional cell playing important roles in human development and tissue wound repairment. Stem cell researches have made breakthrough progress in the twenty-first century, but there still have questions in ethical, cell source, and scale production. In clinical application of stem cells, hematopoietic stem cell transplantation (HSCT) has lasted for half a century. This technology cured tens of thousands of leukemia sufferer and gradually matured. The applications of other stem cells are still in experimental stage and clinic testing stage, but the advancement of the research is rapidly.

According to the different stages of development, the stem cells fall into embryonic stem cells, fetal stem cells, perinatal stem cells, and adult stem cells. Embryonic stem cells are undifferentiated and pluripotent cell lines derived from early embryos. Fetal stem cells are stem cells in human fetal tissues. Adult stem cells are the stem cell residing in the human body after birth, mainly derived from the bone marrow, motivated peripheral blood, and other adult tissues. Perinatal stem cells (PSC) are derived from perinatal tissue source, such as the umbilical cord blood, umbilical cord, placenta, amniotic fluid, and so on. The perinatal stem cell is also known as birth-associated tissue-derived stem cells [1]. The perinatal stem cells represent an intermediate cell type that is different from the embryonic stem cells and adult stem cells. Clinical application of perinatal stem cells began in 1988, the world's first case of cord blood hematopoietic stem cell

Y. Zhang · Z. Li · N. Liu (✉)
School of Medicine, Key Laboratory of Bioactive Materials of Ministry of Education, Nankai University, Tianjin, China
e-mail: zongjinli@nankai.edu.cn; liuna@nankai.edu.cn

Z. C. Han et al. (eds.), *Perinatal Stem Cells*, https://doi.org/10.1007/978-981-13-2703-2_3

transplantation [2]. Perinatal stem cells can also be used as tissue engineering seed cells for the development and application of tissue engineering products. Due to its rich source, no ethical issues, low immunogenicity, easy-to-scale production, and some other advantages, perinatal stem cell sources represent the ideal starting point for regenerative medicine therapeutic applications, such as cardiovascular disease treatment, nervous system disease's treatment, autoimmune disease treatment, and some other diseases. In this chapter we will renew the stemness of perinatal stem cells and compared the differentiations between the perinatal stem cells and other source of stem cells, including embryonic stem cells, iPS cells, and adult stem cells.

3.2 Stemness of ES Cells, iPS Cells, and Adult Stem Cells

Stem cells exhibit stemness, the ability of self-renewal, and differentiate into different cell types in suitable condition (Fig. 3.1). The stemness of the different kinds of stem cells is quite different, which is regulated by various factors. Stemness regulatory network is very complex, which involved in transcription factor expression, histone modification, DNA methylation, and noncoding RNA. At present how to maintain the stemness of stem cells and the regulation mechanism of focused differentiation remains elusive. Scientists will still need more study on this matter to bring cell therapy to the clinic as quickly as possible. Based on the stemness difference, stem cells can be divided into two main kinds: one is the pluripotent stem cells (including embryonic stem cells and induced pluripotent stem cells) and another is multipotent stem cells (mainly the adult stem cells). In the following parts, we will simply introduce the stemness of some kinds of stem cells.

3.2.1 Embryonic Stem Cells

Embryonic stem cells (ESCs) are isolated from the inner cell mass (ICM) of blastocyst stage embryos [3, 4]. In vivo, the inner cell mass develops to form endodermal, mesodermal, and ectodermal tissues of the embryo proper. Embryonic stem cells exhibit the unique property of pluripotency, the ability to form any specialized, differentiated cell types of the organism from which they are derived. In order to maintain pluripotency, embryonic stem cells must proliferate while preventing the stem cell from differentiation, a process known as self-renewal [5]. The pluripotency of ESCs is determined by a well-characterized gene transcription circuitry. The circuitry is assembled by ESC-specific transcription factors, signal-transducing molecules, and epigenetic regulators [6]. Embryonic stem cell lines of almost all species have some general pluripotency factors, such as octamer-binding transcription factor 4 (Oct4), sex-determining region Y-box 2 (Sox2), Nanog, and so on [7]. The transcription factors are highly conservative in almost all species, but the signaling pathway to maintain pluripotency is much different. For

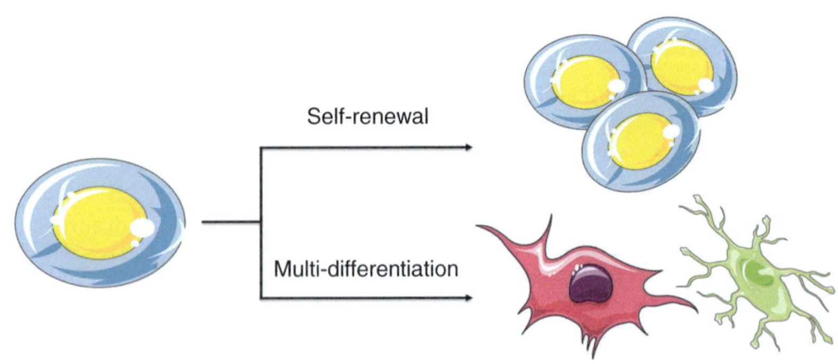

Fig. 3.1 Stemness of stem cells. Stem cell have the ability to self-replicate, which is the self-renewal property of stem cells. Stem cells can also differentiate into several kinds of cell types in the suitable condition. Stemness is the ability of self-renewal and differentiation

example, the leukemia inhibitory factor (LIF) and bone morphogenetic protein 4 (BMP4) are essential to maintain the pluripotency of embryonic stem cell in rodents, such as mouse and rat, but basic fibroblast growth factor (bFGF) and Activin/Nodal are essential in primates, such as monkey and human. Oct4, Sox2, and Nanog are combined to form a complex network to maintain the stemness of the stem cells. Oct4 and Nanog are two transcription factors to maintain stemness of stem cells, which can selectively inhibit the expression of differentiation genes or promote the expression of pluripotency genes through binding to the regulation area of target genes [8]. Embryonic stem cells have the ability to differentiate into any specialized cell type of the human body, and ESC-derived cell types offer great potential for regenerative medicine. However, key to achieve this goal required a strong understanding of stem cell biology, techniques to maintain stem cells, and strategies to make the cell differentiation toward the desired cell type.

3.2.2 iPS Cells

As we reviewed above, embryonic stem cells can give rise to all cell types as we want, so it gains wide attention in the world. But, embryonic stem cells have their own insurmountable defects, such as source limitation and immune rejection. So, people began looking for other alternative method to get pluripotent stem cells. Induced pluripotent stem cell (iPS cell) is a very successful example. In 2006, Yamanaka and his colleagues found that mice somatic cell can be reverted to a pluripotent state by introduction of a small set of transcription factors into it [9]. Human iPS cells were obtained in the following year [10, 11]. The induced pluripotent stem cells are similar to natural ES cells in many aspects, such as the expression of certain stem cell genes, epigenetic modification, self-renewal, and differentiation potential. They have the ability to differentiate into any specialized cell type of the human body. The extended transcriptional network of embryonic stem cells is some key pluripotency markers including Oct4, Sox2, and Nanog. In the last decade, the somatic cells have been reprogrammed to iPS cell through introducing the key pluripotency transcription factors (Oct4, Sox2, Klf4, cMyc), microRNAs, and small molecules [9]. The breakthrough research of iPS cells allow researchers to gain pluripotent stem cells without the ethical controversial of using embryos, providing a new and powerful method to "de-differentiate" cells. Furthermore, tissues derived from iPS cells will be almost perfectly matched to the cell donor, which is an important factor in research of disease modeling and drug screening. IPS cells are expected to help researchers find how to reprogram cells and repair damaged tissues in the human body [12].

3.2.3 Adult Stem Cells

ES cells and iPS cells can give rise to all kinds of cells of all three germ layers, so they have widely potential usage in clinical treatment. But now, we have yet to see any real cures or treatment from these cell sources. The main reason is that ES cells and iPS cells have been shown to form tumors in many studies [13]. Different with ES cells and iPS cells, adult stem cells have already been successfully used in clinical treatment for many years. Adult stem cells (ASCs), also known as somatic stem cells, are found in various regions of the adult organism, such as the bone marrow, skin, eyes, viscera, and brain [14]. Adult stem cells are multipotent stem cells that can differentiate into several cell types of its residing tissue, generally for the purpose of repair. Multipotent cells give rise to cells of specific tissue; hematopoietic stem cells give rise to blood cells, while neural stem cells give rise to neurons and glia. The ability of differentiation and proliferation of adult stem cells is lower than embryonic stem cells [15], but adult stem cells are safer than ES cells and iPS cells when they were used in clinical treatment. The first bone marrow stem cell transplant for leukemia treatment was undergone by Dr. Thomas in

1957 [16]. Hematopoietic stem cell transplantation becomes an effective method to cure some hematologic or non-hematologic diseases. Besides hematopoietic stem cells, mesenchymal stem cell is another non-hematopoietic stem cell in the bone marrow. Experiments have shown that mesenchymal stem cell transplantation can cure kid's osteogenesis imperfecta [17] and get the better effect. Embryonic stem cell research is restricted by its limited source and ethical issues, so research on adult stem cells attracts more attentions, but there still have some deficiencies. Adult stem cells have a lower differentiation and proliferation ability and can't product enough stem cells to be effective in clinical treatment.

3.3 Perinatal Stem Cell

Perinatal stem cells are derived from perinatal tissues, such as umbilical cord blood, umbilical cord, placental blood, placental tissue, and amnion. Normally, two main types of stem cells can be isolated from the perinatal tissues, hematopoietic stem cells and mesenchymal stem cells. The stemness of perinatal stem cells is between the embryonic stem cells and adult stem cells (Fig. 3.2). They are more primitive than adult stem cells. Perinatal stem cells have faster and greater expansion potential than stem cells derived from adults.

3.3.1 Hematopoietic Stem Cells

Hematopoietic stem cell (HSC) is known as multipotent stem cells, present in the hematopoietic tissue in a group of primitive hematopoietic cells. Hematopoietic stem cell is the originator of the blood cells. It has a long history of using the hematopoietic stem cells to cure malignant hematologic disease but the HSCs mainly from the adult bone marrow in the past research. In 1988, the French medical scientist Gluckman firstly applied the umbilical cord blood transplantation (UCBT) to treat a kind of genetic blood disease and made great succeed [18]. After that, the umbilical cord blood has attracted attention and became the mainly source of the hematopoietic stem cells. Functionally, hematopoietic stem cells derived from the umbilical cord blood are capable of producing three lineages of blood cells: erythroid, myeloid, and lymphoid. However, the number of HSCs present in a unit of UCB is usually sufficient for younger children but not for adults [19]. Serikov et al. [20] separated the HSCs from a placental tissue, and the number of it is usually sufficient for two or three adults. In addition, because of the attenuated donor-derived immune response and less risk of graft-versus-host disease (GVHD), potentially larger degree of HLA mismatches at transplantation of perinatal hematopoietic stem cell can be permitted. Due to the hematopoietic stem cells from the placenta which are the early cells, they have higher proliferation ability. Therefore, maybe they can effectively replace hematopoietic stem cells from the umbilical cord blood, peripheral blood, and bone marrow.

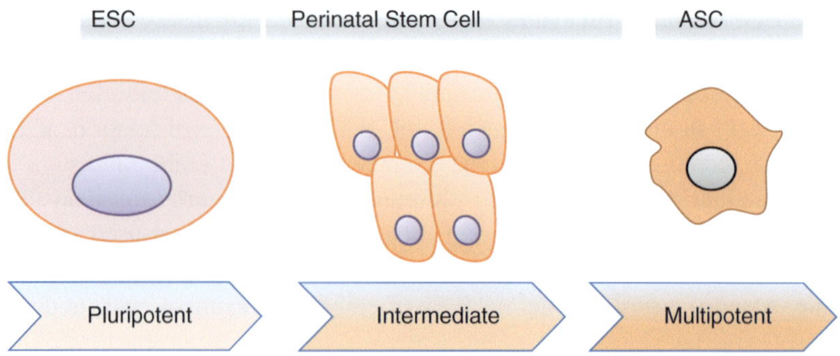

Fig. 3.2 Stemness of perinatal stem cell. *ESC*: embryonic stem cell, *ASC*: adult stem cell

3.3.2 Mesenchymal Stem Cells

Mesenchymal stem cells (MSCs) are pluripotent stem cells which arise from the mesoderm; they have the potential for self-renewal and multi-differentiation. MSCs have many properties that suggest that they are ideal cell for regenerative medicine applications. MSCs can differentiate into all three germ layers and have been shown to be immune privileged, which means that these cells are suitable for allogeneic therapeutic applications [21]. MSCs can be isolated from various tissue types, including the lung, fat, and liver. We used to think the bone marrow is the main source of MSCs, but the number of the bone marrow-derived MSCs is low. Bone marrow-derived MSCs have been shown to senescence around passages 10–12, so bone marrow MSCs can't product enough cells to be effective in treatment [22]. In the view of ethics, medicine, and cell engineering, the MSCs from adult can't be the source of mass production for the routine treatment. Primitive MSCs can be isolated from perinatal tissue source, such as the umbilical cord tissue, umbilical cord blood, placental blood, and placental tissue [23]. The mesenchymal stem cells from perinatal tissues express the typical mesenchymal stem cell markers, such as CD10, CD13, CD29, CD44, CD73, CD90, and CD105. At the same time, they do not express markers of the hematopoietic lineage (CD34 and CD45). MSCs are necessary for the tissue repair and regenerate progress in person's whole life. The origin or source of MSCs may determine their fate and functional characteristics. Understanding the distinct characteristics of each MSC derived from perinatal tissues is important for their optimal therapeutic application.

3.4 Sources for Perinatal Stem Cell Derivation

Gestational tissues, which include the umbilical cord, amniotic fluid, and placenta, are an abundant source of highly multipotent stem cells with potent immunosuppressive activity. These stem cell sources are providing the field of regenerative medicine and giving new hope for the treatment of disease [24]. Gestational tissue has certain advantages as a stem cell source than the "traditional sources," including bone marrow or embryo-derived cells. Such tissue is often discarded after birth and easily available without the destruction of a human embryo [23]. This means that the collection and use of gestational issues effectively avoid the ethical and legal issues. The perinatal stem cells are derived from the amniotic fluid, fetal membranes, amnion, umbilical cord, and placental tissue (Fig. 3.3). In the following part, we will introduce the sources of perinatal stem cells.

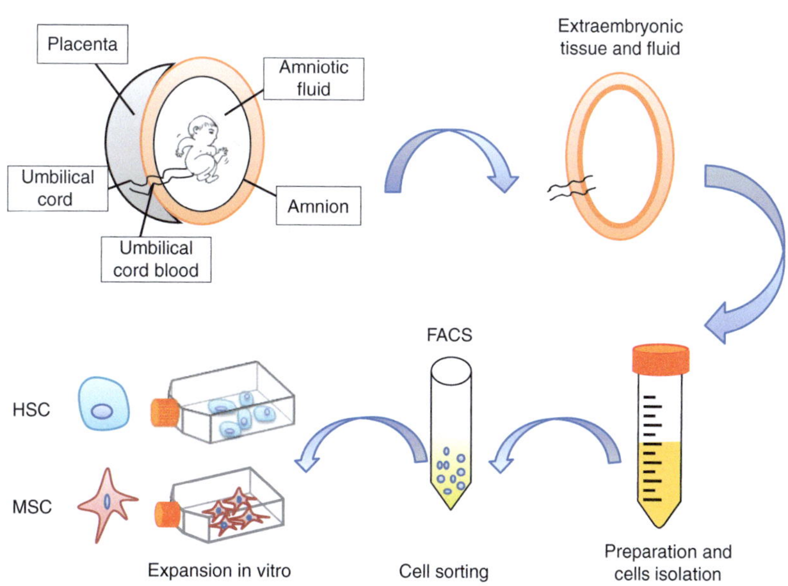

Fig. 3.3 Derivation of perinatal stem cells (MSC and HSC) from perinatal tissues. *MSC*: mesenchymal stem cell [25], *HSC*: hematopoietic stem cell

3.4.1 Umbilical Cord

Traditionally, the bone marrow and mobilized peripheral blood have been the predominant source of autologous and allogeneic hematopoietic stem cells for transplantation to treat leukemia patient. Over the past decades, umbilical cord blood has been established as an alternative source for HSC transplantation. The umbilical cord is the elastic cord that connects the fetus to the placenta during pregnancy. Anatomically, the umbilical cord is coated by the amniotic membrane and contains myxoid connective tissue that consists of two umbilical arteries and one umbilical vein. The vein is coated with a mucous connective proteoglycan-rich matrix that is called Wharton's jelly. Multipotent fibroblast-like mesenchymal cell populations were first identified in the umbilical cord over 10 years. Such cells were originally called umbilical cord matrix stem cells to distinguish them from mesenchymal cells (UCB-MSCs) isolated from umbilical cord blood, as well as from late outgrowth endothelial cells (OECs) and endothelial cells isolated from the umbilical vein (human umbilical vein endothelial cell; HUVEC) [26]. More recently, Wu [27] has found a kind of cell in Wharton's jelly that has a higher multi-differentiation ability; it is the mesenchymal stem cell in Wharton's jelly. The mesenchymal stem cells attract more attention with its unique characteristics, such as widespread source, convenient and harmless to donors. So the mesenchymal stem cells have the widespread application prospect in cell therapy and tissue engineering.

3.4.2 Umbilical Cord Blood

Umbilical cord blood (UCB) is a fetal blood compartment that remains in the placenta and in the attached umbilical cord after birth. UCB contain hematopoietic stem cells, mesenchymal progenitors, endothelial cell precursors, and multipotent/pluripotent lineage stem cells [28]. UCB-derived HSCs are an important part of umbilical cord blood stem cells. UCB-derived HSCs likely originate from hemangioblasts, the precursor of hematopoietic and endothelial progenitors [29]. They have some unique characteristics which are different from bone marrow-derived HSCs and provide for hematopoietic engraftment and regeneration. Umbilical cord blood donations differ from bone marrow donations, in that stem cells can be transplanted without the need to find a match. Thus more patients with leukemia and other diseases can be saved. Umbilical cord blood-derived mesenchymal stem cells (UCB-MSCs) are less common than HSCs. Like AF-MSCs, these cells express transcription factor Oct4. The average amount of UCB-MSCs decreases dramatically with increasing gestational age [30]. MSCs play an important role in skin tissue engineering, immune regulation, and some other aspects.

3.4.3 Amniotic Fluid

Human amniotic fluid is the liquid in the amniotic cavity of the uterus during pregnancy. Throughout the pregnancy, it is essential to maintain the fetal life. Human amniotic fluid can be obtained during amniocentesis at the second trimester. This procedure is already performed in many pregnancies in which the fetus has a congenital abnormality or to determine characteristics such as sex [31]. Amniotic fluid maybe not only is a diagnostic tool but a source of a powerful therapy for a multitude of congenital and adult disorders. Amniotic fluid contains a mixture of different cell types, which was identified as sloughed cells from the fetal amnion membrane and skin, as well as the alimentary, respiratory, and urogenital tracts. In 2002, Prusa [32] has found that the cell in amniotic fluid has some typical characteristics like human stem cell. Then Prusa [33] proposed that the amniotic fluid cell is a kind of stem cell and amniotic fluid stem cells also express the stem cell marker Oct4. Cell culture experiments with these types of cells have demonstrated that they have the ability to differentiate into a variety of cell types, including hematopoietic, adipogenic, chondrocytic, osteogenic, myogenic, endothelial, hepatogenic, pulmonary, neurogenic, and cardiac. It was named as human amniotic fluid-derived stem cells (hAFS). Amniotic fluid also contains a mixture of different cell types. The type and some biological characteristics of amniotic fluid stem cell (AFS) are changed with the gestational age. According to the morphological characters and growth characters, the amniotic fluid cells can be divided into three types: the epithelial cell of E-type, specific amniotic fluid cell of AF-type, and fibroblast of F-type [33]. The epithelial cells of E-type and specific amniotic fluid cell of AF-type are present at the beginning of the cultivation. The specific amniotic fluid cell of AF-type always maintains in the whole cultivation process. The E-type cells tend to decrease rapidly with the extending of incubation time. Amniocentesis obtained from the prenatal diagnosis with genetic testing shows that about 1% of cells can express the cell surface antigen c-Kit (CD117), which is a receptor protein that exist in the surface of various stem cells including embryonic stem cells. Like embryonic stem cells, amniotic fluid stem cells form embryoid bodies in vitro, which stain positive for markers of all three germ layers. About 90% of amniotic fluid stem cells express the Oct4 which is an important transcription factor expressed in the embryonal carcinoma cells, germ cells, and embryonic stem cells. However, unlike embryonic stem cells, amniotic fluid stem cells do not form teratoma when implanted into immune-deficient mice in vivo. This indicates that these cells are in the intermediate stage between embryonic stem cells and adult stem cellsand produce its positive effects. Therefore, the amniotic fluid stem cell is believed to be the safer and available source of the stem cell; they can be applied in regenerative therapies and without ethical issues.

3.4.4 Amnion

The human amnion is a highly elastic, transparent, and extraembryonic tissue that is derived from the epiblast by the end of the first gestational week. It is also called the amniotic membrane. Amnion is a thin film in the innermost of the fetal membranes, which basement membrane is composed of amniotic epithelium (AECs) and filled with the amniotic mesenchymal stem cells. The amnion stem cells mainly include the amniotic epithelial stem cells and amniotic mesenchymal stem cells. Amniotic epithelial cells, uniformly arranged on the basement membrane, are originated from the central region of the epiblast. The amniotic epithelial cells form the innermost layer of the amniotic membrane [34]. The term fetal membrane generally includes 50–70 million amniotic epithelial cells. The amnion stromal cells (ASCs) or amniotic mesenchymal stem cells (AM-MSCs) are located in the compact layer of the stromal matrix which is comprised of interstitial bundles of type I and III collagen [35]; they differentiated from somatopleuric mesodermal cells residing in the end of the epiblast. These cells can be isolated from the stromal matrix of the amniotic membrane by some method, such as enzymatic or mechanical method to remove the amniotic epithelial layer [36]. The stem cells from amnion have the characteristics of easily accessible, not pollution and without ethical issue. The amnion mesenchymal stem cells have a higher proliferative ability combined to the bone marrow stem cells. For this reason, the amnion stem cells become the mainly resource of cells in tissue engineering.

3.4.5 Placental Tissue

The placenta is composed of amniotic membrane, chorion frondosum, and decidua serotina. The fetal surface, a layer close to the fetal, is attached to the chorion. Amniotic membrane is a smooth and transparent thin film without blood vessel, nerve and lymphangion. The maternal surface, called the basal plate, is drilled out a lot of small holes by villus. It consists of a mixture of fetal extravillous trophoblastic cells with mesenchymal stromal cells, natural killer cells, macrophages, and other immune cell types. Chorion is the main part of the placental tissue. It is in the middle layer of the placental tissue. The placenta represents a highly specialized fetomaternal organ for the fetus and the parent to carry out the exchange of the substances and assures the normal growth and development of the fetus during gestation. During the gestational period, placental structures undergo continuous differentiation.

Massive stem cells can be isolated from the human placental tissue, such as amniotic epithelium, amniotic mesenchymal stem cells, chorionic mesenchymal stem cells, chorionic trophoblast stem cells, basal decidual mesenchymal stem cells, hematopoietic stem cells, and so on. The placenta-derived MSCs have some similar characteristics with bone marrow-derived MSCs. Due to the placental MSCs being in a special microenvironment, they also have some differences, e.g., the placental MSCs have higher capability in proliferation, differentiation, and migration. In addition, the expression of some stem cell makers (SOX2, OCT4, SSEA4) in placental MSCs is higher than bone marrow MSCs. Some research shows that the placental MSCs express the special marker frizzled 9 (FZD9, CD349); it provides a simply way to isolate MSCs from the placenta [37]. The placenta has an important role in immune defense. Recently, it was discovered that the placenta has another physiological role, as a hematopoietic organ, in which hematopoietic stem cells (HSCs) develop and expand. So, placenta is a new source for HSCs derivation.

3.5 Stemness of Perinatal Stem Cells

The perinatal tissue is believed to be the natural resources of the stem cell, which contains a variety of stem cell. The perinatal tissue contains sub-totipotent stem cell of embryonic stem cell-like, mesenchymal stem cell, hematopoietic stem cell, neuronal stem cell, epithelial stem cell, vascular stem cell, and so on. The perinatal stem cells have some unique properties, which are different from the embryonic stem cells and adult stem cells. Embryonic stem cells are pluripotent and can give rise to all cell types in the human body. Adult stem cells are multipotent and have the potential to give rise to all cell types of the lineage from which the adult stem cells are derived. The perinatal stem cell is an intermediate cell type that partially combines some multipotent properties of the adult stem cells with pluripotent properties of embryonic stem cells [38]. Almost all mesenchymal stem cells derived from the perinatal issue express a series of embryonic stem cell markers, such as SSEA-3, SSEA-4, Tra-1-60, Tra-1-81, Oct4, Nanog, and so on. Meanwhile, these cells also express markers of mesenchymal stem cells [39]. The hematopoietic stem cells and endothelial cells from perinatal issue whether express the markers of embryonic stem cells have not been reported. It means that perinatal stem cell is a kind of pluripotent stem cells with the potent ability, which is obtained from the embryonic issue during development of embryo. Researchers have recently discovered some signal transduction pathways play an important role in this process by influencing the expression of those transcription factors, such as Jak-STAT, MAPK-ERK, PI3K, WNT, TGFb, bFGF, and so on [40]. The stemness of stem cell is not only related to the above mentioned various stem cell transcription factors but also related to the characteristics of epigenetics, such as distinct miRNA profiles, DNA methylation, and the regulatory mechanism between histone modifications. The study in the stemness of perinatal stem cells is to elucidate the synergetic relationship between cell signal regulatory network mediated by the transcription factors and epigenetics modification [41].

The ability to self-renewal of perinatal HSC is lower than pluripotent stem cells, but the tissue specificity of differentiation is more significant. The hematopoietic stem cell from the umbilical cord has some special biological properties that are different from the hematopoietic stem cell from the bone marrow and peripheral blood [42], with the specific advanced properties of embryonic stem cells. Antigens CD34+ and CD133+ are taken as the important symbol of hematopoietic stem cells, especially the antigen CD34+. Proliferative potential and frequency of repopulating CD34 cells in human umbilical cord blood have been shown to be significantly higher as in the bone marrow or peripheral blood. In preparation for the physiological transition at birth, hematopoiesis of the human fetus greatly increases within the last term of pregnancy. The hematopoietic stem cells from perinatal can differentiate into the immune cells, including specific immunocyte and nonspecific immunocyte. The immunocyte is derived from umbilical cord blood in the state of relatively immature. For this reason, perinatal hematopoietic stem cells contribute to the lower incidence of graft-versus-host disease (GVHD) after transplantation comparable with unrelated stem cells derived from the bone marrow [43].

MSCs can differentiate into many cells in specific condition, such as chondrocyte, lipocyte, neurocyte, osteoblast, muscle cell, and so on. Moreover, MSCs play an important role in hematopoiesis, immunity, inflammatory response, angiogenesis, and some other physiologic functions. In terms of plasticity, mesenchymal stem cells from perinatal issues are multipotent and can be induced to form adipose tissue, bone, cartilage, skeletal muscle cells, cardiomyocyte-like cells, and neural cells [23]. Recent studies show that there are some differences and similarities in biology and function between perinatal mesenchymal stem cell (P-MSC) and bone marrow MSCs. P-MSC has higher proliferation ability, but lower osteogenic differentiation and immunogenicity [44]. Meanwhile, P-MSC has well hematopoietic and immune regulator function, and the mechanism is related to some factors and PEG2 from P-MSC. Human perinatal tissue is typically discarded after birth. So, these tissues can be effectively utilized for research and clinical applications without ethical issues. The P-MSC of the placenta has rich quantity; it has easy-to-scale production and industrialized operation. Therefore, perinatal mesenchymal stem cell is the best available source of stem cell drugs.

3.6 Conclusions and Perspectives

Perinatal tissues are a remarkable source of incredibly valuable stem cells. Perinatal stem cells have immune privileged characteristics, possess a broader multipotent plasticity, and proliferate faster than adult stem cells [45]. Moreover, because perinatal tissues are normally discarded after birth, these cells could be isolated while effectively avoiding ethical issues. The perinatal stem cells have some characteristics different from the embryonic stem cells and adult stem cells. The perinatal stem cells represent an intermediate cell type which has recently been described to combine qualities of both adult stem cell and embryonic stem cell and possess a broad multipotent plasticity. Although adult stem cells have achieved success clinically, its disadvantage is also very obvious. First, the adult stem cells possess a lower proliferation and differentiation ability and can't meet the required quantity in clinic. Second, it is harm to the donor when we obtained it from the donor. Third, adult stem cells isolated from the adult issues, so the risk of carrying the harmful microorganisms such as virus is higher. The perinatal stem cells have some advantages compared with adult stem cells. First, perinatal tissues are normally discarded after birth, which provide perfect resource for perinatal stem cell derivation while effectively avoiding ethical issues. Second, the perinatal stem cells represent an intermediate cell type which can differentiate into three layers: the ectoderm, mesoderm, and endoderm. Third, the proliferation ability of the perinatal stem cells is lower than embryonic stem cells, but higher than adult stem cells. Fourth, it is harmless to the donor. Fifth, due to the placental barrier, the placenta has a lower rate of carrying out the virus or some other harmful microbes. Sixth, perinatal stem cells, especially perinatal MSCs, have low immunogenicity and decrease the chances of immune rejection. The perinatal stem cells have those unique properties compared with other kinds of stem cells. The perinatal stem cells have been clinically succeeded; it presents the great therapeutic potential in clinical practice.

Acknowledgments This work was supported by grants from the National Natural Science Foundation of China (31771636, 81671734, and 81501528), National Key R&D Plan (2017YFA0103201 and 2011DAV00088), Tianjin Natural Science Foundation (18JCYBJC24400).

References

1. Witkowska-Zimny M, Wrobel E. Perinatal sources of mesenchymal stem cells: Wharton's jelly, amnion and chorion. Cell Mol Biol Lett. 2011;16(3):493–514.
2. Kurtzberg J, et al. Untying the Gordian knot: policies, practices, and ethical issues related to banking of umbilical cord blood. J Clin Invest. 2005;115(10):2592–7.
3. Martin GR. Isolation of a pluripotent cell line from early mouse embryos cultured in medium conditioned by teratocarcinoma stem cells. Proc Natl Acad Sci U S A. 1981;78(12):7634–8.
4. Thomson JA, et al. Embryonic stem cell lines derived from human blastocysts. Science. 1998;282(5391):1145–7.
5. Kingham E, Oreffo RO. Embryonic and induced pluripotent stem cells: understanding, creating, and exploiting the nano-niche for regenerative medicine. ACS Nano. 2013;7(3):1867–81.
6. Hadjimichael C, et al. Common stemness regulators of embryonic and cancer stem cells. World J Stem Cells. 2015;7(9):1150–84.
7. Fong H, et al. Regulation of self-renewal and pluripotency by Sox2 in human embryonic stem cells. Stem Cells. 2008;26(8):1931–8.
8. Rodda DJ, et al. Transcriptional regulation of Nanog by OCT4 and SOX2. J Biol Chem. 2005;280(26):24731–7.
9. Takahashi K, Yamanaka S. Induction of pluripotent stem cells from mouse embryonic and adult fibroblast cultures by defined factors. Cell. 2006;126(4):663–76.
10. Takahashi K, et al. Induction of pluripotent stem cells from adult human fibroblasts by defined factors. Cell. 2007;131(5):861–72.
11. Yu J, et al. Induced pluripotent stem cell lines derived from human somatic cells. Science. 2007;318(5858):1917–20.
12. Ye L, et al. Induced pluripotent stem cells and their potential for basic and clinical sciences. Curr Cardiol Rev. 2013;9(1):63–72.
13. Knoepfler PS. Deconstructing stem cell tumorigenicity: a roadmap to safe regenerative medicine. Stem Cells. 2009;27(5):1050–6.
14. Mariano ED, et al. Adult stem cells in neural repair: current options, limitations and perspectives. World J Stem Cells. 2015;7(2):477–82.
15. Clarke D, Frisen J. Differentiation potential of adult stem cells. Curr Opin Genet Dev. 2001;11(5):575–80.
16. Macdermot HE. The papers of Dr. Thomas Simpson (1833-1918). Can Med Assoc J. 1957;77(3):266–7.
17. Horwitz EM, et al. Transplantability and therapeutic effects of bone marrow-derived mesenchymal cells in children with osteogenesis imperfecta. Nat Med. 1999;5(3):309–13.
18. de Witte T, et al. Haematopoietic stem cell transplantation for patients with myelo-dysplastic syndromes and secondary acute myeloid leukaemias: a report on behalf of the Chronic Leukaemia Working Party of the European Group for Blood and Marrow Transplantation (EBMT). Br J Haematol. 2000;110(3):620–30.
19. Barcena A, et al. Human placenta and chorion: potential additional sources of hematopoietic stem cells for transplantation. Transfusion. 2011;51(Suppl 4):94S–105S.
20. Serikov V, et al. Human term placenta as a source of hematopoietic cells. Exp Biol Med (Maywood). 2009;234(7):813–23.
21. Weiss ML, et al. Immune properties of human umbilical cord Wharton's jelly-derived cells. Stem Cells. 2008;26(11):2865–74.
22. Karahuseyinoglu S, et al. Biology of stem cells in human umbilical cord stroma: in situ and in vitro surveys. Stem Cells. 2007;25(2):319–31.
23. Parolini O, et al. Concise review: isolation and characterization of cells from human term placenta: outcome of the first international workshop on placenta derived stem cells. Stem Cells. 2008; 26(2):300–11.
24. De Coppi P, et al. Isolation of amniotic stem cell lines with potential for therapy. Nat Biotechnol. 2007;25(1):100–6.
25. Si JW, Wang XD, Shen SG. Perinatal stem cells: a promising cell resource for tissue engineering of craniofacial bone. World J Stem Cells. 2015;7(1):149–59.
26. Baudin B, et al. A protocol for isolation and culture of human umbilical vein endothelial cells. Nat Protoc. 2007;2(3):481–5.
27. Wu S, et al. Microvesicles derived from human umbilical cord Wharton's jelly mesenchymal stem cells attenuate bladder tumor cell growth in vitro and in vivo. PLoS One. 2013;8(4):e61366.
28. Park DH, et al. Transplantation of umbilical cord blood stem cells for treating spinal cord injury. Stem Cell Rev. 2011;7(1):181–94.
29. Gucciardo L, et al. Fetal mesenchymal stem cells: isolation, properties and potential use in perinatology and regenerative medicine. BJOG. 2009;116(2):166–72.
30. Chen Y, et al. Mesenchymal stem cells: a promising candidate in regenerative medicine. Int J Biochem Cell Biol. 2008;40(5):815–20.
31. Kamath-Rayne BD, et al. Amniotic fluid: the use of high-dimensional biology to understand fetal well-being. Reprod Sci. 2014;21(1):6–19.
32. Prusa AR, Hengstschlager M. Amniotic fluid cells and human stem cell research: a new connection. Med Sci Monit. 2002;8(11):RA253–7.
33. Prusa AR, et al. Oct-4-expressing cells in human amniotic fluid: a new source for stem cell research? Hum Reprod. 2003;18(7):1489–93.
34. Niknejad H, et al. Properties of the amniotic membrane for potential use in tissue engineering. Eur Cell Mater. 2008;15:88–99.
35. Ilancheran S, et al. Stem cells derived from human fetal membranes display multilineage differentiation potential. Biol Reprod. 2007;77(3):577–88.
36. Miki T, Strom SC. Amnion-derived pluripotent/multipotent stem cells. Stem Cell Rev. 2006;2(2):133–42.

37. Battula VL, et al. Prospective isolation and characterization of mesenchymal stem cells from human placenta using a frizzled-9-specific monoclonal antibody. Differentiation. 2008;76(4):326–36.
38. Pappa KI, Anagnou NP. Novel sources of fetal stem cells: where do they fit on the developmental continuum? Regen Med. 2009;4(3):423–33.
39. Miao Z, et al. Isolation of mesenchymal stem cells from human placenta: comparison with human bone marrow mesenchymal stem cells. Cell Biol Int. 2006;30(9):681–7.
40. Chang F, et al. Signal transduction mediated by the Ras/Raf/MEK/ERK pathway from cytokine receptors to transcription factors: potential targeting for therapeutic intervention. Leukemia. 2003;17(7):1263–93.
41. Gifford CA, et al. Transcriptional and epigenetic dynamics during specification of human embryonic stem cells. Cell. 2013;153(5):1149–63.
42. Aljitawi OS. Ex vivo expansion of umbilical cord blood: where are we? Int J Hematol. 2012;95(4):371–9.
43. Rocha V, et al. Clinical use of umbilical cord blood hematopoietic stem cells. Biol Blood Marrow Transplant. 2006;12(1 Suppl 1):34–41.
44. Moll G, et al. Different Procoagulant activity of therapeutic Mesenchymal stromal cells derived from bone marrow and placental decidua. Stem Cells Dev. 2015;24(19):2269–79.
45. Marcus AJ, Woodbury D. Fetal stem cells from extraembryonic tissues: do not discard. J Cell Mol Med. 2008;12(3):730–42.

4 Quality Assessment of Umbilical Cord Tissue in Newborn Stem Cell Banking

Katherine Stewart Brown

Abstract

Umbilical cord tissue is rich in mesenchymal stromal cells and other stem and progenitor cell populations with potential therapeutic and research value. The umbilical cord tissue, like umbilical cord blood, can be collected in a noninvasive manner following delivery. Given the potential utilization of mesenchymal stromal cells, from a variety of perinatal and adult tissues, in regenerative medicine applications, a growing portion of private newborn stem cell banks offer cryopreservation of umbilical cord tissue, alongside cord blood, for future cell-based applications. Identification of consensus standards for objectively qualifying an umbilical cord tissue-based unit prior to advancing to clinical grade expansion will establish comparability in products prepared with different processing or cryopreservation methods and help standardize the industry. We summarize here the application of a nondestructive, systematic evaluation of umbilical cord tissue health that can be applied prior to processing or manipulation in order to establish comparability among umbilical cord tissue specimens. This or similar approaches are useful in product development efforts and quality control programs for newborn stem cell banking.

K. S. Brown (✉)
Cbr Systems, Inc., South San Francisco, CA, USA
e-mail: kbrown@cordblood.com

4.1 Newborn Stem Cell Banking

In the fall of 1988, the first successful transplant utilizing cord blood as a hematopoietic stem cell graft was performed for a boy with Fanconi's anemia [1]. Demonstration of the capacity of cord blood to reconstitute a patient's blood and immune system, in conjunction with the confirmation that cord blood hematopoietic stem and progenitor cells could be cryopreserved for future clinical use, led to the development of the cord blood banking industry in the 1990s [2–4]. The industry includes both public banks, which store cord blood units for use in an unrelated recipient, and private or family banks, which store cord blood for future potential use by the donor (autologous) or a first- or second-degree relative (allogeneic).

The first exploration of feasibility of the clinical application of autologous mesenchymal stromal cell (MSC) isolation and infusion was reported more than 20 years ago [5]. Since that time there has been a considerable amount of evidence suggesting their safety across a wide variety of conditions in both allogeneic and autologous settings, and preclinical studies suggest potential therapeutic benefits due primarily to trophic and immune modulatory effects; it is estimated that over 450 clinical trials are now evaluating MSCs as part of a therapeutic intervention, and MSCs are widely recognized to hold great promise in regenerative medicine [6].

MSCs can be found in many different tissues, including the bone marrow, adipose tissue, dental

Z. C. Han et al. (eds.), *Perinatal Stem Cells*, https://doi.org/10.1007/978-981-13-2703-2_4

pulp, and perinatal tissues such as umbilical cord blood, umbilical cord tissue, and placental tissue [7]. Interestingly, several groups have observed that MSC frequency and function are negatively impacted by increased age and certain chronic disease conditions [8, 9]. The enhanced proliferative capacity of MSCs isolated from newborn tissues, in combination with the reduced risk of exposure to virus and environmental toxins compared to aged adult tissues, makes the cryopreservation of newborn tissues especially attractive [10]. Umbilical cord tissue is collected following cord blood collection and does not impact cord blood unit quality, supporting the concept that multiple newborn tissues can be acquired from the same donor [11]. Given the interest and anticipated utilization of MSCs in a variety of clinical settings and the aforementioned advantages of tissue acquisition and cryopreservation of biologically young material, numerous private cord blood banks, and more recently some public banks, have or are establishing programs for umbilical cord tissue cryopreservation.

4.2 Umbilical Cord Tissue Banking

As many groups have reported, MSCs can be isolated from several locations in the umbilical cord: Wharton's jelly, which comprises the majority of the volume of the umbilical cord, and the perivascular, intravascular, and subamniotic regions (see, e.g., [12] and summarized in [7, 13]). While cells isolated from the perivascular region are phenotypically distinct from the MSCs isolated from the surrounding Wharton's jelly [14], these regions are relatively indiscrete, and there is insufficient preclinical evidence to determine if cells from a particular anatomical region should be prioritized for clinical applications [15]. In addition to MSCs, other cells with potential therapeutic value can be isolated from the vessels and amniotic membrane [16–19].

The most common approaches to isolating cells from umbilical cord tissue are enzymatic digestion and outgrowth from an explant established by segmenting the tissue into small pieces [7, 13]. Although enzymatic digestion is used by a number of facilities to isolate cells prior to cryopreservation, there remains a lack of standardization in both enzymes and duration of incubation times. Evidence suggests that suboptimal enzymatic digestion conditions may lead to cellular damage that some have postulated result in a concomitant negative impact on cellular function [7, 20, 21]. In contrast, isolation of cells by outgrowth from an explant avoids exposure to potentially detrimental agents and, importantly, can be performed on both freshly collected umbilical cord tissue and thawed, previously cryopreserved cord tissue. In fact umbilical cord tissue can be segmented, submerged in a DMSO-based cryoprotectant, cryopreserved, and subsequently thawed without compromising the ability to isolate cells by explant outgrowth nor any discernable functional impact to the cells themselves [21–24]. Lastly, several groups have recently reported on the use of tools to facilitate a nonenzymatic, mechanically based approach including benchtop devices for tissue dissociation (gentleMACS™ Dissociator; Miltenyi Biotec Inc.) or single-use disposable devices for solid tissue processing (AC:Px™; AuxoCell Laboratories Inc.) [25, 26]. The basic approaches summarized above can be further modified, and variations on these approaches have been summarized previously [7, 13]. Although the initial product of the aforementioned processes is heterogeneous, the predominant stem cell population isolated by these methods meets the criteria for defining MSCs [15, 26–28]. Greater purity can be achieved with the addition of a selection step, such as cell sorting or culture in a MSC supportive media.

With an estimated four million cord blood units stored worldwide across private and public banks, cord blood banking is arguably a driver in the larger field of cell and biobanking [2]. The establishment and continued refinement in the cord blood banking industry of standardized approaches for collection, processing, storage, and quality control help ensure the safety of cell therapy products stored with the intent of future clinical use. In the context of products stored for intended clinical applications, quality control can be defined as "Operational techniques and activities to monitor and eliminate the causes of unsatisfactory performance at any stage of a process" [29]. A particular advantage of incorporating umbilical cord tissue storage under the umbrella of cord blood banking is that product workflow can be modeled off of cord blood storage steps and existing processes and documentation can often be translated directly or adapted to accommodate the new product.

4.3 Quality Assessment of Umbilical Cord Tissue

Of the more than 100 private cord blood banks offering storage of an umbilical cord tissue product, close to half cryopreserve the tissue as a composite material while the remaining store a cellular product [30]. Utilization of an enzyme digestion or device for tissue dissociation to isolate umbilical cord cells pre-cryopreservation results in a single cell suspension, and the product is amenable to assessment of cell content and viability, as well as a quality monitoring workflow similar to what has been previously established for cord blood banking. As such, quality assessments, including cell dose or viability thresholds, can be readily applied.

Isolation of umbilical cord tissue cells by explant is robust to process variables and well suited to the remote collection model common among private banks [20]. One of the challenges inherent to this approach though is quality assessments reported in the literature are generally limited to cell yield resulting from explant outgrowth. We developed a scoring system that incorporates adherence of cells to culture plastic and evidence of cell proliferation in order to generate additional metrics, beyond cell yield, to utilize in quality assessments. Applying the aforementioned normalized scoring matrix to tissue explants plated in a uniform manner allows for a systematic, quantitative measurement of explant outgrowth that correlates with measurements of metabolic activity and cell yield and confirms previously qualitative evidence of biological variability between donors [11]. Once such a system is established internally, it can be utilized in statistical analysis to understand biological variability, and it simplifies the execution of processes relevant to quality control in a biobank setting, such as regular process control monitoring and evaluation or validation of proposed process changes. Specifically, this or a similar, quantitative assay can be used to establish baseline quality and biological variability of cord tissue units at receipt and, importantly, prior to any manipulation or destruction of the incoming material. This allows for a more accurate assessment of whether a proposed process change positively or negatively affects umbilical cord tissue quality because the contribution of potentially confounding variables has been identified and, if necessary, data can be normalized to an internal control. To illustrate, using a score ratio to normalize for biological variability in input material, we demonstrated that cryopreservation of umbilical cord tissue as a composite material does not compromise the ability to isolate cells by explant outgrowth [11].

There are technical aspects of the explant assay which, if not addressed, can contribute to variability in cell yield, complicating the standardization of quality assessments. Any variation of an explant assay that is used to gather quantitative information must take steps to ensure uniformity of the individual explant cultures, including reducing improper adherence of explants to the culture dish, an occurrence that can reduce cell recovery and confound results in comparative analysis. At our institution we have addressed this by adding a short, dry incubation time after plating and prior to media addition to help ensure tissue adherence, and explants are plated in a predefined, gridded pattern to monitor the location of explants and the resulting expected colonies. The Nagamura-Inoue group took a different approach, describing the use of a stainless steel mesh (Cellamigo®; Tsubakimoto Chain Co., Japan) that, when placed over tissue explants, ensures continued contact with the culture plastic helping to improve cell yield [31]. Both approaches are effective in reducing the potential of the small tissue fragments to lose contact with cell culture plastic thereby compromising the consistency of the explant assay.

As with any product development effort, it is important to examine external variables and internal manufacturing and storage steps for their potential to affect process yield or capability. For instance, some have reported challenges in isolating cells from umbilical cord tissue stored in a serum-containing cryoprotectant (see, for instance, [32]). However, a comparison of cryoprotectant formulations suggests that storage of umbilical cord tissue in a serum- and xeno-free cryoprotectant produces superior cell yield upon thawing and explant culture compared to serum-containing solutions [24]. Similarly, utilizing a serum- and xeno-free cryoprotectant (CryoStor® CS10, BioLife Solutions), we have found no evidence that cell yield from explant outgrowth is compromised when isolating cells from thawed, previously cryopreserved cord tissue [11].

4.4 Incorporation of Quality Assessment into Umbilical Cord Tissue Banking

The quality of the starting material and the steps involved in preparation for storage help define the utility of the post-thaw product [33]. We propose there is benefit in determining the quality of cord tissue intended for cryopreservation even if solely for the purpose of developing standard operating procedures, executing validations, and establishing metrics for quality control and process monitoring programs. These of course are part of the large picture of unit quality, which included programs for infectious disease marker testing and sterility monitoring.

Recognizing that there are areas of commonalities, some members of the cord blood banking community have advocated for establishing globally harmonized standards for umbilical cord blood and tissue products. Standards are challenging to implement at the time of storage given different approaches to quality assessment of umbilical cord tissue units cryopreserved as a cellular product and those stored as a composite material. However, standardization could be implemented after thawing of a cell suspension or, in the case of whole tissue cryopreservation, recovery of cells by explant outgrowth from thawed material. In practice, a panel of agreed upon assays could confirm that a product meets specified acceptance criteria, regardless of the cryopreservation approach, before moving to ex vivo expansion required to prepare clinically relevant cell doses. In this setting standardized assays performed prior to clinical grade expansion would complement, not replace, appropriate post-expansion critical quality parameters for identity, purity, potency, and sterility of a product. Further, as an industry, newborn stem cell banks have not yet established standard panels to routinely examine factors relevant to potential therapeutic benefits of MSCs such as immunomodulatory or angiogenic properties. Incorporation of functional assays will be an important consideration as clinical trials and the potential utilization of stored umbilical cord tissue material advances.

4.5 Regulatory Considerations

Products stored for intended clinical applications must comply with applicable standards and regulations. In the United States, cell- and tissue-based products are regulated by the Food and Drug Administration (FDA) utilizing a risk-based approach first outlined in 1997 and later codified under Title 21 of the Code of Federal Regulations (CFR), Part 1271 for human cells, tissues, and cellular- and tissue-based products (HCT/Ps) [34]. Additionally, the particular regulatory pathway for an HCT/P is based on classification as either a 351 or 361 product under the Public Health Services Act [34]. Briefly, a cellular therapy product which is (1) minimally manipulated, (2) intended for homolgous use, (3) not combined with a device or drug, except for sterilizing, preserving, or strage agents, and (4) has a systemic effect or is dependent on the metabolic actvity of cells and is for autologous use, allogeneic use in a first- or second- degree relative, or reproductive use presents less risk and is designated a "361 product", while a product that fails to meet the above criteria is desgignated as a "351 product" (additional details in [34]). Those HCT/Ps that meet the definition of a 361 have a less onerous path to clinical application in the United States as, for example, premarket approval is not required, and the regulatory focus is on protecting recipients against the transmission of infectious diseases [35]. In contrast, 351 products are considered biological drugs and regulated as such, including requiring an Investigational New Drug application. This is of relevance to umbilical cord tissue banking and the various methods being explored for preparing the tissue or cellular component for cryopreservation. The FDA defines minimal manipulation as:

1. For structural tissue, processing that does not alter the original relevant characteristics of the tissue relating to the tissue's utility for reconstruction, repair, or replacement
2. For cells or nonstructural tissues, processing that does not alter the relevant biological characteristics of cells or tissues

(21 CRF 1271.3(f)). The FDA recently published draft guidance documents which, in part, are intended to clarify their position on these definitions and other topics related to the regulation of HCT/Ps (see, e.g., Minimal Manipulation of Human Cells, Tissues, and Cellular- and Tissue-Based Products (2015)). Based on the draft guidances and examples contained therein, it seems logical that the application of collagenase or extensive mechanical dissociation to isolate cells in efforts to facilitate storage of a cell suspension would result in a product considered to be more than minimal manipulated as it alters the original characteristics of the umbilical tissue itself. Isolation of cells from thawed, previously cryopreserved umbilical cord tissue by explant outgrowth may face the same regulatory pathway. Although the thawed tissue itself is likely not more than minimally manipulated, there is the potential that the culturing of cells during explant isolation may result in a product that FDA considers to be more than minimally manipulated. In September of 2016, the FDA held a 2-day public hearing to elicit feedback from the public on the draft guidances and published the finalized guidance documents the following year. This was an unusual step for the FDA, and many stakeholders appreciated the opportunity for additional dialog with the regulatory agency.

4.6 Conclusion

As others have noted, consistency in product manufacturing of MSCs is critical, perhaps especially so in the setting of private banking, where every unit is inherently unique due to biological variability and reference standards may be of particular use [36]. The approaches discussed here are worth considering regardless of whether a newborn stem cell bank offers storage of umbilical cord tissue as a single cell suspension or as a composite material. The health and quality of the umbilical cord tissue prior to any manipulation are rarely quantified; one can argue that in product development, the impact of processing step variables, without a baseline assessment of incoming quality, is somewhat incomplete. Irrespective of the processing method adopted, incorporating an assay to determine health at receipt by the banking facility allows for reliable and consistent quantification, and thereby determination of baseline equivalence, of the quality of umbilical cord tissue units prior to manipulation and without destruction of the product. This information can be used to develop an internal reference standard with the potential to enhance comparability in product development efforts and quality control processes, as well as to provide a frame of reference useful in inter-facility collaborations.

Institutions that have or are in the process of establishing an umbilical cord tissue product will identify approaches appropriate for their organization and define optimal conditions for processing and cryopreservation. It is incumbent upon the institution to address technical challenges associated with the chosen procedural method in order to demonstrate reproducible product consistency and quality throughout transport, processing, and storage. This should be done in conjunction with the development of standard operating procedures and execution of validations with appropriate and specified acceptance criteria, in accordance with regulatory requirements and industry best practice.

References

1. Gluckman E, Broxmeyer HA, Auerbach AD, et al. Hematopoietic reconstitution in a patient with Fanconi's anemia by means of umbilical-cord blood from an HLA-identical sibling. N Engl J Med. 1989;321:1174–8. https://doi.org/10.1056/NEJM198910263211707.
2. Ballen KK, Verter F, Kurtzberg J. Umbilical cord blood donation: public or private? Bone Marrow Transplant. 2015;50:1271. https://doi.org/10.1038/bmt.2015.124.
3. Broxmeyer HE, Lee M-R, Hangoc G, et al. Hematopoietic stem/progenitor cells, generation of induced pluripotent stem cells, and isolation of endothelial progenitors from 21- to 23.5-year cryopreserved cord blood. Blood. 2011;117:4773–7. https://doi.org/10.1182/blood-2011-01-330514.
4. Broxmeyer HE, Douglas GW, Hangoc G, et al. Human umbilical cord blood as a potential source of transplantable hematopoietic stem/progenitor cells. Stem Cells. 1989;86:3828–32. https://doi.org/10.1634/stemcells.21-2-217.
5. Lazarus HM, Haynesworth SE, Gerson SL, et al. Ex vivo expansion and subsequent infusion of human bone marrow-derived stromal progenitor cells (mesenchymal progenitor cells): implications for therapeutic use. Bone Marrow Transplant. 1995;16:557–64.
6. Caplan AI, Hariri R. Body management: mesenchymal stem cells control the internal regenerator. Stem Cells Transl Med. 2015;4:695. https://doi.org/10.5966/sctm.2014-0291.
7. Batsali AK, Kastrinaki MC. Mesenchymal stem cells derived from Wharton's jelly of the umbilical cord: biological properties and emerging clinical applications. Curr Stem Cell Res Ther. 2013;8:144–55.
8. Choudhery MS, Badowski M, Muise A, Pierce J. Donor age negatively impacts adipose tissue-derived mesenchymal stem cell expansion and differentiation. J Transl Med. 2014;12:8.
9. Caplan AI. Why are MSCs therapeutic? New data: new insight. J Pathol. 2009;217:318–24.
10. Capelli C, Gotti E, Morigi M, et al. Minimally manipulated whole human umbilical cord is a rich source of clinical-grade human mesenchymal stromal cells expanded in human platelet lysate. Cytotherapy. 2011;13:786. https://doi.org/10.3109/14653249.2011.563294.
11. Skiles ML, Brown KS, Tatz W, et al. Quantitative analysis of mesenchymal stromal cell isolation from umbilical cord tissue cryopreserved as a composite material at a stem cell bank. Cytotherapy. 2018;20(4):564–75.
12. Ishige I, Nagamura-Inoue T, Honda MJ, et al. Comparison of mesenchymal stem cells derived from arterial, venous, and Wharton's jelly explants of human umbilical cord. J Transl Med. 2009;90:261–9. https://doi.org/10.1007/s12185-009-0377-3.
13. Bongso A, Fong C-Y. The therapeutic potential, challenges and future clinical directions of stem cells from the Wharton's jelly of the human umbilical cord. Stem Cell Rev Rep. 2013;9:226–40. https://doi.org/10.1007/s12015-012-9418-z.
14. Sarugaser R, Lickorish D, Baksh D, et al. Human umbilical cord perivascular (HUCPV) cells: a source of mesenchymal progenitors. Stem Cells. 2005;23:220–9. https://doi.org/10.1634/stemcells.2004-0166.
15. Maslova O, Novak M, Kruzliak P. Umbilical cord tissue-derived cells as therapeutic agents. Stem Cells Int. 2015;2015:150609. https://doi.org/10.1155/2015/150609.
16. Reza HM, Ng B-Y, Gimeno FL, et al. Umbilical cord lining stem cells as a novel and promising source for ocular surface regeneration. Stem Cell Rev Rep. 2011;7:935. https://doi.org/10.1007/s12015-011-9245-7.
17. Reza HM, Ng B-Y, Phan TT, et al. Characterization of a novel umbilical cord lining cell with CD227 positivity and unique pattern of p63 expression and function. Stem Cell Rev Rep. 2010;7:624. https://doi.org/10.1007/s12015-010-9214-6.
18. Hayward CJ, Fradette J, Galbraith T, et al. Harvesting the potential of the human umbilical cord: isolation and characterisation of four cell types for tissue engineering applications. Cells Tissues Organs. 2013;197:37–54. https://doi.org/10.1159/000341254.
19. Lim IJ, Phan TT. Epithelial and mesenchymal stem cells from the umbilical cord lining membrane. Cell Transplant. 2014;23:497. https://doi.org/10.3727/096368914x678346.
20. Iftimia-Mander A, Hourd P, Dainty R, Thomas RJ. Mesenchymal stem cell isolation from human umbilical cord tissue: understanding and minimizing variability in cell yield for process optimization. Biopreserv Biobank. 2013;11:291–8. https://doi.org/10.1089/bio.2013.0027.
21. Nagamura T, He H. Umbilical cord-derived mesenchymal stem cells: their advantages and potential clinical utility. World J Stem Cells. 2014;6:195–202.
22. Choudhery M, Badowski M, Muise A, Harris D. Utility of cryopreserved umbilical cord tissue for regenerative medicine. Curr Stem Cell Res Ther. 2013;8:370–2. https://doi.org/10.2174/1574888x11308050004.
23. Friedman R, Betancur M, Boissel L, et al. Umbilical cord mesenchymal stem cells: adjuvants for human cell transplantation. Biol Blood Marrow Transplant. 2007;13:1477. https://doi.org/10.1016/j.bbmt.2007.08.048.
24. Shimazu T, Mori Y, Takahashi A, et al. Serum- and xeno-free cryopreservation of human umbilical cord tissue as mesenchymal stromal cell source. Cytotherapy. 2015;17:593–600. https://doi.org/10.1016/j.jcyt.2015.03.604.
25. Taghizadeh R. Perinatal cells and the landscape of ancillary cord blood banking services. [Webinar]. Recording of May 12, 2016 AABB Cellular Therapies Webinar; 2016.

26. Emnett RJ, Kaul A, Babic A, et al. Evaluation of tissue homogenization to support the generation of GMP-compliant mesenchymal stromal cells from the umbilical cord. Stem Cells Int. 2016;2016:1. https://doi.org/10.1155/2016/3274054.
27. Mack A, Brown KS, Faust E, et al. GMP-compatible iPSC derivation from human umbilical cord blood and tissue across multiple donors. ISSCR Abstract W201; 2016.
28. Dominici M, Le Blanc K, Mueller I. Minimal criteria for defining multipotent mesenchymal stromal cells. The International Society for Cellular Therapy position statement. Cytotherapy. 2006;8:315–7.
29. Cardwell L, Sugrue MW. Quality and process control. In: Areman EM, Loper K, editors. Cellular therapy: principles, methods, and regulations. 2nd ed. Bethesda: AABB Press; 2016. p. 85–104.
30. Verter F, Couto PS. Survey of how cord blood banks process cord tissue. ICBS abstracts. Transfusion. 2016;56:1A–12A. https://doi.org/10.1111/trf.13686.
31. Mori Y, Ohshimo J, Shimazu T, et al. Improved explant method to isolate umbilical cord-derived mesenchymal stem cells and their immunosuppressive properties. Tissue Eng Part C Methods. 2015;21:367–72. https://doi.org/10.1089/ten.tec.2014.0385.
32. Chatzistamatiou TK, Papassavas AC. Optimizing isolation culture and freezing methods to preserve Wharton's jelly's mesenchymal stem cell (MSC) properties: an MSC banking protocol validation for the Hellenic Cord Blood Bank. Transfusion. 2014;54:3108–20.
33. Akel S. Cell banking: process development and cell preservation. Translational regenerative medicine. 2015. https://doi.org/10.1016/c2012-0-06956-6.
34. Harvath L. A brief history of the US FDA regulation of human cells and tissues. In: Areman EM, Loper K, editors. Cellular therapy: principles, methods, and regulations. 2nd ed. Bethesda: AABB Press; 2016. p. 2–7.
35. Eisenstein M. Regulation: rewriting the regenerative rulebook. Nature. 2016;540:S64. https://doi.org/10.1038/540s63a.
36. Viswanathan S, Keating A, Deans R, et al. Soliciting strategies for developing cell-based reference materials to advance mesenchymal stromal cell research and clinical translation. Stem Cells Dev. 2014;23:1157. https://doi.org/10.1089/scd.2013.0591.

5 Safety and Genetic Stability of Cultured Perinatal Mesenchymal Stem Cells

Youwei Wang

Abstract

Perinatal mesenchymal stem cells are considered as a promising tool to treat many diseases in clinical studies. Being different with chemical or protein drugs, perinatal stem cells were used as living cells. No method is well established to evaluate the safety issues of perinatal mesenchymal stem cells. This is a challenge not only for patients and physicians but also for manufacturers and administrators. Safety issues of perinatal stem cells could come from perinatal stem cells themselves, which include, but not limited to, cytokine storm, unwanted differentiation, and tumorigenesis. Other safety concerns of perinatal stem cells are related to processing and quality control when processing or manufacturing mesenchymal stem cells. After in vitro expansion, which is inevitable for clinical application, perinatal mesenchymal stem cells may change their biological characteristics. So it is critical to answer whether in vitro culture decreases the clinical value of perinatal mesenchymal stem cells or increases any risk for the clinical application. Preclinical studies and clinical data, quality control, and potential risk factors for manufacturing and clinical application will be discussed in this chapter.

Y. Wang (✉)
Cell Therapy Lab, James Cancer Hospital, Ohio State University, Columbus, OH, USA

Perinatal tissues are an important source of mesenchymal stem cells. Unlike bone marrow derived mesenchymal stem cells which need to take bone marrow from health donors, perinatal mesenchymal stem cells usually come from umbilical cord, placenta or amniotic fluid, which are discarded tissues after birth. This makes collection of perinatal mesenchymal stem cells easier, safer, noninvasive, and painless. Perinatal-derived mesenchymal stem cells showed better proliferation and in vitro life span; thus more cells can be gained from one donor than adipose tissue or bone marrow-derived mesenchymal stem cells. Additionally, the frequency of primary mesenchymal stem cells in perinatal tissues is higher than adult tissues [1]. Nevertheless, the mount of primary mesenchymal stem cells, even from perinatal tissues, is not enough for clinical application. So in vitro expansion is necessary for clinical application of mesenchymal stem cells. Fortunately, perinatal mesenchymal stem cells can be expanded in vitro, without losing multipotency and the ability of immuneregulation, which was evidenced by many preclinical or clinical studies. However, the safety of perinatal mesenchymal stem cells, in vitro cultured perinatal mesenchymal stem cells in particular, are always critical for the clinical applications, and more attention should be paid. In this chapter, we will discuss recently studies focused on the safety evaluation which should be respected for perinatal stem cell-based cell therapies.

Z. C. Han et al. (eds.), *Perinatal Stem Cells*, https://doi.org/10.1007/978-981-13-2703-2_5

5.1 Safety Analysis of Perinatal Mesenchymal Stem Cells in Preclinical Studies

Preclinical safety study should be performed before perinatal mesenchymal stem cells are used in clinical trials. Perinatal mesenchymal stem cells belong to mesenchymal stem cells. The safety issues of mesenchymal stem cells were evaluated in many in vivo studies. Considering the difference between human and other animals, the safety evaluation of human cells should be performed in different species. Actually, safety issues of mesenchymal stem cells were investigated in many different animal models, including mice, rats, cats, canines, rabbits, horses, and monkeys. Mesenchymal stem cells used in all of these studies, most of them are derived from perinatal tissues, and are expanded in vitro.

No mesenchymal stem cell-related adverse event was observed in the studies using rodent animal models, which includes mice and rats. To reduce immune rejection between different species, human mesenchymal stem cell safety studies were performed in immune-deficient mice. Human mesenchymal stem cells can survive in immune-deficient mice as long as 8 months. No malignant transformation or any adverse event to the recipients was found, even when very high dose (2.5×10^8 cells/kg body weight) of mesenchymal stem cells was administrated [2]. In most of clinical trials, patients received $1–2 \times 10^6$ cells/kg body weight mesenchymal stem cells [3] or less [4]. In diabetic nude rat model, intramuscular injection of placenta mesenchymal stem cells seems safe and alleviates critical limb ischemia [5]. Besides intravenous injection, safety of Wharton's jelly-derived mesenchymal stem cells was evaluated in retinal degeneration rat model. Wharton's jelly mesenchymal stem cells were subretinally injected in Royal College of Surgeons rats. Wharton's jelly mesenchymal stem cells do not migrate systemically. No tumorigenesis was observed in this study, which means that subretinal injection of Wharton's jelly mesenchymal stem cells was safe [6].

Safety of intraperitoneal injection of mesenchymal stem cells was evaluated in cats. No severe adverse effects were observed in mesenchymal stem cell-treated cats. Only two cats treated with mesenchymal stem cells appear lethargic and less interactive. Both of them recovered without any treatment in 3 days [7]. Safety study in equine showed that umbilical cord blood mesenchymal stem cell treatment promotes recovery from tendon and ligament injuries. No adverse effect related to the treatment was observed in this study [8]. In rabbit models, umbilical cord-derived mesenchymal stem cells were evaluated in osteochondral defect disease or anterior cruciate ligament reconstruction. Umbilical cord mesenchymal stem cells promote cartilage repair and are beneficial for anterior cruciate ligament reconstruction. No immune rejection or other server adverse event was observed in these rabbit model-based preclinical studies [9, 10]. Similar data were also observed in minipig model [11]. In a canine model focused on mesenchymal stem cell-related bone repair, transplantation of mesenchymal stem cells promotes new bone formation without any nonspecific differentiation or migration [12].

A toxicity study of human umbilical cord mesenchymal stem cells distinguished itself by using healthy cynomolgus monkeys, which possess a close evolutionary relationship to human [13]. The mesenchymal stem cells are isolated, expanded, and prepared in a clean room facility following the GMP (good manufacturing practice) guidance. The safety issues are evaluated in a GLP (good laboratory practice) laboratory. Two doses, 2×10^6 and 2×10^7 cells/kg body weight, were evaluated in this study. Toxicity analyses include body weight, body temperature, urinalysis, lymphoproliferation, blood cell counts, biochemistry analyses and other safety related issues. No mesenchymal stem cell-related adverse effects were observed in this toxicity study in cynomolgus monkeys. Another question is how long transplanted mesenchymal stem cells can survive in the recipients and whether they can differentiate to other functional cells. In principle, allologous cells cannot survive in immune-sufficient recipients. Whether the clinical effects of mesenchymal stem cells come from secreting cytokine or differentiating into functional cells is a controversial question. Whereas, in this study, concentration of mesenchymal stem cells in blood culminate at 30 min after transplantation and become almost undetectable in 8 h. Two weeks after transplantation, all tissues of the cynomolgus monkeys are negative for transplanted mesenchymal stem cells, which was measured by RT-PCR targeting human genetic sequence. Together, preclinical studies showed that perinatal mesenchymal stem cells are safe and worth for further clinical trial.

5.2 Safety Evaluation of Perinatal Mesenchymal Stem Cells in Clinical Trials

Perinatal mesenchymal stem cells were used in many clinical trials for treating diverse diseases. Few adverse events were reported in these clinical trials of perinatal mesenchymal stem cells. In a clinical trial where placenta-derived mesenchymal stem cells were used to treat type 2 diabetes, daily insulin requirement was decreased after placenta mesenchymal stem cell transplantation. No adverse event was observed. Furthermore, renal and cardio function was improved in some patients [3]. Similar data were observed in a clinical study using Wharton's jelly-derived mesenchymal stem cells to treat type 2 diabetes [14]. Data from these clinical trials support that perinatal mesenchymal stem cells are safe for treating diabetes patients.

As perinatal mesenchymal stem cells show great immunomodulation activities in in vitro assays [1] and animal experiments [15], many clinical trials of perinatal mesenchymal stem cells focus on autoimmune diseases. In a clinical study, where 21 severe aplastic anemia patients received haploidentical hematopoietic stem cell transplantation without T cell depletion, co-transplantation of third-party donor-derived umbilical cord mesenchymal stem cells reduces the risk of graft failure and severe GvHD (graft-versus-host disease). Although some hematopoietic stem cell transplantation-related issues were observed, such as GvHD, no umbilical cord mesenchymal stem cell transplantation-related adverse event was found in this clinical trial [4]. Another study enrolled 17 severe aplastic anemia patients got similar data [16]. Together, these clinical investigations support that co-transplantation of umbilical cord mesenchymal stem cells is not only safe but also reduces the risk of graft failure and incidence of severe GvHD. In a clinical study where long-term safety of treating refractory systemic lupus erythematosus patients with umbilical cord mesenchymal stem cells was evaluated, only one of seven patients had mild, dizzy, and warm sensation after mesenchymal stem cell administration. These symptoms disappeared without any treatment in 10 min. No other adverse event was observed in this 6-year follow-up study. No change of serum tumor markers was found before and 6 years after mesenchymal stem cell transplantation [17]. In a multicenter clinical trial, 40 systemic lupus erythematosus patients were recruited and treated with umbilical cord mesenchymal stem cells. Data from the four clinical centers showed satisfactory clinical response to umbilical cord mesenchymal stem cell treatment in systemic lupus erythematosus patients. No umbilical cord mesenchymal stem cell transplantation-related adverse event was observed in this multicenter clinical study [18]. Clinical studies also revealed that transplantation of umbilical cord mesenchymal stem cells made clinical improvement of inflammatory bowel disease. Insomnia, low fever, and feeling hot in the face were reported in some of the patients who received 1×10^6 cells/kg body weight. Without any specific treatment, all of these symptoms restored spontaneously. So, based on these data, umbilical cord mesenchymal stem cells are safe to treat inflammatory bowel disease patients [19].

In a clinical trial, which treated decompensated liver cirrhosis patients with umbilical cord mesenchymal stem cells, liver function was improved, while ascites was reduced. Within 6 h after administration, mild fever was observed in 4 of 30 patients who received umbilical cord mesenchymal stem cells transfusion. This symptom recovered within 12 h without any specific treatment. No other adverse effect was reported in this clinical trial [20]. For the acute-on-chronic liver failure patients, umbilical cord mesenchymal stem cell treatment significantly improves overall survival rate and liver functions. Umbilical cord mesenchymal stem cell transplantation-related mild fever was also observed in these clinical trials. It is not observed in the control group which did not received umbilical cord mesenchymal stem cell treatment [21, 22]. These data show that umbilical cord mesenchymal stem cells are safe for treating decompensated liver cirrhosis and acute-on-chronic liver failure patients. Together, increasing data of clinical studies support that perinatal mesenchymal stem cells are safe for clinical applications.

5.3 Potential Safety Concerns and Genetic Stability

Clinical grade mesenchymal stem cells are subject to microbiological contamination; thus effective procedures should be established to maintain sterility of mesenchymal stem cells, which is extremely important for the safety issues of clinical application. Manufacturing and quality control should start from donor screening through final product administration. All of the manufacture processing, examination, equipment, environment, and personnel of the manufacturer should be compliant with GMP or GTP (good tissue practice) requirements. Actually, most GTP requirements are covered by GMP. The regulation and control of manufacturing and quality will be discussed in other chapters.

So far, no universal standards for in vitro expansion of clinical grade mesenchymal stem cells were established, and in many countries adequately controlled bovine serum is allowed in cell therapy which includes mesenchymal stem cells. Nevertheless, it does not mean that we can disregard the safety issue of bovine serum. Usually, bovine serum will be added to the medium when culturing mesenchymal stem cells in vitro. Serum is a putative source of prion or virus transmission and ill-defined, which could be considered as an obstacle to clinical application of mesenchymal stem cells. Although rigorous quality control can minimize the risk of contamination by pathogens, it can not exclude the unknown or undetectable zoonotic disease completely. The bath-to-bath variability of serum, which can be amplified in final products of mesenchymal stem cells, tremendously impairs the manufacturing control. Some studies replace serum with substances derived from bovine serum, such as bovine serum albumin. Bovine serum albumin is much safer than serum, and the variation of bovine serum albumin between different batches is much smaller than that of bovine serum. However, medium containing bovine serum albumin is not animal-derived component-free either. Other studies use human AB serum or substances derived from human blood, platelet lysates in particular, to replace serum in in vitro expansion of mesenchymal stem cells. Using human serum or substances derived from human is prohibited in some countries (e.g., China). As the risk of contamination with pathogens, which may transmit much easier from person to person than from bovine to human, can not be completely eliminated by donor screening or testing for communicable diseases, human serum or its extract should not be considered as a safer source than bovine serum. Besides, the preparation of platelet lysates needs to pool buffy coat or thrombocyte derived from different donors, any undetected pathogen from one donor will contaminate the medium and the mesenchymal stem cells. So serum-free, chemical-defined medium for mesenchymal stem cells is preferred, which will conquer all the drawback of bovine serum.

Serum-free medium-derived mesenchymal stem cells may possess some different biological characteristics, compared with mesenchymal stem cells cultured in serum-containing medium. Usually, serum-free medium will not change the multilineage differentiation potential of mesenchymal stem cells. Some studies found that serum-free medium may change the cell surface markers of mesenchymal stem cells [23, 24]. However, there is no evidence to support that change of surface markers lead to decreased clinical potential [25]. Serum-free medium-derived mesenchymal stem cells may gain different epigenetic status compared with serum-containing medium-derived mesenchymal stem cells [26]. A study focused on canine and equine mesenchymal stem cells found that serum-free medium may affect the expression of PGE2 (prostaglandin E2) and the immunoregulating ability [27]. In serum-free medium, human umbilical cord mesenchymal stem cells may change their biological characteristics, growth rate, and in vitro lifespan in particular. However, they can still repress IFN-gamma production by activated T cells, being similar with serum-containing medium-derived mesenchymal stem cells [25]. Based on the current data, serum-free medium improved the safety of mesenchymal stem cells for clinical application without impairing the efficacy. There are different commercial available serum-free media for mesenchymal stem cells. They may be prepared based on different recipes and consisted of different compounds. Further investigation is needed to evaluate the inconsistent data derived from different serum-free medium for mesenchymal stem cells. After that, we can figure out which serum-free medium is the best choice for expanding mesenchymal stem cells for clinical application. Besides mesenchymal stem cell culture medium, bovine serum is also added in cryoprotectants. Some studies investigated the serum-free cryopreservation for mesenchymal stem cells. Comparing serum-free medium, serum-free cryoprotectants work better and may not change the biological characteristics or cause any safety issue. Now, many commercial serum-free cryoprotectants are available. They improved the safety of clinical grade mesenchymal stem cells without changing their biological characteristics which include multilineage potential, surface markers, and immunoregulation ability.

Genetic stability is a critical safety issue for cell therapy products, including in vitro expanded stem cells. Genetic instability is considered as a hallmark of cancer, and in vitro expanded stem cells with genetic mutation could gain growth advantage and may cause cancer after transplantation. In in vitro expansion, karyotypic instability was found in embryonic stem cells by traditional karyotyping [28–31], which means that the mutation in embryonic stem cell chromosome is huge and easy to detected. Considering the ethical issues, iPSCs (induced pluripotent stem cells) are a great candidate for replacing embryonic stem cells. However, the studies focusing on genetic stability of iPSCs reveal that genomic instability is also observed in the generation and in vitro culture of iPSCs [32]. Although many studies were devoted to improve the genomic stability of iPSCs, so far maintaining genomic stability when erasing the identity of a terminal-differentiated cell to acquire pluripotency may be a mission impossible [33].

The frequency and size of genomic mutation in the culture of mesenchymal stem cells are much smaller than those of embryonic stem cells and iPSCs. Rare chromosomal change is observed in in vitro expanded mesenchymal stem cells [34]. Usually, the genomic instability of mesenchymal stem cells consists of copy number variations (CNVs) and single-nucleotide changes (SNCs), which can be detected by aCGH (array-based comparative genomic hybridization) or whole genome sequencing separately. Using aCGH, CNVs are observed at late passage (P30) in seven of nine clone mesenchymal stem cells. The other two clones do not show CNVs at P30. Transcriptome analysis revealed that genes involved in cell cycle control and DNA damage response were downregulated in the clones showing poor genomic stability; meanwhile, these pathways were upregulated in the clones with good genomic stability [34]. Another study based on whole genome sequencing found that SNCs, which preexist in uncultured mononuclear cells with low frequency, will reach 17–36% in late passage (P13) [35]. Another more important question is the biological significance of the mutations in in vitro expanded mesenchymal stem cells. A study focused on umbilical cord-derived mesenchymal stem cells just addressed this question. In this study, even umbilical cord mesenchymal stem cells with the worst genomic stability cannot form tumor in immune-deficient mice nor become immortal. All of the umbilical cord mesenchymal stem cells, with or without genomic mutation, will senesce and stop proliferating after a period of in vitro expansion, which varies between 20 and 45 passages.

Umbilical cord mesenchymal stem cells derived from different donors differ greatly in genomic stability. If the genomic stability can be evaluated before transplantation, it should improve the safety of mesenchymal stem cells. Actually, we can freeze perinatal mesenchymal stem cells at early passage in master cell banks and then take one vial of them to do long-term in vitro culture. The genomic stability can be evaluated at late passage by aCGH or sequencing. The perinatal mesenchymal stem cell clones with no or few genetic mutations should be a better candidate for clinical application, compared with the perinatal mesenchymal stem cell clone with huge genetic aberration at late passage. So far there is not any study about the threshold or standard about genetic stability for mesenchymal stem cells' clinical application. Based on previous study, we may use 10 CNVs as a temporary threshold. Mesenchymal stem cells with less than 10 CNVs can be considered as good in genomic stability. This is not a reasonable standard; however, it is better than none and helps us to exclude mesenchymal stem cells with huge genetic mutation in in vitro culture from clinical transplantation.

Murine bone marrow mesenchymal stem cells can proceed to a malignant transformation in in vitro culture. Transformed murine mesenchymal stem cells gained increased c-myc expression and generate fibrosarcomas in vivo. Karyotyping showed that chromosomal abnormalities were accumulated in transformed murine mesenchymal stem cells [36]. In a study focused on cynomolgus macaques, spontaneous transformation was also observed in in vitro cultured bone marrow-derived mesenchymal stem cells. Transformed cynomolgus macaque bone marrow mesenchymal stem cells lost multiple differentiation potentials, and telomerase activity increased significantly in transformed bone marrow-derived mesenchymal stem cells. Subcutaneous tumors were observed in NOD/SCID mice inoculated with transformed bone marrow-derived mesenchymal stem cells [37, 38].

Many studies investigated malignant transformation of human mesenchymal stem cells. First, spontaneous transformation was reported in adipose-derived mesenchymal stem cells. However, short tandem repeat (STR) analysis revealed that the so-called transformed mesenchymal stem cells are contaminating tumor cell lines which are manipulated in the lab [39–43]. Recently, malignant transformation was observed in the human bone marrow and liver-derived mesenchymal stem cells, which was confirmed by STR analysis. So the possibility of transformation of human mesenchymal stem cells can not be completely eliminated. However, the chance should be very low. As in at least three individual studies, which investigate mesenchymal stem cells derived from more than 50 donors, malignant transformation of mesenchymal stem cells in in vitro expansion was not observed in these studies [34, 38, 40]. Taking all the reports together, the malignant transformation of human mesenchymal stem cells in in vitro expansion, if it exists, seems to be a little probability event. However, we can not ignore the possibility of malignant transformation. If we take the multiple-hit hypothesis [44], adult mesenchymal stem cells may accumulate more mutation before in vitro expansion, comparing with perinatal mesenchymal stem cells. In this point, perinatal mesenchymal stem cells should be a better candidate for clinical application. Whether a mesenchymal stem cell clone undergoes malignant transformation seems dependent on the original culture, or the donor, which is supported by the finding that umbilical cord mesenchymal stem cells derived from different donors possessed variant genomic stability. This implies that we can choose the cells with good genomic stability and without malignant transformation in in vitro culture. For example, we can build a master cell bank, in which mesenchymal stem cells were expanded and frozen at any early passage. One vial of mesenchymal stem cells in master cell bank will be cultured till senescence. The in vitro life span varies among mesenchymal stem cells derived from different tissues. For umbilical cord mesenchymal stem cells, they will get senescence and stop proliferating at 20–40 passages. So if the in vitro life span of a umbilical cord mesenchymal stem cell clone exceeds too much of 40 passages, it probably undergoes malignant transformation. Besides, the genomic stability of umbilical cord mesenchymal stem cells can also be evaluated by aCGH between mesenchymal stem cells in master cell bank and after long-term in vitro culture. Only the cells with good genomic stability and without malignant transformation are safe for clinical application. Then we can manufacture cell therapy products using these master cell banks.

Just like any other drugs, mesenchymal stem cell-based cell therapy should have its own indications and contraindications. Cancer may be a major contraindication for mesenchymal stem cell-based cell therapy, as many studies have showed that mesenchymal stem cells can promote tumor cell growth and metastasis. Mesenchymal stem cells are found to localize in tumor stroma or tumor environment, and they can promote tumor cell survival and increase drug resistance [45], which is observed at least in breast carcinomas [46], osteosarcoma [47], head and neck squamous cell carcinoma [48], hepatic tumors [49], pancreatic tumor [50], ovarian cancer [51], melanoma [52], and colon cancer [53]. In animal model, co-injection tumor cells with umbilical cord mesenchymal stem cells will promote tumor cell growth and metastasis [54].

Tumor environment, conditional medium of tumor cells, or some cytokines, e.g., TGF-beta (transforming growth factor-beta) and SDF-1 (stromal-derived factor-1), which are enriched in tumor environment, can recruit mesenchymal stem cells to primary tumors. Then tumor environment educates mesenchymal stem cells to promote tumor progression [47, 55, 56].

Many studies revealed numerous mechanisms by which mesenchymal stem cells promote tumor development. Mesenchymal stem cells can differentiate into vascular endothelial cells and are involved in angiogenesis and then promote tumor growth [57]. Besides, in a lymphoma xenograft model, direct cell-cell contact interactions are shown to be critical for the promotion of tumor cell proliferation by mesenchymal stem cells [58]. Some evidence also showed that mesenchymal stem cells may also stimulate epithelial to mesenchymal transition, which makes cancer more invasive [59]. In vitro and in vivo studies demonstrated that mesenchymal stem cell-derived exosomes can promote tumor growth. Through activating extracellular signal-regulated kinase pathway, mesenchymal stem cell-derived exosomes elevate vascular endothelial growth factor expression and then promote tumor progression [60].

5.4 Future Studies for Safer Clinical Application of Perinatal Mesenchymal Stem Cells

Although perinatal mesenchymal stem cells showed promising results in many clinical trials, the detailed mechanisms for that is not well studied or complete clear. If the mechanism for mesenchymal stem cell-based therapies was not deciphered, we can not maximize the therapeutic efficacy nor minimize the potential adverse effects. So one of the most important further investigations should be exploring the mechanisms for perinatal mesenchymal stem cell-based therapies in different diseases. Perinatal mesenchymal stem cells used in most clinical trials are cultured in serum-containing medium, and the manufacturing procedure contains many hands-on steps, which makes the cell products bear the risk of contamination and impairs quality control. Some studies tried to prepare mesenchymal stem cells with automatic equipment or serum-free medium. However, it has not been used in clinical trials. So far, few clinical trials screened the donors for perinatal mesenchymal stem cells. Growing evidence implied that the variation among different donors cannot be ignored. Perinatal mesenchymal stem cells derived from the donors with better genomic stability and potency will benefit the safety and effectiveness of clinical trials.

References

1. Lu LL, et al. Isolation and characterization of human umbilical cord mesenchymal stem cells with hematopoiesis-supportive function and other potentials. Haematologica. 2006;91(8):1017–26.
2. Ra JC, et al. Safety of intravenous infusion of human adipose tissue-derived mesenchymal stem

cells in animals and humans. Stem Cells Dev. 2011;20(8):1297–308.
3. Jiang R, et al. Transplantation of placenta-derived mesenchymal stem cells in type 2 diabetes: a pilot study. Front Med. 2011;5(1):94–100.
4. Wu Y, et al. Cotransplantation of haploidentical hematopoietic and umbilical cord mesenchymal stem cells for severe aplastic anemia: successful engraftment and mild GVHD. Stem Cell Res. 2014;12(1):132–8.
5. Liang L, et al. Transplantation of human placenta derived mesenchymal stem cell alleviates critical limb ischemia in diabetic nude rat. Cell Transplant. 2017;26(1):45–61.
6. Leow SN, et al. Safety and efficacy of human Wharton's jelly-derived mesenchymal stem cells therapy for retinal degeneration. PLoS One. 2015;10(6):e0128973.
7. Parys M, et al. Safety of intraperitoneal injection of adipose tissue-derived autologous mesenchymal stem cells in cats. J Vet Intern Med. 2016;30(1):157–63.
8. Van Loon VJ, et al. Clinical follow-up of horses treated with allogeneic equine mesenchymal stem cells derived from umbilical cord blood for different tendon and ligament disorders. Vet Q. 2014;34(2):92–7.
9. Park YB, et al. Effect of transplanting various concentrations of a composite of human umbilical cord blood-derived mesenchymal stem cells and hyaluronic acid hydrogel on articular cartilage repair in a rabbit model. PLoS One. 2016;11(11):e0165446.
10. Jang KM, et al. Efficacy and safety of human umbilical cord blood-derived mesenchymal stem cells in anterior cruciate ligament reconstruction of a rabbit model: new strategy to enhance tendon graft healing. Arthroscopy. 2015;31(8):1530–9.
11. Park YB, et al. Cartilage regeneration in osteoarthritic patients by a composite of allogeneic umbilical cord blood-derived Mesenchymal stem cells and hyaluronate hydrogel: results from a clinical trial for safety and proof-of-concept with 7 years of extended follow-up. Stem Cells Transl Med. 2017;6(2):613–21.
12. De Kok IJ, et al. Evaluation of mesenchymal stem cells following implantation in alveolar sockets: a canine safety study. Int J Oral Maxillofac Implants. 2005;20(4):511–8.
13. Wang Y, et al. A toxicity study of multiple-administration human umbilical cord mesenchymal stem cells in cynomolgus monkeys. Stem Cells Dev. 2012;21(9):1401–8.
14. Hu J, et al. Long term effect and safety of Wharton's jelly-derived mesenchymal stem cells on type 2 diabetes. Exp Ther Med. 2016;12(3):1857–66.
15. Tisato V, et al. Mesenchymal stem cells of cord blood origin are effective at preventing but not treating graft-versus-host disease. Leukemia. 2007;21(9):1992–9.
16. Li XH, et al. Reduced intensity conditioning, combined transplantation of haploidentical hematopoietic stem cells and mesenchymal stem cells in patients with severe aplastic anemia. PLoS One. 2014;9(3):e89666.
17. Wang D, et al. Long-term safety of umbilical cord mesenchymal stem cells transplantation for systemic lupus erythematosus: a 6-year follow-up study. Clin Exp Med. 2017;17(3):333–40.
18. Wang D, et al. Umbilical cord mesenchymal stem cell transplantation in active and refractory systemic lupus erythematosus: a multicenter clinical study. Arthritis Res Ther. 2014;16(2):R79.
19. Liang J, et al. Allogeneic mesenchymal stem cell transplantation in seven patients with refractory inflammatory bowel disease. Gut. 2012;61(3):468–9.
20. Zhang Z, et al. Human umbilical cord mesenchymal stem cells improve liver function and ascites in decompensated liver cirrhosis patients. J Gastroenterol Hepatol. 2012;27(Suppl 2):112–20.
21. Li YH, et al. Umbilical cord-derived mesenchymal stem cell transplantation in hepatitis B virus related acute-on-chronic liver failure treated with plasma exchange and Entecavir: a 24-month prospective study. Stem Cell Rev. 2016;12(6):645–53.
22. Shi M, et al. Human mesenchymal stem cell transfusion is safe and improves liver function in acute-on-chronic liver failure patients. Stem Cells Transl Med. 2012;1(10):725–31.
23. Chen G, et al. Monitoring the biology stability of human umbilical cord-derived mesenchymal stem cells during long-term culture in serum-free medium. Cell Tissue Bank. 2014;15(4):513–21.
24. Mark P, et al. Human mesenchymal stem cells display reduced expression of CD105 after culture in serum-free medium. Stem Cells Int. 2013;2013:698076.
25. Wang Y, et al. Human mesenchymal stem cells possess different biological characteristics but do not change their therapeutic potential when cultured in serum free medium. Stem Cell Res Ther. 2014;5(6):132.
26. Zhu Y, et al. Alteration of histone acetylation pattern during long-term serum-free culture conditions of human fetal placental mesenchymal stem cells. PLoS One. 2015;10(2):e0117068.
27. Clark KC, et al. Canine and equine mesenchymal stem cells grown in serum free media have altered immunophenotype. Stem Cell Rev. 2016;12(2):245–56.
28. Buzzard JJ, et al. Karyotype of human ES cells during extended culture. Nat Biotechnol. 2004;22(4):381–2. author reply 382.
29. Draper JS, et al. Recurrent gain of chromosomes 17q and 12 in cultured human embryonic stem cells. Nat Biotechnol. 2004;22(1):53–4.
30. Lefort N, et al. Human embryonic stem cells reveal recurrent genomic instability at 20q11.21. Nat Biotechnol. 2008;26(12):1364–6.
31. Spits C, et al. Recurrent chromosomal abnormalities in human embryonic stem cells. Nat Biotechnol. 2008;26(12):1361–3.
32. Ronen D, Benvenisty N. Genomic stability in reprogramming. Curr Opin Genet Dev. 2012;22(5):444–9.
33. von Joest M, Bua Aguin S, Li H. Genomic stability during cellular reprogramming: mission impossible? Mutat Res. 2016;788:12–6.
34. Wang Y, et al. Long-term cultured mesenchymal stem cells frequently develop genomic mutations but do not

undergo malignant transformation. Cell Death Dis. 2013;4:e950.
35. Cai J, et al. Whole-genome sequencing identifies genetic variances in culture-expanded human mesenchymal stem cells. Stem Cell Rep. 2014;3(2):227–33.
36. Miura M, et al. Accumulated chromosomal instability in murine bone marrow mesenchymal stem cells leads to malignant transformation. Stem Cells. 2006;24(4):1095–103.
37. Ren Z, et al. Spontaneous transformation of adult mesenchymal stem cells from cynomolgus macaques in vitro. Exp Cell Res. 2011;317(20):2950–7.
38. Ren Z, Zhang YA, Chen Z. Spontaneous transformation of cynomolgus mesenchymal stem cells in vitro: further confirmation by short tandem repeat analysis. Exp Cell Res. 2012;318(5):435–40.
39. Pan Q, et al. Detection of spontaneous tumorigenic transformation during culture expansion of human mesenchymal stromal cells. Exp Biol Med (Maywood). 2014;239(1):105–15.
40. Bernardo ME, et al. Human bone marrow derived mesenchymal stem cells do not undergo transformation after long-term in vitro culture and do not exhibit telomere maintenance mechanisms. Cancer Res. 2007;67(19):9142–9.
41. de la Fuente R, et al. Retraction: spontaneous human adult stem cell transformation. Cancer Res. 2010;70(16):6682.
42. Rubio D, et al. Spontaneous human adult stem cell transformation. Cancer Res. 2005;65(8):3035–9.
43. Garcia S, et al. Pitfalls in spontaneous in vitro transformation of human mesenchymal stem cells. Exp Cell Res. 2010;316(9):1648–50.
44. Knudson AG Jr. Mutation and cancer: statistical study of retinoblastoma. Proc Natl Acad Sci U S A. 1971;68(4):820–3.
45. Tu B, et al. Mesenchymal stem cells promote osteosarcoma cell survival and drug resistance through activation of STAT3. Oncotarget. 2016;7(30):48296–308.
46. Karnoub AE, et al. Mesenchymal stem cells within tumour stroma promote breast cancer metastasis. Nature. 2007;449(7162):557–63.
47. Xu WT, et al. Human mesenchymal stem cells (hMSCs) target osteosarcoma and promote its growth and pulmonary metastasis. Cancer Lett. 2009;281(1):32–41.
48. De Boeck A, et al. Resident and bone marrow-derived mesenchymal stem cells in head and neck squamous cell carcinoma. Oral Oncol. 2010;46(5):336–42.
49. Li MH, et al. Study of bone mesenchymal stem cells tropism for hepatic tumors and effect on the form of tumor stromal. Zhonghua Yi Xue Za Zhi. 2010;90(5):349–54.
50. Mathew E, et al. Mesenchymal stem cells promote pancreatic tumor growth by inducing alternative polarization of macrophages. Neoplasia. 2016;18(3):142–51.
51. Coffman LG, et al. Human carcinoma-associated mesenchymal stem cells promote ovarian cancer chemotherapy resistance via a BMP4/HH signaling loop. Oncotarget. 2016;7(6):6916–32.
52. Peppicelli S, et al. Extracellular acidity strengthens mesenchymal stem cells to promote melanoma progression. Cell Cycle. 2015;14(19):3088–100.
53. Shinagawa K, et al. Mesenchymal stem cells enhance growth and metastasis of colon cancer. Int J Cancer. 2010;127(10):2323–33.
54. Yang X, et al. Human umbilical cord mesenchymal stem cells promote carcinoma growth and lymph node metastasis when co-injected with esophageal carcinoma cells in nude mice. Cancer Cell Int. 2014;14(1):93.
55. Barcellos-de-Souza P, et al. Mesenchymal stem cells are recruited and activated into carcinoma-associated fibroblasts by prostate cancer microenvironment-derived TGF-beta1. Stem Cells. 2016;34(10):2536–47.
56. Mishra PJ, et al. Carcinoma-associated fibroblast-like differentiation of human mesenchymal stem cells. Cancer Res. 2008;68(11):4331–9.
57. Zhang K, et al. Bone marrow mesenchymal stem cells induce angiogenesis and promote bladder cancer growth in a rabbit model. Urol Int. 2010;84(1):94–9.
58. Roorda BD, et al. Mesenchymal stem cells contribute to tumor cell proliferation by direct cell-cell contact interactions. Cancer Investig. 2010;28(5):526–34.
59. Martin FT, et al. Potential role of mesenchymal stem cells (MSCs) in the breast tumour microenvironment: stimulation of epithelial to mesenchymal transition (EMT). Breast Cancer Res Treat. 2010;124(2):317–26.
60. Zhu W, et al. Exosomes derived from human bone marrow mesenchymal stem cells promote tumor growth in vivo. Cancer Lett. 2012;315(1):28–37.

Therapeutic Application of Perinatal Mesenchymal Stem Cells in Nervous System Diseases

6

Wenbin Liao

Abstract

The nervous system diseases or disorders are caused by the malfunction and abnormalities in the nerves, spinal cord, or brain, resulting in paralysis, seizures, confusion, or distorted perception. The currently available regimens to restore the impaired or lost neurologic functions are limited and not effective. Thus it is urgent to find novel strategies to address the unmet medical need. Stem cell therapy provides a promising strategy in the treatment of nervous system diseases. Perinatal mesenchymal stem cells (MSC) are ideal cell candidates for the cell-based therapy for nervous system diseases, and recent advances have showed that perinatal MSC transplantation is safe and effective in both preclinical and clinical studies in several nervous system diseases including stroke, spinal cord injury, Parkinson's disease, and Alzheimer's disease.

6.1 Introduction

The nervous system is a complex organ that includes neurons, astrocytes, oligodendrocytes, and other nonneural cells types. Diseases of the nervous system, including degenerative diseases, cerebrovascular diseases, and injury, affect millions of people of all ages. For the majority of the nervous system diseases, it is a big challenge to restore the lost neural function and prevent the degenerating neural cells. Therefore, there are huge unmet medical needs for the treatment of diseases in nervous system. The paradigm that lost neural cells are not irreversible has been changed because of the discovery of neurogenesis in the adult brain [1, 2]. Adult neurogenic niches such as subventricular zone (SVZ) and dentate gyrus (DG) harbor a pool of endogenous neural stem cells (NSC) that could potentially be exploited for repair and/or regeneration in injured or diseased conditions [3, 4]. Indeed, activation or enhancement of endogenous neurogenesis after injury or other insults in the nervous system has been frequently reported ever since [5–9].

Therefore, exogenous delivery of additional neural stem cells to replace the damaged neural cell and tissue has been explored extensively in a variety of nervous system diseases [10–14] including stroke, Parkinson's disease, spinal cord injury, multiple sclerosis, etc. As expected, numerous studies showed the therapeutic effects of the neural stem cells for the nervous system diseases. However, the correlation of therapeutic effects with cell replacement or regeneration by the transplanted stem cells seems not to be supported by the evidence in vivo. The low efficiency of cell homing, survival, and engraftment and more importantly the wiring with the host neural networks are related to the difficulty for cell replacement by the exogenously transplanted

W. Liao (✉)
Baylx, Inc., Lake Forest, CA, USA

Z. C. Han et al. (eds.), *Perinatal Stem Cells*, https://doi.org/10.1007/978-981-13-2703-2_6

cells [14, 15]. In contrary, the mechanisms underlying the improved outcome after neural stem cell transplantation in neural disorders were thought to be related to the anti-inflammation activities [16] as well as the ability to induce an enhanced endogenous neurogenesis, angiogenesis, synaptogenesis, and neural tissue remodeling by transplanted neural stem cells. These indirect functions of neural stem cells mainly depend on their paracrine effects by secreting a variety of cytokines with versatile biological activities.

The notion that cell-based therapy for nervous system diseases is not necessarily involving the direct cell replacement has been recognized by many researchers in the fields. Mesenchymal stem cells are thought to be able to differentiate into neural cell types and thus had been originally tested for their regenerative effects in nervous diseases. The accumulated data result in a similar conclusion to that for neural stem cell-based studies. Mesenchymal stem cells also act via indirect mechanisms for the positive effects in nervous diseases such as stroke, spinal cord injury, Parkinson's disease, etc. The in vivo differentiation of MSC toward neural cells is extremely rare, and the cell replacement pathway by MSC-based therapies is not supported.

Compared to neural stem cells, mesenchymal stem cells possess several advantages. The most important one is the rich cell source to obtain MSC. MSC exist in almost all tissue and organs [17, 18], and among all of the tissues, perinatal tissues stand out as the most abundant and ideal source in terms of advantages including, but not limited to, no ethical issues, robust cell proliferation, ease to standardize, and scale-up for industrial manufacture. In contrast, NSC are originally obtained from fetal or adult neural tissues thus facing donor shortage and substantial ethical issues. Recent advances allow the derivation of NSC from embryonic stem cells (ESC) or induced pluripotent stem cells (iPSC), however, there are still difficulties to obtain pure NSCs from differentiation, and the contamination of pluripotent cells could lead to tumorigenesis. Another advantage of MSC over NSC is that MSC are more potent in secreting of biological factors especially those with activities in immunomodulation, anti-inflammation, and angiogenesis. In addition, MSC are also able to produce neurotrophic cytokines with neuroprotective effects. And last but not least, MSC have an established safe profile as supported by numerous human clinical studies [19, 20], while several reports have raised the cautions of in situ mass formation for NSC-based studies [11, 12, 21].

In nervous system, our hope for cell or tissue replacement to regenerate the tissue and restore the lost neural functions may still await complete understanding of the complex organ system as well as new technology to make ex vivo neural cells or tissue organs that are capable of integrating into the host. As the current cell-based therapies for nervous system still mainly rely on the indirect mechanisms of the transplanted cells, MSC may be the ideal cell candidate for protecting the damaged neural tissues and restoring the functions in neural disorders. In this chapter, we will discuss the current status of both preclinical and clinical studies testing perinatal MSC in the treatment of nervous system diseases.

6.2 Preclinical Studies of Perinatal Mesenchymal Stem Cell Therapy for Nervous System Diseases

6.2.1 Perinatal Mesenchymal Stem Cells for Stroke

Stroke is caused by the disruption of blood flow to the brain. According to World Heart Association, 15 million people worldwide suffer a stroke every year. Stroke is the second leading cause of disability, after dementia. In the United States, stroke is the leading cause of disability and the fourth leading cause of death of all diseases. The only FDA-approved drug, tissue plasminogen activator (tPA), has only been beneficial to less than 5% of ischemic stroke patients due to the its limitations such as limited therapeutic window (4.5 h from disease onset to tPA administration) and a risk of hemorrhagic transformation associated with the treatment [22, 23]. In addition, to date no effective treatment is proved to promote neuronal function recovery.

Therefore, innovative strategies have been explored to address the significant unmet clinical need for stroke to increase the therapeutic window specifically targeting the restorative phase poststroke. In the latest decades, stem cell-based therapies have become one of the most promising approaches to restore function in stroke. Not surprisingly, perinatal MSC have been widely investigated in the treatment of stroke.

6.2.1.1 Umbilical Cord Mesenchymal Stem Cells (UC-MSC)

Umbilical cord tissue is the most common origin for MSC among all of the perinatal MSC studies in stroke. A number of research groups, including ours, have reported the benefits of human UC-MSC implantation in stroke animals. The most common animal model for stroke is the transient rodent middle cerebral artery occlusion (MCAO) model. In 2008, the result of the first experimental study on UC-MSC in stroke was reported by Koh et al. [24]. Before the in vivo transplantation, they showed that 20 days after the induction of in vitro neuronal differentiation, about 77.4% of the hUC-MSC showed neuronal cell morphology and were positive for neuronal cell markers, but functional neuronal-type channels were not detected by electrophysiological analysis. In in vivo study, intracerebral implantation of the hUC-MSC into immunosuppressed ischemic stroke rats improved neurobehavioral function and reduced infarct volume as compared to control rats. In addition, most of the implanted hUC-MSC homed to the damaged hemisphere. Although some of these cells expressed detectable levels of neuron-specific markers, they failed to become functionally active neuronal cells; the authors concluded the improvement in neurobehavioral function and the reduction of infarct volume might be related to the neuroprotective effects of hUC-MSC rather than the formation of a new network between host neurons and the implanted hUC-MSC [24].

In the following year (2009), our group also published a work about human UC-MSC in the treatment of rat stroke [25, 26]. 2X105 UC-MSC were injected cerebrally to rats 2 h after MCAO to evaluate their fate and function in vivo. We found that the transplanted UC-MSC survived for at least 5 weeks in rat brain, and the UC-MSC treatment significantly reduced injury volume and neurologic functional deficits of rats after stroke. More importantly, UC-MSC were found to widely incorporate into cerebral vasculature and substantially increased vascular density by increasing the vascular endothelial growth factor (VEGF) and basic fibroblast growth factor (bFGF) expression in ipsilateral hemisphere of stroke. These data suggested the benefits of UC-MSC transplantation in stroke rat might be mediated by their ability to promote angiogenesis. In parallel, the therapeutic effects of UC-MSC in a hemorrhagic stroke rat model had been reported [25, 26]; the underlying mechanisms were thought to correlate with anti-inflammation and increased angiogenesis. In a rat model of embolic stroke, Zhang et al. [27] demonstrated intravenous infusion of hUC-MSC significantly improved neurological functional recovery without reducing infarct volume compared to vehicle-treated control rats. The underlying mechanisms were likely related to the significantly enhanced synaptogenesis and vessel density in the ischemic boundary zone (IBZ), as well as increased endogenous neural progenitor cell proliferation in the subventricular zone (SVZ) after hUC-MSC treatment. Similar results have been reported by many other groups. Shams et al. [28] tested if coadministration of ASA with hUC-MSC cells could bring better outcome improvement in stroke rats as compared to those with ASA and hUC-MSC alone. The results suggested that the learning and memory disturbance in the ASA and UC-MSC-treated animals was less pronounced than ischemic animals. However, the addition of ASA to hUC-MSC did not raise the outcome higher than administration of hUC-MSC alone.

Take the advantage of MR imaging system, Jiang et al. [29] monitored the white matter and vascular reorganization over a 12-week period, and their data suggested that MRI could be a useful tool to measure structural recovery after stroke. MR imaging can detect the brain reorganization resulting from the effect of hUC-MSC therapy, and it has found a positive correlation between the MRI measurements and brain structural changes and functional behavioral tests after stroke. MRI ventricular volumes provided the most sensitive index in monitoring brain remodeling and treatment effects and highly correlated with histological and functional measurements. A recent study [30] investigated the UC-MSC biodistribution of intravenously administered indium-111 (In-111) oxine-labeled cells in a rat model of MCAO using single photon emission computed tomography (SPECT). In addition to a significant increase in vascular and synaptophysin density in stroke areas of rats that received In-111-labeled hUC-MSC, they found that most of the UC-MSC trafficked to the lungs immediately following IV administration with a drastic decrease 3 days postinjection. There are about 1% of the total administrated UC-MSC at the site of injury in stroke areas of animals. Therefore, regarding the limited number of cells homed to or surviving in the injured nerve system after stroke, it is reasonable to speculate that cell replacement mechanism is not the main explanation for the benefit of UC-MSC in stroke. The paracrine effects of UC-MSC by secreting of a variety of biologically potent cytokines, which increase the angiogenesis, synaptogenesis, endogenous neural stem/progenitor cell activation, etc., probably weigh more for the therapeutic functions, regardless of the routes and timing of UC-MSC administration [31–33]. In agreement with this notion, Zhou et al. [34] found that hUC-MSC transplantation improved the recovery of neuronal function in a hypoxic-ischemic brain damage (HIBD) rat model, and the benefit may be related to the secreted IL-8 by UC-MSC, which enhanced angiogenesis in the hippocampus via the JNK pathway. Zhou et al. [34] provided evidence that neurons after oxygen glucose deprivation in the brain of stroke mouse triggered the VEGF-A secretion from hUC-MSC by Notch1 signaling and that contributes to the angiogenic effect of hUC-MSC transplantation in stroked brain.

6.2.1.2 Umbilical Cord Blood Mesenchymal Stem Cells (UCB-MSC)

It is well known that umbilical cord blood is a rich source for hematopoietic stem cells (HSC) and UCB-HSC in transplantation for blood diseases has been an important alternative to bone marrow (BM) HSC transplantation. UCB also contain non-hematopoietic stem cells including MSC, endothelial progenitor cells, etc. A number of researchers have tested the potential of UCB-MSC in restoring neurological functions in stroke animals.

Chung et al. [35] examined the effects of hUCB-MSC delivered through the basilar artery in a canine thromboembolic brain ischemia model. Cerebral ischemia was induced through occlusion of the middle cerebral artery by injecting thrombus emboli into ten beagles. The results showed that transplanted cells had differentiated into neurons and astrocytes and were positive for von Willebrand factor (vWF) and neuroprotective factors, such as brain-derived neurotrophic factor (BDNF) and vascular endothelial growth factor (VEGF), at 4 weeks after the transplantation. The transplantation of UCB-MSC was able to reduce the infarction lesion volume and induce earlier recovery from the neurological deficit. In a rat stroke model, Lim et al. [36] determined the therapeutic potential of intrathecal or intravenous administration of MSC. The rats receiving hUCB-MSC intrathecally had a significantly higher number of migrated cells within the ischemic area when compared with animals receiving cells intravenously, although animals that received hUCB-MSC in both groups had similar significantly improved motor function when compared with untreated control animals. Consistently, in a permanent MCAO rat model, Kim et al. [37, 38] evaluated the outcome improvement of rat after severe brain injury and UCB-MSC intervention. All of the abnormalities including increased brain infarct volume as measured by MRI, impaired functional tests, and histologic abnormalities in the penumbra were significantly improved by MSC transplantation 6 h after MCAO.

Park et al. [39, 40] administered three types of donor-derived human hUCB-MSC intracerebrally into the ischemic stroke rat model. Functional and pathological assessments revealed a significant better neurological performance of rats after hUCB-MSC treatment with an accumulation of neuronal progenitor cells, angiogenic cytokines, and growth factors that can promote tissue regeneration and repairs in regions adjacent to ischemic site. The similar pattern of neurogenic and angiogenic profiles of the three types of hUCB-MSC suggested that therapeutic effect of hUCB-MSC in the ischemic brain is independent of the type of donor. In a separate study [39, 40], the same group demonstrated that hUCB-MSC transplantation resulted in a greater number of newly generated cells and angiogenic and tissue repair factors and a lower number of inflammatory events in the penumbra zone which may be associated with the increased expression of thrombospondin1, pantraxin3, and vascular endothelial growth factor under hypoxic conditions than under normoxic conditions.

In a recent study, Hocum Stone et al. [41] addressed another mechanism of action and compared the therapeutic efficacy of high- versus low-passaged hUCB-MSC in MCAO rat model. There were no adverse reactions detected, and both behavioral and histological analyses had shown that the administration of these cells reduced the infarct volume as well as improves the functional outcome of these rats following stroke for both high- and low-passaged hUCB-MSC. Flow cytometry also revealed a restoration of normal levels of macrophages and T cells and microglia in the brain following treatment. The data suggest that hUCB-MSC may act by both inhibiting immune cell migration into the brain from the periphery and by inhibition of immune cell activation within the brain.

These studies suggest that hUCB-MSC have a beneficial effect on cerebral ischemia, which may also be explained by their potency in enhancing angiogenesis, neurogenesis, and anti-inflammatory events after the ischemia injury in stroke.

6.2.1.3 Amniotic Membrane or Fluid Mesenchymal Stem Cells (AM/AF-MSC)

Amnion-derived cells have also been suggested as a potent graft source for cell therapy in stroke [42]. Indeed, several researchers have tested AF or AM-MSC in animal model of stroke. Rehni et al. [43] compared the effects of intracerebroventricular administration of amniotic fluid-derived stem cells to that of embryonic neuronal stem cells in stroke animals. After a serial of motor, somatosensory, and memory assessments, the authors concluded that transplantation of amniotic fluid stem cells was able to benefit the ischemic brain which was comparable to the neuroprotective effect of embryonic neuronal stem cells. Liu et al. [44] reported the AM-MSC survived and migrated to the ischemic area of rats which significantly ameliorate behavioral dysfunction and reduce infarct volume of ischemic rats. Glial cell line-derived neurotrophic factor (GDNF) overexpressed AM-MSC could further enhance the improvement of rats after stroke.

Tajiri et al. [45] isolated viable rat AF-MSC and demonstrated their therapeutic benefits via intravenous transplantation in a rodent model of ischemic stroke. They found that there were more proliferating cells in the dentate gyrus (DG) and the SVZ of AF-MSC-transplanted stroke animals compared to vehicle-infused stroke animals, suggesting a possible enhancement of endogenous repair mechanisms by AF-MSC.

6.2.1.4 Placenta Mesenchymal Stem Cells (P-MSC)

Similar to other perinatal MSC, the placenta-derived MSC (P-MSC) has also been reported to be able to accelerate the neurological function recovery of stroke animals. Yarygin et al. [46] demonstrated that intravenous infusion of P-MSC resulted in a substantial reduction of the infarct volume in the brain. Although the labeled P-MSC accumulated around the ischemic region, in the hippocampus, and in the SVZ of both hemispheres, only few human MSC adjacent to the ischemic focus expressed markers of astrocytes or neurons. The therapeutic effects of P-MSC in stroke rats were thought to be related to the activation of the host neural stem and progenitor cells and their migration into the ischemic tissue and adjacent areas, rather than cell replacement of damaged rat neurons and glial cells by transplanted human cells.

Chen et al. [47, 48] tested whether P-MSC infusion could improve neurological functional recovery in a rat stroke model and investigated the potential mechanisms underlying the MSC-induced neuroprotective effect. They demonstrated that P-MSC infusion significantly accelerated functional recovery and dramatically decreased lesion volume, TUNEL, and cleaved caspase three-positive cell number in the ischemic brain. P-MSC treatment also significantly augmented the expression of HGF, BDNF, and VEGF levels in the IBZ of the ischemic brain compared to controls, suggesting a potent paracrine activity of P-MSC in promoting angiogenesis and neurogenesis and inhibiting neural apoptosis in CNS injury after stroke.

There was an interesting finding from Kranz et al. [49] when they investigated the effect of intravenously infused P-MSC in stroke animals. They compared the effects of MSC from the human maternal and fetal placenta and also at different doses in focal ischemia rat. The results found that double infusion of P-MSC was superior to single transplantation in the functional tests and benefits for all outcome parameters were observed only for maternally derived MSC.

6.2.2 Perinatal Mesenchymal Stem Cells for Spinal Cord Injury (SCI)

Spinal cord injury (SCI) is caused by traumatic injuries or disease to the spinal cord with immediate consequences due to loss of motor, sensory, and autonomic nervous system functions, as well as later problems including muscle wasting, chronic pain, urinary infections, and pressure sores. Spinal cord injury affects millions of people worldwide and typically has lifelong consequences and thus is a huge healthcare burden for many countries. Although there is no effective therapeutic option currently available, recent advances have shown that stem cell transplantation strategies hold promise to enhance functional recovery after SCI.

Perinatal MSC have frequently been explored for the treatment of SCI. Cui et al. [50] investigated the effects of human UCB-MSC transplantation on the functional restoration of spinal cord injury (SCI). A SCI model was established using the modified Allen's method, and the rats that were injected with UCB-MSC at the injury site were compared to saline control using a series of locomotor rating systems. The data suggested transplantation of UCB-MSC promoted the recovery of the damaged function of spinal cord nerves in SCI rats. Lim et al. [51] determined the effects of allogenic UCB-MSC with or without recombinant methionyl human granulocyte colony-stimulating factor (hGCSF) on a canine lumbar spinal cord injury. The functional Olby scores in the groups with the UCB-MSC and UCB + hGCSF were significantly higher than in control groups from 2 weeks after the transplantation; however, there were no significant differences between the UCB-MSC and UCBG groups. Significant improvement in the nerve conduction velocity and structural consistency of the nerve cell bodies were also found in MSC-treated dogs. Park et al. [52, 53] also demonstrated UCB-MSC-promoted functional recovery of contused spinal cord in rats. In addition, in MSC-transplanted group, TUNEL-positive cells were decreased, and BrdU-positive neural precursor and oligo-lineage cells were significantly increased rats compared with control group, suggesting endogenous cell proliferation and oligogenesis contribute to MSC-promoted functional recovery following SCI. A group of researchers [54] compared the effects of MSC from fat, bone marrow, Wharton's jelly, and umbilical cord blood for treating spinal cord injuries in dogs. The results showed there were no significant differences in functional recovery among the MSC groups, but UCB-MSC induced more nerve regeneration and anti-inflammation activity in SCI dogs.

Most SCI patients suffer from long-lasting neuropathic pain, which is caused by damage or diseases affecting the central or peripheral nervous system. MSC are also reported to alleviate the neuropathic pain following SCI. Yousefifard et al. [55] compared the effects of alleviating functional deficits and neuropathic pain between BM-MSC and UC-MSC transplantations. Intrathecal injection of UC-MSC led to improving functional recovery, allodynia, and hyperalgesia in a similar extent to that of BM-MSC. The survival rate of UC-MSC was significantly higher than BM-MSC, and transplantation of UC-MSC brought about better results than

BM-MSC in windup of wide dynamic range neurons. In agreement with the results of UC-MSC, both UCB-MSC and AM-MSC transplantation also could suppress mechanical allodynia [56], and this effect was thought to be closely associated with the modulation of spinal cord microglia activity and NMDA receptor phosphorylation.

6.2.3 Perinatal Mesenchymal Stem Cells for Parkinson's Disease (PD)

Parkinson's disease (PD) is the second most frequent neurodegenerative disease after Alzheimer's disease, which is caused by degeneration of dopamine (DA) neurons in the substantia nigra pars compacta (SNc) of the midbrain. Mesenchymal stem cells have been thought to be able to differentiate into dopamine-secreting cells [57, 58]. Another report also showed that conditioned medium from human amniotic epithelial cells was able to induce the differentiation of UCB-MSC into dopaminergic neuron-like cells [59, 60]. After transplantation into PD animal models [61, 62], amnion-derived cells can survive for several months and differentiate into TH-positive cells in PD rats. The grafts significantly ameliorated apomorphine-induced turns, attenuated the loss of TH-positive cells and prevented the fall of DA and its metabolites 3, 4-dihydroxyphenylacetic acid (DOPAC) and homovanillic acid (HVA) in PD rats. Similar results were also reported with P-MSC [52, 53] and UC-MSC. For examples, Shetty et al. [63] found that after dopaminergic differentiation, UC-MSC expressed significant higher levels of tyrosine hydroxyalse and Nurr1 compared to BM-MSC. The in vivo efficacy results of the differentiated and undifferentiated cell types suggest differentiated UC-MSC are potentially better for clinical use than the undifferentiated cells, if they are able to locate at the site of action in adequate numbers. Zhao et al. [64] transplanted hUC-MSC, and hUC-MSC-derived DA-like neurons were transplanted into the striatum and SN of a rat model of PD that is induced by 6-hydroxydopamine (6-OHDA).The results suggested that hUC-MSC-derived DA-like neurons appear to be favorable candidates for cell replacement therapy in PD, and heat shock protein 60 (Hsp60) could be a positive indicator for behavioral recovery.

When hepatocyte growth factor (HGF) was overexpressed in UC-MSC [65], cultural supernatant from HGF-UC-MSC could promote regeneration of damaged PD cells at higher efficacy than the supernatant from hUC-MSC alone. Yan et al. [66] transduced Lmx1α and NTN, two genes important for DA neuron functions, into UC-MSC, and transplantation of these genetically modified UC-MSC in the PD monkey models ameliorated the functional deficits by the behavioral measures.

Different from MSC for stroke treatment, it seems more studies showed that MSC after induction toward DA-like phenotype performed better than unmodified native MSC in terms of the outcome in PD animal models. This observation may be explained by the unique pathogenesis of PD that the specific cell type dopamine-producing cells are reduced and dopamine needs to be supplemented.

6.2.4 Perinatal Mesenchymal Stem Cells for Alzheimer's Disease

Alzheimer's disease (AD) is a chronic neurodegenerative disease, mostly affecting medial temporal lobe and associative neocortical structures. AD is the most common cause of dementia and increases with age. As aging is becoming a global issue, it is urgent to search for new therapies to overcome this disorder. AD pathogenesis remains unclear, and current therapeutical strategies have shown limited effectiveness. Accumulating evidence has supported a role for neuroinflammation and immune system dysregulation in AD [67, 68]. Given their potent immunomodulatory and anti-inflammatory functions [69–71], MSC have become an increasingly attractive option to potentially treat and slow down the progression of AD.

The modulation of neuroinflammation by perinatal MSC has been reported by many groups in AD models. Lee et al. [72] showed that co-culture of human UCB-MSC with hippocampus neurons treated with amyloid-β (Aβ) dramatically decreased the apoptosis of the neurons. Moreover, in an acute Aβ-induced AD model, UCB-MSC treatment downregulated the markers of glial activation, oxidative stress, and apoptosis levels in the brain, which was accompanied by cognitive rescue with restoration of learning/memory function. In a following study using a APP/PS1 transgenic AD mouse model, similar results were found that UCB-MSC transplantation drastically reduced the Aβ deposition, β-secretase 1 (BACE-1) levels, and tau hyperphosphorylation in the brain. These effects were associated with reversal of disease-associated microglial neuroinflammation, as evidenced by decreased microglia-induced proinflammatory cytokines, elevated alternatively activated microglia, and increased anti-inflammatory cytokines. Consistently, Boutajangout et al. [73] observed significant improvements on cognitive and sensorimotor tasks in human UC-MSC-treated APP/PS1 mice, which probably resulted from a reduction of the amyloid beta burden in the cortex and the hippocampus by UC-MSC treatment. In the same animal model, Yang et al. [59, 60] induced human UC-MSC into neuron-like cells (NC) and transplanted the differentiated NC cells into AD mice. Decreased Aβ deposition and improved memory in AD mice were detected, which was thought to be associated with activating M2-like microglia and modulating neuroinflammation.

In two recent studies [37, 38, 74], soluble factors secreted by UCB-MSC that were able to enhance neurogenesis and reduce neuroinflammation in AD models were identified. The placenta and amniotic MSC were also reported to mediate neuroprotection by regulating neuronal death, neurogenesis, glia cell activation in hippocampus, and altering cytokine expression in mouse AD model [75–77]. AM-MSC secreted high levels of transforming growth factor-β (TGF-β) under in vitro inflammatory environment conditions and showed significant long-lasting improvement in AD pathology and memory function via immunomodulatory and paracrine mechanisms [76].

6.2.5 Perinatal Mesenchymal Stem Cells for Other Nervous System Diseases

As there are limited or no effective therapies for the vast majority of diseases in the nervous system, like other stem cells, perinatal MSC-based therapies have also been attempted in many other neural diseases besides those discussed above.

Selim et al. [78] illustrated that intravenous administration of placenta MSC significantly ameliorated the disease course, decreasing brain inflammation and degenerating neurons in rat experimental autoimmune encephalomyelitis (EAE), an animal model of human multiple sclerosis. MSC treatment reduced axonal damage and increased oligodendrocyte precursors and upregulated the expression of neurotrophic factors including brain-derived neurotrophic factor (BDNF), nerve growth factor (NGF), and neurotrophin 3 (NTF3) in the brain. UC-MSC were also reported to prevent the onset of EAE and induce an improved outcome in established EAE in rats [79]. Similar results were observed in mouse EAE [80–83], where intracerebrally transplanted placenta MSC or UC-MSC significantly decreased the disease severity, perivascular immune cell infiltrations, demyelination, and axonal injury in the spinal cord and increased the survival of EAE animals when compared to saline, which might be related to the reduction of the inflammation by anti-inflammatory protein, TNF-α-stimulated gene/protein 6 (TSG-6).

In an experimental autoimmune myasthenia gravis (EAMG) induced by subcutaneous injection of synthetic analogue of acetylcholine receptor (AchR) [84], intravenously injected human UC-MSC specifically are home to spleen tissue and significantly improved the functional deficits of EAMG mice. In addition, AchR antibody level was dramatically decreased in UC-MSC-treated group when compared with untreated control, which was related to the ability of hMSC to inhibit the proliferation of AchR-specific lymphocyte.

Perinatal MSC were also shown promises in the treatment of motor neuron diseases. In a mouse model of amyotrophic lateral sclerosis

(ALS), an adult-onset progressive neurodegenerative disease involving degeneration of motor neurons in the central nervous system, Sun et al. [85] evaluated the long-term effects of intravenous administration of human AM-MSC into a hSOD1(G93A) mouse model. hAM-SCs were able to home the spinal cord and significantly retarded disease progression, extended survival, improved motor function, prevented motor neuron loss, and decreased neuroinflammation, as compared with the treatment with PBS, in ALS mice.

MSC could also treat traumatic brain injury (TBI), a form of nondegenerative, acquired damage to the brain from a sudden external mechanical force. When injected in a weight-drop model of TBI in rats [86, 87], human UC-MSC were able to migrate to the injured site by SDF1/CXCR4 signaling. UC-MSC transplantation reduced TBI-induced brain edema and lesion volume, leading to improved neurologic function, memory and cognition. A recent study demonstrated the similar results by human AM-MSC transplantation in mouse TBI model [88]. Additionally, they found that treatment with AM-MSC or conditioned medium showed comparable protective effects in terms of neuronal rescue, promotion of M2 microglia polarization and induction of trophic factors, indicating the human MSC exerted their functions by paracrine mechanism.

Additional animal models that have been used to support the protective or restorative effects of perinatal MSC include Huntington's disease (HD) [89], sciatic or optic nerve injury [47, 48, 90], and experimental spina bifida [91].

6.3 Clinical Studies of Perinatal MSC in Nervous System Diseases

Liu et al. [81–83] studied clinical effect and safety of UC-MSC in treating spinal cord injury (SCI) by intrathecal injection. Treatment was effective in 13 of 22 patients. In most patients with effective outcomes, motor or sensory functions, or both, were improved, and bowel and bladder control ability was improved. No short- or long-term treatment-related adverse events occurred in the study. A similar study was conducted to evaluate [92] the UC-MSC transplantation in the treatment for sequelae of thoracolumbar spinal cord injury in 34 patients. Patients receiving UC-MSC exhibited accelerated neurological functional recovery as shown by an increase in maximum urinary flow rate and bladder capacity, as well as a decrease in residue urine volume and maximum detrusor pressure. This efficacy is superior to that of rehabilitation therapy and self-healing. The same group also compared the clinical efficacies of two transplantation approaches for UC-MSC in treating SCI: open surgical exploration or CT-guided local stem cell transplantation. Twenty seven subjects were divided into three groups: open surgical transplantation group, CT-guided transplantation group, and control without stem cell transplantation. The results showed CT-guided local stem cell transplantation exhibited the best clinical outcome [93].

Four patients with stroke (three with ischemic and one with hemorrhagic stroke) in the middle cerebral artery territory were recruited in a study to test the feasibility of delivering UC-MSC via catheter to the proximal end of the lesion artery [94]. One single dose of 2×10^7 UC-MSC was administrated to each patient. No major accidents were observed. Functional assessments demonstrated two of the three ischemic stroke patients improved muscle strength, while the hemorrhagic stroke patient failed to gain improvement.

In a clinical transplantation of UC-MSC in multiple sclerosis patients [95], 23 patients were enrolled, and 13 of them were given UC-MSC intravenously plus traditional anti-inflammatory treatment, whereas the control patients received the anti-inflammatory treatment only. The results found that overall symptoms of the UC-MSC-treated patients improved compared to patients in the control group. UC-MSC treatment also decreased the relapse occurrence of MS patients. A shift from Th1 to Th2 immunity may explain the benefits in UC-MSC-treated patients. Infusions of placenta MSC in relapsing-remitting multiple sclerosis and secondary progressive multiple sclerosis were proved safe in a multicenter, randomized, double-blind, placebo-controlled, two-dose-ranging study [96], but the efficacy of placenta MSC in this setting still awaits the data from further larger trials.

Lv et al. [97] reported the results of transplantation of human cord blood mononuclear cells (CB-MNCs) and UC-MSC in children's autism. Thirty-seven subjects diagnosed with autism were enrolled into this study, and 23 subjects received 4 stem cell infusions through intravenous and intrathecal injections. Assessment results of the Childhood Autism Rating Scale (CARS), Clinical Global Impression (CGI) scale, and Aberrant Behavior Checklist (ABC) combination of CB-MNC and UC-MSC showed larger therapeutic effects than the CB-MNC transplantation alone. As a following study, another 20 subjects were enrolled to receive combination of CB-MNC and UC-MSC transplantation [98]. Transplantation of CB-MNC and UC-MSC could increase HGF, BDNF, and NGF levels in the CSF of patients with autism.

A recent study by Miao et al. [99] reported the safety and efficacy results of intrathecally administrated UC-MSC by lumbar puncture in various neurological conditions in a total of 100 patients. All patients were followed up for more than 1 year after the treatment. Mild to moderate side effects (headache, low-grade fever, low back pain, and lower limb pain) were observed in 22 (22%) patients, and functional indices were improved in 47 patients (47%): 12 patients with spinal cord injury, 11 patients with cerebral palsy, 9 patients with post-traumatic brain syndrome, 9 patients with post-brain infarction syndrome, 3 patients with spinocerebellar ataxias, and 3 patients with motor neuron disease. The results support the intrathecal administration of UC-MSC is a safe and effective approach to treat neurological disorders. Further analysis suggested that the beneficial effects of UC-MSC might act via inhibition of mitogen-activated protein kinase pathway-mediated apoptosis [100].

To date, there are limited clinical studies that have been done for evaluating the safety and efficacy of perinatal MSC. Nevertheless, the results from these studies support that perinatal MSC share a similar safe profile to MSC derived from the bone marrow or other tissues [20]. Also patients receiving perinatal MSC have a trend to a better recovery as compared to the placebo or untreated control. Therefore, larger randomized controlled trials (RCT) are warranted to further elucidate the efficacy of perinatal MSC in nervous system diseases.

6.4 Mechanism of Action of Perinatal MSC for Their Therapeutic Effects in Nervous System Diseases

Like other tissue-derived MSC, it is still controversial that perinatal MSC are able to differentiate into bona fide functional neural cells such as neuron, astrocyte, and oligodendrocyte. In nervous system diseases, the expectation that transplanted cells would differentiate in situ into desired cell types and replace the lost or damaged neural cell and tissues has barely been satisfied with numerous attempts. The difficulty comes not only from the requirement of in vivo precise differentiation at the right time and location but also from the needs of sufficient homing, engraftment, and survival of transplanted cells, as well as their integration into the host neural network. Surprisingly, it seems that MSC transplantation, via either local or systemic injection, is able to elicit a significant functional improvement in a variety of nervous system diseases, in the absence of substantial cell replacement contributions. An acceptable concept is that MSC act through indirect manners such as paracrine mechanisms to affect both neural cells and nonneural cells, which leads to one or more of the following: enhanced endogenous neurogenesis, angiogenesis, synaptogenesis and neural cell and tissue remodeling, and reduced neural cell death, reduced neuroinflammation, and systemic inflammation (Fig. 6.1) after transplantation in nervous system diseases. In respect to the complexity of the pathogenesis of the different diseases in nervous system, it is reasonable to acknowledge that MSC probably take advantages of different mechanisms to benefit the injured or diseased neural tissues in different diseases or even in the different stages of a disease. MSC are living therapeutics and are thought to

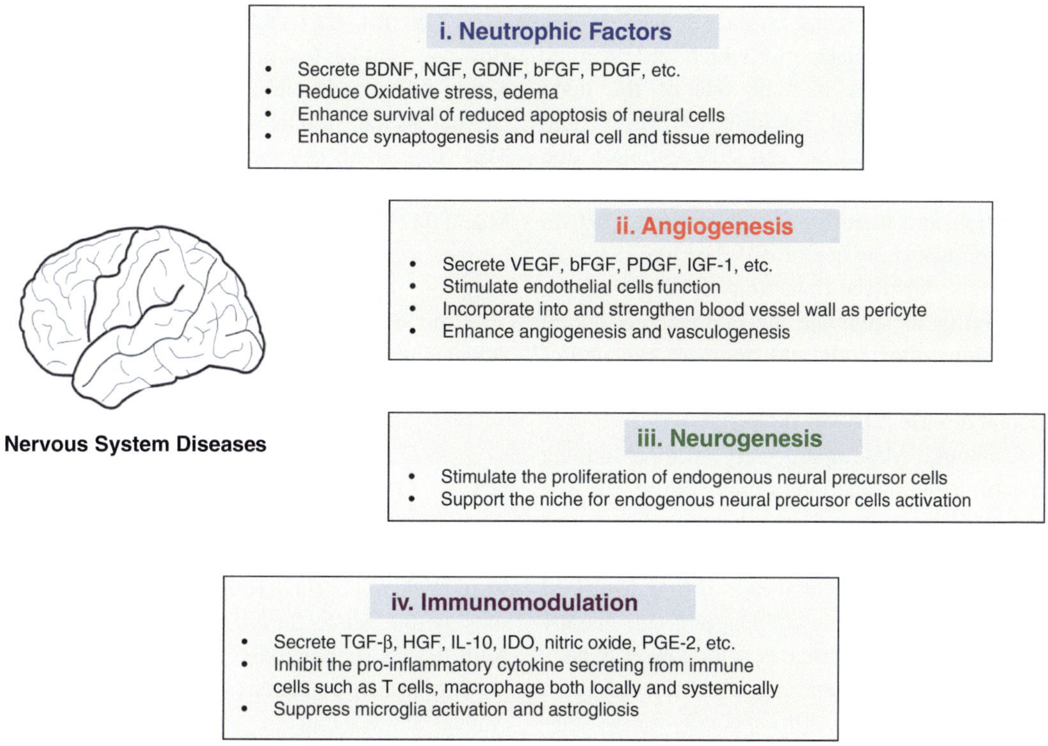

Fig. 6.1 Summary of proposed mechanisms of perinatal mesenchymal stromal cells (MSC)-based therapies for nervous system diseases. Important potential mechanisms of action for MSC action in nervous system diseases include one or more of the following: (1) secretion of neurotrophic factors, (2) promotion of angiogenesis, (3) stimulation of neurogenesis, and (4) modulation of immune responses. Note that some of the factors secreted by MSC may have multifactorial functions such as bFGF have effects on both neurogenesis and angiogenesis. In addition, the mechanisms of MSC action may vary depending on the specific disease of nervous system. *BDNF* brain-derived neurotrophic factor, *PDGF* platelet-derived growth factor, *bFGF* basic fibroblast growth factor, *TGF-β* transforming growth factor beta, *IGF-1* insulin-like growth factor 1, *NGF* nerve growth factor, *HGF* hepatocyte growth factor, *IDO* indoleamine 2,3-dioxygenase, *PGE-2* prostaglandin E 2, *VEGF* vascular endothelial growth factor, *IL-10* interleukin 10

be able to interpret the complex in vivo microenvironment and respond accordingly. This is supported by the evidence that MSC can switch their functions depending on the different environments of inflammation [101].

Indeed, MSC have been reported to adopt different actions in different animal models of nervous diseases. For example, in ischemic stroke conditions, MSC were able to enhance the angiogenesis by incorporating into the vascular system and also secreting angiogenic cytokine such as VEGF and bFGF [25, 26, 39, 40, 47, 48], in order to compensate the reduced blood supply after ischemia. In diseases where inflammation is the main cause such as multiple sclerosis and Alzheimer's disease, MSC may take a major role to inhibit the neuroinflammation and regulate the immune cells [37, 38, 59, 60, 74, 80–83]. Additionally, in neural autoimmune disease, loss of self-tolerance is one of the main contributors to the dysfunctional immune system. Like in autoimmune diseases of other tissue organs [102, 103], MSC are also able to restore the immune balance in neural autoimmune disease by regulating T-regulatory cell, which is crucial in maintaining the self-tolerance. UC-MSC have been shown to reverse the suppressive deficiency of T-regulatory cells from peripheral blood of patients with multiple sclerosis in a co-culture [104]. MSC also suppress the proliferation of activated T cells using contact-dependent and paracrine mechanisms, in which indoleamine 2,3-dioxygenase 1 may be one of the main effector molecules. In addition, UC-MSC could interfere with dendritic cell differentiation and maturation, thus directly affecting antigen presentation and therefore T-cell priming [79]. In another aspect, as neural cell damage or loss is commonly involved in nervous system diseases, the ability to stimulate the endogenous neural stem/progenitor cells and augment the neurogenesis is another mechanism of MSC actions in neural diseases [16, 46–48, 52, 53].

Although MSC mainly rely on their indirect mechanisms to function, it may still be true that the more transplanted cells home, survive, and engraft in the nervous system, the better will be for the therapeutic outcomes in neural disorders. Unfortunately, one big challenge of cell therapy for nervous system diseases is the low efficiency of homing, survival, and engraftment in the target site of injury [25, 26, 30], especially for systemically transplanted MSC. The blood-brain barrier presents the imperative obstacle for systemically transplanted MSC to migrate to the neural parenchyma [105, 106]. Nevertheless, the timing, routes, and dosage of cell administration may all account for the ultimate cell fate in vivo [107]. Kholodenko et al. [108] investigated penetration of placenta MSC through the brain-blood barrier and their distribution in the brain of experimental stroke animals using two models of stroke. The results demonstrated that the efficiency of MSC crossing the blood-brain barrier, the number of MSC attaining the ischemic focus and neurogenic zones, and the time of death of transplanted MSC largely depended on the degree and nature of injury to the central nervous system, thus should be taken into account when planning the experiments for evaluation of the effects of cell therapy on the models of neurological diseases and in clinical studies in the field of regenerative neurology.

Modification of MSC surface markers has also been applied to increase their homing. MSC express a series of adhesion molecules (e.g., VLA-4, VCAM-1) and chemokine receptors (e.g., CXCR4, CCR2) [81–83, 87, 109, 110]. However, unlike leukocytes that are equipped with a full set of homing molecules for migration across the endothelial barrier to target sites, MSC lack some key factors such as P-selectin glycoprotein ligand 1 (PSGL-1), LFA-1, and Mac-1 [111]. Therefore, modification of MSC by increasing of the existed homing molecules and more importantly expressing the nonexisted ones has been tested and proved to effectively enhance the homing ability of MSC to nervous system [112–114].

6.5 Conclusion and Future Perspectives

In appreciation of the advantages of perinatal tissues as the source for MSC compared to other adult tissues such as bone marrow and adipose, we have seen a tremendous increase of new preclinical studies that are using umbilical cord-, placenta-, or amnion-derived MSC to prove the concept of their therapeutic potentials in neural disorders. In agreement with the results from studies investigating other tissue-derived MSC, perinatal MSC show comparable or, in many cases, even more potent therapeutic effects in animal models of a variety of nervous system diseases. Moreover, resembling their counterparts in other tissues, perinatal MSC also act through multifactorial mechanisms for the beneficial outcomes.

In contrast, there are limited reports about clinical studies using perinatal MSC to date, which may be due to the complexity and time-consuming nature of clinical trial itself, the relatively low number of physicians that have recognized the invaluable source of perinatal tissues for MSC, or the limited number of GMP facilities that are able to manufacture clinical grade MSC from perinatal tissues. Nevertheless, it is optimistic to foresee that more and more results of clinical studies will come out in the soon future, as we have seen an increasing number of registered clinical trials (e.g., from www.clinicaltrials.gov) worldwide.

References

1. Eriksson PS, Perfilieva E, Bjork-Eriksson T, Alborn AM, Nordborg C, Peterson DA, Gage FH. Neurogenesis in the adult human hippocampus. Nat Med. 1998;4(11):1313–7.
2. Gross CG. Neurogenesis in the adult brain: death of a dogma. Nat Rev Neurosci. 2000;1(1):67–73.
3. Eriksson PS. Neurogenesis and its implications for regeneration in the adult brain. J Rehabil Med. 2003;(41 Suppl):17–9.
4. Gage FH, Kempermann G, Palmer TD, Peterson DA, Ray J. Multipotent progenitor cells in the adult dentate gyrus. J Neurobiol. 1998;36(2):249–66.
5. Bingham B, Liu D, Wood A, Cho S. Ischemia-stimulated neurogenesis is regulated by proliferation, migration, differentiation and caspase activation of hippocampal precursor cells. Brain Res. 2005;1058(1–2):167–77.
6. Ibrahim S, Hu W, Wang X, Gao X, He C, Chen J. Traumatic brain injury causes aberrant migration of adult-born neurons in the hippocampus. Sci Rep. 2016;6:21793.
7. Kokaia Z, Lindvall O. Neurogenesis after ischaemic brain insults. Curr Opin Neurobiol. 2003;13(1):127–32.
8. Rusanescu G, Mao J. Peripheral nerve injury induces adult brain neurogenesis and remodelling. J Cell Mol Med. 2017;21(2):299–314.
9. Yang H, Lu P, McKay HM, Bernot T, Keirstead H, Steward O, Gage FH, Edgerton VR, Tuszynski MH. Endogenous neurogenesis replaces oligodendrocytes and astrocytes after primate spinal cord injury. J Neurosci. 2006;26(8):2157–66.
10. Akiyama Y, Honmou O, Kato T, Uede T, Hashi K, Kocsis JD. Transplantation of clonal neural precursor cells derived from adult human brain establishes functional peripheral myelin in the rat spinal cord. Exp Neurol. 2001;167(1):27–39.
11. Chen KS, McIntyre JC, Lieberman AP, Martens JR, Patil PG. Human spinal autografts of olfactory epithelial stem cells recapitulate donor site histology, maintaining proliferative and differentiation capacity many years after transplantation. Acta Neuropathol. 2016;131(4):639–40.
12. Chen L, Zhang G, Gu Y, Guo X. Meta-analysis and systematic review of neural stem cells therapy for experimental ischemia stroke in preclinical studies. Sci Rep. 2016;6:32291.
13. Garitaonandia I, Gonzalez R, Christiansen-Weber T, Abramihina T, Poustovoitov M, Noskov A, Sherman G, Semechkin A, Snyder E, Kern R. Neural stem cell tumorigenicity and biodistribution assessment for phase I clinical trial in Parkinson's disease. Sci Rep. 2016;6:34478.
14. Zhang C, Cao J, Li X, Xu H, Wang W, Wang L, Zhao X, Li W, Jiao J, Hu B, Zhou Q, Zhao T. Treatment of multiple sclerosis by transplantation of neural stem cells derived from induced pluripotent stem cells. Sci China Life Sci. 2016;59(9):950–7.
15. Ramos-Gomez M, Martinez-Serrano A. Tracking of iron-labeled human neural stem cells by magnetic resonance imaging in cell replacement therapy for Parkinson's disease. Neural Regen Res. 2016;11(1):49–52.
16. Cheng Z, Zhu W, Cao K, Wu F, Li J, Wang G, Li H, Lu M, Ren Y, He X. Anti-inflammatory mechanism of neural stem cell transplantation in spinal cord injury. Int J Mol Sci. 2016;17(9).
17. Crisan M, Yap S, Casteilla L, Chen CW, Corselli M, Park TS, Andriolo G, Sun B, Zheng B, Zhang L, Norotte C, Teng PN, Traas J, Schugar R, Deasy BM, Badylak S, Buhring HJ, Giacobino JP, Lazzari L, Huard J, Peault B. A perivascular origin for mesenchymal stem cells in multiple human organs. Cell Stem Cell. 2008;3(3):301–13.
18. Murray IR, West CC, Hardy WR, James AW, Park TS, Nguyen A, Tawonsawatruk T, Lazzari L, Soo C, Peault B. Natural history of mesenchymal stem cells, from vessel walls to culture vessels. Cell Mol Life Sci. 2014;71(8):1353–74.
19. Centeno CJ, Al-Sayegh H, Freeman MD, Smith J, Murrell WD, Bubnov R. A multi-center analysis of adverse events among two thousand, three hundred and seventy two adult patients undergoing adult autologous stem cell therapy for orthopaedic conditions. Int Orthop. 2016;40(8):1755–65.
20. Lalu MM, McIntyre L, Pugliese C, Fergusson D, Winston BW, Marshall JC, Granton J, Stewart DJ, Canadian Critical Care Trials Group. Safety of cell therapy with mesenchymal stromal cells (SafeCell): a systematic review and meta-analysis of clinical trials. PLoS One. 2012;7(10):e47559.
21. Dlouhy BJ, Awe O, Rao RC, Kirby PA, Hitchon PW. Autograft-derived spinal cord mass following olfactory mucosal cell transplantation in a spinal cord injury patient: case report. J Neurosurg Spine. 2014;21(4):618–22.
22. Graham GD. Tissue plasminogen activator for acute ischemic stroke in clinical practice: a meta-analysis of safety data. Stroke. 2003;34(12):2847–50.
23. Seet RC, Rabinstein AA. Symptomatic intracranial hemorrhage following intravenous thrombolysis for acute ischemic stroke: a critical review of case definitions. Cerebrovasc Dis. 2012;34(2):106–14.
24. Koh SH, Kim KS, Choi MR, Jung KH, Park KS, Chai YG, Roh W, Hwang SJ, Ko HJ, Huh YM, Kim HT, Kim SH. Implantation of human umbilical cord-derived mesenchymal stem cells as a neuroprotective therapy for ischemic stroke in rats. Brain Res. 2008;1229:233–48.
25. Liao W, Xie J, Zhong J, Liu Y, Du L, Zhou B, Xu J, Liu P, Yang S, Wang J, Han Z, Han ZC. Therapeutic effect of human umbilical cord multipotent mesenchymal stromal cells in a rat model of stroke. Transplantation. 2009;87(3):350–9.
26. Liao W, Zhong J, Yu J, Xie J, Liu Y, Du L, Yang S, Liu P, Xu J, Wang J, Han Z, Han ZC. Therapeutic

benefit of human umbilical cord derived mesenchymal stromal cells in intracerebral hemorrhage rat: implications of anti-inflammation and angiogenesis. Cell Physiol Biochem. 2009;24(3–4):307–16.
27. Zhang L, Yi L, Chopp M, Kramer BC, Romanko M, Gosiewska A, Hong K. Intravenous administration of human umbilical tissue-derived cells improves neurological function in aged rats after embolic stroke. Cell Transplant. 2013;22(9):1569–76.
28. Shams ara A, Sheibani V, Esmaeilpour K, Eslaminejad T, Nematollahi-Mahani SN. Coadministration of the human umbilical cord matrix-derived mesenchymal cells and aspirin alters postischemic brain injury in rats. J Stroke Cerebrovasc Dis. 2015;24(9):2005–16.
29. Jiang Q, Thiffault C, Kramer BC, Ding GL, Zhang L, Nejad-Davarani SP, Li L, Arbab AS, Lu M, Navia B, Victor SJ, Hong K, Li QJ, Wang SY, Li Y, Chopp M. MRI detects brain reorganization after human umbilical tissue-derived cells (hUTC) treatment of stroke in rat. PLoS One. 2012;7(8):e42845.
30. Arbab AS, Thiffault C, Navia B, Victor SJ, Hong K, Zhang L, Jiang Q, Varma NR, Iskander A, Chopp M. Tracking of In-111-labeled human umbilical tissue-derived cells (hUTC) in a rat model of cerebral ischemia using SPECT imaging. BMC Med Imaging. 2012;12:33.
31. Shehadah A, Chen J, Kramer B, Zacharek A, Cui Y, Roberts C, Lu M, Chopp M. Efficacy of single and multiple injections of human umbilical tissue-derived cells following experimental stroke in rats. PLoS One. 2013;8(1):e54083.
32. Zhang L, Li Y, Romanko M, Kramer BC, Gosiewska A, Chopp M, Hong K. Different routes of administration of human umbilical tissue-derived cells improve functional recovery in the rat after focal cerebral ischemia. Brain Res. 2012;1489:104–12.
33. Zhang L, Li Y, Zhang C, Chopp M, Gosiewska A, Hong K. Delayed administration of human umbilical tissue-derived cells improved neurological functional recovery in a rodent model of focal ischemia. Stroke. 2011;42(5):1437–44.
34. Zhou X, Gu J, Gu Y, He M, Bi Y, Chen J, Li T. Human umbilical cord-derived mesenchymal stem cells improve learning and memory function in hypoxic-ischemic brain-damaged rats via an IL-8-mediated secretion mechanism rather than differentiation pattern induction. Cell Physiol Biochem. 2015;35(6):2383–401.
35. Chung DJ, Choi CB, Lee SH, Kang EH, Lee JH, Hwang SH, Han H, Lee JH, Choe BY, Lee SY, Kim HY. Intraarterially delivered human umbilical cord blood-derived mesenchymal stem cells in canine cerebral ischemia. J Neurosci Res. 2009;87(16):3554–67.
36. Lim JY, Jeong CH, Jun JA, Kim SM, Ryu CH, Hou Y, Oh W, Chang JW, Jeun SS. Therapeutic effects of human umbilical cord blood-derived mesenchymal stem cells after intrathecal administration by lumbar puncture in a rat model of cerebral ischemia. Stem Cell Res Ther. 2011;2(5):38.
37. Kim ES, Ahn SY, Im GH, Sung DK, Park YR, Choi SH, Choi SJ, Chang YS, Oh W, Lee JH, Park WS. Human umbilical cord blood-derived mesenchymal stem cell transplantation attenuates severe brain injury by permanent middle cerebral artery occlusion in newborn rats. Pediatr Res. 2012;72(3):277–84.
38. Kim JY, Kim DH, Kim JH, Lee D, Jeon HB, Kwon SJ, Kim SM, Yoo YJ, Lee EH, Choi SJ, Seo SW, Lee JI, Na DL, Yang YS, Oh W, Chang JW. Soluble intracellular adhesion molecule-1 secreted by human umbilical cord blood-derived mesenchymal stem cell reduces amyloid-beta plaques. Cell Death Differ. 2012;19(4):680–91.
39. Park HW, Chang JW, Yang YS, Oh W, Hwang JH, Kim DG, Paek SH. The effect of donor-dependent administration of human umbilical cord blood-derived mesenchymal stem cells following focal cerebral ischemia in rats. Exp Neurobiol. 2015;24(4):358–65.
40. Park HW, Moon HE, Kim HS, Paek SL, Kim Y, Chang JW, Yang YS, Kim K, Oh W, Hwang JH, Kim JW, Kim DG, Paek SH. Human umbilical cord blood-derived mesenchymal stem cells improve functional recovery through thrombospondin1, pantraxin3, and vascular endothelial growth factor in the ischemic rat brain. J Neurosci Res. 2015;93(12):1814–25.
41. Hocum Stone LL, Xiao F, Rotschafer J, Nan Z, Juliano M, Sanberg CD, Sanberg PR, Kuzmin-Nichols N, Grande A, Cheeran MC, Low WC. Amelioration of ischemic brain injury in rats with human umbilical cord blood stem cells: mechanisms of action. Cell Transplant. 2016;25(8):1473–88.
42. Yu SJ, Soncini M, Kaneko Y, Hess DC, Parolini O, Borlongan CV. Amnion: a potent graft source for cell therapy in stroke. Cell Transplant. 2009;18(2):111–8.
43. Rehni AK, Singh N, Jaggi AS, Singh M. Amniotic fluid derived stem cells ameliorate focal cerebral ischaemia-reperfusion injury induced behavioural deficits in mice. Behav Brain Res. 2007;183(1):95–100.
44. Liu T, Wu J, Huang Q, Hou Y, Jiang Z, Zang S, Guo L. Human amniotic epithelial cells ameliorate behavioral dysfunction and reduce infarct size in the rat middle cerebral artery occlusion model. Shock. 2008;29(5):603–11.
45. Tajiri N, Acosta S, Glover LE, Bickford PC, Jacotte Simancas A, Yasuhara T, Date I, Solomita MA, Antonucci I, Stuppia L, Kaneko Y, Borlongan CV. Intravenous grafts of amniotic fluid-derived stem cells induce endogenous cell proliferation and attenuate behavioral deficits in ischemic stroke rats. PLoS One. 2012;7(8):e43779.
46. Yarygin KN, Kholodenko IV, Konieva AA, Burunova VV, Tairova RT, Gubsky LV, Cheglakov IB, Pirogov YA, Yarygin VN, Skvortsova VI. Mechanisms of positive effects of transplantation of human placental mesenchymal stem cells on recovery of rats after experimental ischemic stroke. Bull Exp Biol Med. 2009;148(6):862–8.

47. Chen J, Shehadah A, Pal A, Zacharek A, Cui X, Cui Y, Roberts C, Lu M, Zeitlin A, Hariri R, Chopp M. Neuroprotective effect of human placenta-derived cell treatment of stroke in rats. Cell Transplant. 2013;22(5):871–9.
48. Chen M, Xiang Z, Cai J. The anti-apoptotic and neuro-protective effects of human umbilical cord blood mesenchymal stem cells (hUCB-MSCs) on acute optic nerve injury is transient. Brain Res. 2013;1532:63–75.
49. Kranz A, Wagner DC, Kamprad M, Scholz M, Schmidt UR, Nitzsche F, Aberman Z, Emmrich F, Riegelsberger UM, Boltze J. Transplantation of placenta-derived mesenchymal stromal cells upon experimental stroke in rats. Brain Res. 2010;1315:128–36.
50. Cui B, Li E, Yang B, Wang B. Human umbilical cord blood-derived mesenchymal stem cell transplantation for the treatment of spinal cord injury. Exp Ther Med. 2014;7(5):1233–6.
51. Lim JH, Byeon YE, Ryu HH, Jeong YH, Lee YW, Kim WH, Kang KS, Kweon OK. Transplantation of canine umbilical cord blood-derived mesenchymal stem cells in experimentally induced spinal cord injured dogs. J Vet Sci. 2007;8(3):275–82.
52. Park S, Kim E, Koh SE, Maeng S, Lee WD, Lim J, Shim I, Lee YJ. Dopaminergic differentiation of neural progenitors derived from placental mesenchymal stem cells in the brains of Parkinson's disease model rats and alleviation of asymmetric rotational behavior. Brain Res. 2012;1466:158–66.
53. Park SI, Lim JY, Jeong CH, Kim SM, Jun JA, Jeun SS, Oh WI. Human umbilical cord blood-derived mesenchymal stem cell therapy promotes functional recovery of contused rat spinal cord through enhancement of endogenous cell proliferation and oligogenesis. J Biomed Biotechnol. 2012;2012:362473.
54. Ryu HH, Kang BJ, Park SS, Kim Y, Sung GJ, Woo HM, Kim WH, Kweon OK. Comparison of mesenchymal stem cells derived from fat, bone marrow, Wharton's jelly, and umbilical cord blood for treating spinal cord injuries in dogs. J Vet Med Sci. 2012;74(12):1617–30.
55. Yousefifard M, Nasirinezhad F, Shardi Manaheji H, Janzadeh A, Hosseini M, Keshavarz M. Human bone marrow-derived and umbilical cord-derived mesenchymal stem cells for alleviating neuropathic pain in a spinal cord injury model. Stem Cell Res Ther. 2016;7:36.
56. Roh DH, Seo MS, Choi HS, Park SB, Han HJ, Beitz AJ, Kang KS, Lee JH. Transplantation of human umbilical cord blood or amniotic epithelial stem cells alleviates mechanical allodynia after spinal cord injury in rats. Cell Transplant. 2013;22(9):1577–90.
57. Borkowska P, Fila-Danilow A, Paul-Samojedny M, Kowalczyk M, Hart J, Ryszawy J, Kowalski J. Differentiation of adult rat mesenchymal stem cells to GABAergic, dopaminergic and cholinergic neurons. Pharmacol Rep. 2015;67(2):179–86.
58. Chun SY, Soker S, Jang YJ, Kwon TG, Yoo ES. Differentiation of human dental pulp stem cells into dopaminergic neuron-like cells in vitro. J Korean Med Sci. 2016;31(2):171–7.
59. Yang H, Xie Z, Wei L, Yang H, Yang S, Zhu Z, Wang P, Zhao C, Bi J. Human umbilical cord mesenchymal stem cell-derived neuron-like cells rescue memory deficits and reduce amyloid-beta deposition in an AbetaPP/PS1 transgenic mouse model. Stem Cell Res Ther. 2013;4(4):76.
60. Yang S, Sun HM, Yan JH, Xue H, Wu B, Dong F, Li WS, Ji FQ, Zhou DS. Conditioned medium from human amniotic epithelial cells may induce the differentiation of human umbilical cord blood mesenchymal stem cells into dopaminergic neuron-like cells. J Neurosci Res. 2013;91(7):978–86.
61. Kakishita K, Nakao N, Sakuragawa N, Itakura T. Implantation of human amniotic epithelial cells prevents the degeneration of nigral dopamine neurons in rats with 6-hydroxydopamine lesions. Brain Res. 2003;980(1):48–56.
62. Yang X, Song L, Wu N, Liu Z, Xue S, Hui G. An experimental study on intracerebroventricular transplantation of human amniotic epithelial cells in a rat model of Parkinson's disease. Neurol Res. 2010;32(10):1054–9.
63. Shetty P, Thakur AM, Viswanathan C. Dopaminergic cells, derived from a high efficiency differentiation protocol from umbilical cord derived mesenchymal stem cells, alleviate symptoms in a Parkinson's disease rodent model. Cell Biol Int. 2013;37(2):167–80.
64. Zhao C, Li H, Zhao XJ, Liu ZX, Zhou P, Liu Y, Feng MJ. Heat shock protein 60 affects behavioral improvement in a rat model of Parkinson's disease grafted with human umbilical cord mesenchymal stem cell-derived dopaminergic-like neurons. Neurochem Res. 2016;41(6):1238–49.
65. Liu XS, Li JF, Wang SS, Wang YT, Zhang YZ, Yin HL, Geng S, Gong HC, Han B, Wang YL. Human umbilical cord mesenchymal stem cells infected with adenovirus expressing HGF promote regeneration of damaged neuron cells in a Parkinson's disease model. Biomed Res Int. 2014;2014:909657.
66. Yan M, Sun M, Zhou Y, Wang W, He Z, Tang D, Lu S, Wang X, Li S, Wang W, Li H. Conversion of human umbilical cord mesenchymal stem cells in Wharton's jelly to dopamine neurons mediated by the Lmx1a and neurturin in vitro: potential therapeutic application for Parkinson's disease in a rhesus monkey model. PLoS One. 2013;8(5):e64000.
67. Guerriero F, Sgarlata C, Francis M, Maurizi N, Faragli A, Perna S, Rondanelli M, Rollone M, Ricevuti G. Neuroinflammation, immune system and Alzheimer disease: searching for the missing link. Aging Clin Exp Res. 2017;29(5):821–31.
68. Minter MR, Taylor JM, Crack PJ. The contribution of neuroinflammation to amyloid toxicity in Alzheimer's disease. J Neurochem. 2016;136(3):457–74.
69. Gao F, Chiu SM, Motan DA, Zhang Z, Chen L, Ji HL, Tse HF, Fu QL, Lian Q. Mesenchymal stem cells and immunomodulation: current status and future prospects. Cell Death Dis. 2016;7:e2062.

70. Prockop DJ, Oh JY. Mesenchymal stem/stromal cells (MSCs): role as guardians of inflammation. Mol Ther. 2012;20(1):14–20.
71. Uccelli A, de Rosbo NK. The immunomodulatory function of mesenchymal stem cells: mode of action and pathways. Ann N Y Acad Sci. 2015;1351:114–26.
72. Lee HJ, Lee JK, Lee H, Shin JW, Carter JE, Sakamoto T, Jin HK, Bae JS. The therapeutic potential of human umbilical cord blood-derived mesenchymal stem cells in Alzheimer's disease. Neurosci Lett. 2010;481(1):30–5.
73. Boutajangout A, Noorwali A, Atta H, Wisniewski T. Human umbilical cord stem cell xenografts improve cognitive decline and reduce the amyloid burden in a mouse model of Alzheimer's disease. Curr Alzheimer Res. 2017;14(1):104–11.
74. Kim DH, Lee D, Chang EH, Kim JH, Hwang JW, Kim JY, Kyung JW, Kim SH, Oh JS, Shim SM, Na DL, Oh W, Chang JW. GDF-15 secreted from human umbilical cord blood mesenchymal stem cells delivered through the cerebrospinal fluid promotes hippocampal neurogenesis and synaptic activity in an Alzheimer's disease model. Stem Cells Dev. 2015;24(20):2378–90.
75. Jiao H, Shi K, Zhang W, Yang L, Yang L, Guan F, Yang B. Therapeutic potential of human amniotic membrane-derived mesenchymal stem cells in APP transgenic mice. Oncol Lett. 2016;12(3):1877–83.
76. Kim KS, Kim HS, Park JM, Kim HW, Park MK, Lee HS, Lim DS, Lee TH, Chopp M, Moon J. Long-term immunomodulatory effect of amniotic stem cells in an Alzheimer's disease model. Neurobiol Aging. 2013;34(10):2408–20.
77. Yun HM, Kim HS, Park KR, Shin JM, Kang AR, il Lee K, Song S, Kim YB, Han SB, Chung HM, Hong JT. Placenta-derived mesenchymal stem cells improve memory dysfunction in an Abeta1-42-infused mouse model of Alzheimer's disease. Cell Death Dis. 2013;4:e958.
78. Selim AO, Selim SA, Shalaby SM, Mosaad H, Saber T. Neuroprotective effects of placenta-derived mesenchymal stromal cells in a rat model of experimental autoimmune encephalomyelitis. Cytotherapy. 2016;18(9):1100–13.
79. Donders R, Vanheusden M, Bogie JF, Ravanidis S, Thewissen K, Stinissen P, Gyselaers W, Hendriks JJ, Hellings N. Human Wharton's jelly-derived stem cells display immunomodulatory properties and transiently improve rat experimental autoimmune encephalomyelitis. Cell Transplant. 2015;24(10):2077–98.
80. Fisher-Shoval Y, Barhum Y, Sadan O, Yust-Katz S, Ben-Zur T, Lev N, Benkler C, Hod M, Melamed E, Offen D. Transplantation of placenta-derived mesenchymal stem cells in the EAE mouse model of MS. J Mol Neurosci. 2012;48(1):176–84.
81. Liu J, Han D, Wang Z, Xue M, Zhu L, Yan H, Zheng X, Guo Z, Wang H. Clinical analysis of the treatment of spinal cord injury with umbilical cord mesenchymal stem cells. Cytotherapy. 2013;15(2):185–91.
82. Liu L, Eckert MA, Riazifar H, Kang DK, Agalliu D, Zhao W. From blood to the brain: can systemically transplanted mesenchymal stem cells cross the blood-brain barrier? Stem Cells Int. 2013;2013:435093.
83. Liu R, Zhang Z, Lu Z, Borlongan C, Pan J, Chen J, Qian L, Liu Z, Zhu L, Zhang J, Xu Y. Human umbilical cord stem cells ameliorate experimental autoimmune encephalomyelitis by regulating immunoinflammation and remyelination. Stem Cells Dev. 2013;22(7):1053–62.
84. Yu J, Zheng C, Ren X, Li J, Liu M, Zhang L, Liang L, Du W, Han ZC. Intravenous administration of bone marrow mesenchymal stem cells benefits experimental autoimmune myasthenia gravis mice through an immunomodulatory action. Scand J Immunol. 2010;72(3):242–9.
85. Sun H, Hou Z, Yang H, Meng M, Li P, Zou Q, Yang L, Chen Y, Chai H, Zhong H, Yang ZZ, Zhao J, Lai L, Jiang X, Xiao Z. Multiple systemic transplantations of human amniotic mesenchymal stem cells exert therapeutic effects in an ALS mouse model. Cell Tissue Res. 2014;357(3):571–82.
86. Cheng T, Yang B, Li D, Ma S, Tian Y, Qu R, Zhang W, Zhang Y, Hu K, Guan F, Wang J. Wharton's Jelly transplantation improves neurologic function in a rat model of traumatic brain injury. Cell Mol Neurobiol. 2015;35(5):641–9.
87. Ma J, Liu N, Yi B, Zhang X, Gao BB, Zhang Y, Xu R, Li X, Dai Y. Transplanted hUCB-MSCs migrated to the damaged area by SDF-1/CXCR4 signaling to promote functional recovery after traumatic brain injury in rats. Neurol Res. 2015;37(1):50–6.
88. Pischiutta F, Brunelli L, Romele P, Silini A, Sammali E, Paracchini L, Marchini S, Talamini L, Bigini P, Boncoraglio GB, Pastorelli R, De Simoni MG, Parolini O, Zanier ER. Protection of brain injury by amniotic mesenchymal stromal cell-secreted metabolites. Crit Care Med. 2016;44(11):e1118–31.
89. Fink KD, Rossignol J, Crane AT, Davis KK, Bombard MC, Bavar AM, Clerc S, Lowrance SA, Song C, Lescaudron L, Dunbar GL. Transplantation of umbilical cord-derived mesenchymal stem cells into the striata of R6/2 mice: behavioral and neuropathological analysis. Stem Cell Res Ther. 2013;4(5):130.
90. Yang DY, Sheu ML, Su HL, Cheng FC, Chen YJ, Chen CJ, Chiu WT, Yiin JJ, Sheehan J, Pan HC. Dual regeneration of muscle and nerve by intravenous administration of human amniotic fluid-derived mesenchymal stem cells regulated by stromal cell-derived factor-1alpha in a sciatic nerve injury model. J Neurosurg. 2012;116(6):1357–67.
91. Dionigi B, Ahmed A, Brazzo J 3rd, Connors JP, Zurakowski D, Fauza DO. Partial or complete coverage of experimental spina bifida by simple intra-amniotic injection of concentrated amniotic mesenchymal stem cells. J Pediatr Surg. 2015;50(1):69–73.
92. Cheng H, Liu X, Hua R, Dai G, Wang X, Gao J, An Y. Clinical observation of umbilical cord mesenchymal stem cell transplantation in treatment

for sequelae of thoracolumbar spinal cord injury. J Transl Med. 2014;12:253.
93. Dai G, Liu X, Zhang Z, Wang X, Li M, Cheng H, Hua R, Shi J, Wang R, Qin C, Gao J, An Y. Comparative analysis of curative effect of CT-guided stem cell transplantation and open surgical transplantation for sequelae of spinal cord injury. J Transl Med. 2013;11:315.
94. Jiang Y, Zhu W, Zhu J, Wu L, Xu G, Liu X. Feasibility of delivering mesenchymal stem cells via catheter to the proximal end of the lesion artery in patients with stroke in the territory of the middle cerebral artery. Cell Transplant. 2013;22(12):2291–8.
95. Li JF, Zhang DJ, Geng T, Chen L, Huang H, Yin HL, Zhang YZ, Lou JY, Cao B, Wang YL. The potential of human umbilical cord-derived mesenchymal stem cells as a novel cellular therapy for multiple sclerosis. Cell Transplant. 2014;23(Suppl 1):S113–22.
96. Lublin FD, Bowen JD, Huddlestone J, Kremenchutzky M, Carpenter A, Corboy JR, Freedman MS, Krupp L, Paulo C, Hariri RJ, Fischkoff SA. Human placenta-derived cells (PDA-001) for the treatment of adults with multiple sclerosis: a randomized, placebo-controlled, multiple-dose study. Mult Scler Relat Disord. 2014;3(6):696–704.
97. Lv YT, Zhang Y, Liu M, Qiuwaxi JN, Ashwood P, Cho SC, Huan Y, Ge RC, Chen XW, Wang ZJ, Kim BJ, Hu X. Transplantation of human cord blood mononuclear cells and umbilical cord-derived mesenchymal stem cells in autism. J Transl Med. 2013;11:196.
98. Li Q, Chen CF, Wang DY, Lu YT, Huan Y, Fang SX, Han Y, Ge RC, Chen XW. Changes in growth factor levels in the cerebrospinal fluid of autism patients after transplantation of human umbilical cord blood mononuclear cells and umbilical cord-derived mesenchymal stem cells. Genet Mol Res. 2016;15(2).
99. Miao X, Wu X, Shi W. Umbilical cord mesenchymal stem cells in neurological disorders: a clinical study. Indian J Biochem Biophys. 2015;52(2):140–6.
100. Zhang R, Chen H, Zheng Z, Liu Q, Xu L. Umbilical cord-derived mesenchymal stem cell therapy for neurological disorders via inhibition of mitogen-activated protein kinase pathway-mediated apoptosis. Mol Med Rep. 2015;11(3):1807–12.
101. Bernardo ME, Fibbe WE. Mesenchymal stromal cells: sensors and switchers of inflammation. Cell Stem Cell. 2013;13(4):392–402.
102. Chao YH, Wu HP, Wu KH, Tsai YG, Peng CT, Lin KC, Chao WR, Lee MS, Fu YC. An increase in CD3+CD4+CD25+ regulatory T cells after administration of umbilical cord-derived mesenchymal stem cells during sepsis. PLoS One. 2014;9(10):e110338.
103. Wang L, Zhang Y, Li H, Hong J, Chen X, Li M, Bai W, Wang J, Liu Y, Wu M. Clinical observation of employment of umbilical cord derived mesenchymal stem cell for juvenile idiopathic arthritis therapy. Stem Cells Int. 2016;2016:9165267.
104. Yang H, Sun J, Wang F, Li Y, Bi J, Qu T. Umbilical cord-derived mesenchymal stem cells reversed the suppressive deficiency of T regulatory cells from peripheral blood of patients with multiple sclerosis in a co-culture - a preliminary study. Oncotarget. 2016;7(45):72537–45.
105. Lee NK, Yang J, Chang EH, Park SE, Lee J, Choi SJ, Oh W, Chang JW, Na DL. Intra-arterially delivered mesenchymal stem cells are not detected in the brain parenchyma in an Alzheimer's disease mouse model. PLoS One. 2016;11(5):e0155912.
106. Park SE, Lee NK, Lee J, Hwang JW, Choi SJ, Hwang H, Hyung B, Chang JW, Na DL. Distribution of human umbilical cord blood-derived mesenchymal stem cells in the Alzheimer's disease transgenic mouse after a single intravenous injection. Neuroreport. 2016;27(4):235–41.
107. Violatto MB, Santangelo C, Capelli C, Frapolli R, Ferrari R, Sitia L, Tortarolo M, Talamini L, Previdi S, Moscatelli D, Salmona M, Introna M, Bendotti C, Bigini P. Longitudinal tracking of triple labeled umbilical cord derived mesenchymal stromal cells in a mouse model of Amyotrophic Lateral Sclerosis. Stem Cell Res. 2015;15(1):243–53.
108. Kholodenko IV, Yarygin KN, Gubsky LV, Konieva AA, Tairova RT, Povarova OV, Kholodenko RV, Burunova VV, Yarygin VN, Skvortsova VI. Intravenous xenotransplantation of human placental mesenchymal stem cells to rats: comparative analysis of homing in rat brain in two models of experimental ischemic stroke. Bull Exp Biol Med. 2012;154(1):118–23.
109. Ruster B, Gottig S, Ludwig RJ, Bistrian R, Muller S, Seifried E, Gille J, Henschler R. Mesenchymal stem cells display coordinated rolling and adhesion behavior on endothelial cells. Blood. 2006;108(12):3938–44.
110. Steingen C, Brenig F, Baumgartner L, Schmidt J, Schmidt A, Bloch W. Characterization of key mechanisms in transmigration and invasion of mesenchymal stem cells. J Mol Cell Cardiol. 2008;44(6):1072–84.
111. da Silva Meirelles L, Caplan AI, Nardi NB. In search of the in vivo identity of mesenchymal stem cells. Stem Cells. 2008;26(9):2287–99.
112. Liao W, Pham V, Liu L, Riazifar M, Pone EJ, Zhang SX, Ma F, Lu M, Walsh CM, Zhao W. Mesenchymal stem cells engineered to express selectin ligands and IL-10 exert enhanced therapeutic efficacy in murine experimental autoimmune encephalomyelitis. Biomaterials. 2016;77:87–97.
113. Shi W, Huang CJ, Xu XD, Jin GH, Huang RQ, Huang JF, Chen YN, Ju SQ, Wang Y, Shi YW, Qin JB, Zhang YQ, Liu QQ, Wang XB, Zhang XH, Chen J. Transplantation of RADA16-BDNF peptide scaffold with human umbilical cord mesenchymal stem cells forced with CXCR4 and activated astrocytes for repair of traumatic brain injury. Acta Biomater. 2016;45:247–61.
114. Yu X, Chen D, Zhang Y, Wu X, Huang Z, Zhou H, Zhang Y, Zhang Z. Overexpression of CXCR4 in mesenchymal stem cells promotes migration, neuroprotection and angiogenesis in a rat model of stroke. J Neurol Sci. 2012;316(1–2):141–9.

Umbilical Cord Blood as a Source of Novel Reagents and Therapeutics

7

Paolo Rebulla, Sergi Querol, and Alejandro Madrigal

7.1 Introduction: Why Developing Novel Uses of Umbilical Cord Blood?

The newborn's blood left in the placenta after delivery (usually termed "placental blood" or "umbilical cord blood," UCB) contains a large number of therapeutically valuable hemopoietic stem and progenitor cells. UCB can be donated to the community for solidarity purposes by informed, consenting parents [1, 2]. After collection by venipuncture of umbilical cord vessels by trained midwives or other health operators, UCB is transported at controlled temperature to UCB banks, where it is processed, cryopreserved, and stored in liquid nitrogen for future use in hemopoietic transplant procedures performed in patients suffering from severe blood diseases.

The therapeutic efficacy of UCB transplant is documented by several large clinical trials performed with both family-related and family-unrelated UCB donors and by registry data [3]. Following the implementation of not-for-profit cord blood banking programs in the early 1990s, a global inventory of allogeneic UCB units currently exceeding 700,000 units has been made available to transplant centers by 139 public cord blood banks, which have facilitated the performance of more than 30,000 UCB transplants [3]. Accreditation programs are available which certify conformance of cord blood banking technical and operational procedures with national regulations and internationally recognized quality standards [4].

As compared to transplants performed with the traditional sources of therapeutic hemopoietic stem cells, represented by the bone marrow or mobilized peripheral blood from fully matched family-related and family-unrelated donors, UCB transplant offers the important advantages of more permissible donor-to-recipient compatibility for the human leukocyte antigen (HLA) system, prompt availability and lower frequency, and severity of graft-versus-host disease [3]. Despite these advantages, UCB transplant is currently hampered by higher transplant-related mortality and cost caused by longer hospital patient stay due to delayed neutrophil and platelet reconstitution. Although it is hoped that delayed reconstitution could be overcome by novel drugs facilitating UCB stem cell homing into the bone marrow [5] and pretransplant patient conditioning procedures [6], significant economic support is and will be needed for long-term UCB cryopreservation in technologically advanced and expensive facilities.

P. Rebulla (✉)
Department of Transfusion Medicine and Hematology, Foundation Ca' Granda Ospedale Maggiore Policlinico, Milan, Italy
e-mail: paolo.rebulla@policlinico.mi.it

S. Querol
Barcelona Cord Blood Bank, Programa Concordia, Banc Sang i Teixits, Barcelona, Spain

A. Madrigal
Anthony Nolan Research Institute, Royal Free and University College, London, UK

Z. C. Han et al. (eds.), *Perinatal Stem Cells*, https://doi.org/10.1007/978-981-13-2703-2_7

The above scenario has become even more critical in the last few years, due to a progressive decrease in the number of UCB units released and a parallel reduction of cost recovery by the public banks which followed an increased favor of clinicians for hemopoietic stem cell transplants performed using HLA haplo-identical donors, usually represented by readily available and highly motivated patients' parents and siblings [7, 8]. Although conclusive evidence comparing the long-term clinical outcomes of UCB versus HLA haplo-identical transplant is still lacking [9, 10], concern has been expressed by some administrators on the long-term sustainability of public UCB banking [11].

Another element of potential difficulty for public UCB banking programs is the evidence that banked UCB units unable to provide a minimum cell dose per kg of the recipient body weight have been and will probably be infrequently requested by transplant centers, due to their poor clinical outcome. To limit the negative economic impact of such units with low cell count, several public banks have recently chosen to discard donated UCB units containing less than 1.2–1.5×10^9 nucleated cells. Although the current appropriateness of choosing high cell count thresholds for banking is hardly disputable, it is possible that a desirable hemopoietic transplant inventory expansion with only units with very high cell counts could be negatively impacted by the recent concern expressed by obstetric societies on the practice of immediate or very early clamping of the umbilical cord after delivery [12, 13] and by the World Health Organization recommendation to implement routine procedures of delayed cord clamping [14], as the latter cause smaller newborn's blood volume and number of cells left in the placenta [15, 16]. In this regard, it was encouraging that the National Swedish Cord Blood Bank reported in 2016 that implementation of a change from early cord clamping (within 15 s after birth) to delayed clamping at 60 s after birth was associated with a modest reduction of the collected CB volume from 119 ± 28 mL to 111 ± 32 mL [17]. However, opposite conclusions were drawn by the Canadian Blood Services' Cord Blood Bank, which reported that "delayed clamping greatly diminishes the volume and TNC [total nucleated cell] threshold of units collected for a public cord blood bank" [18]. Additional studies are needed to show if the findings of the study performed in Sweden, which used a dedicated staff of six trained midwives working full time with day-time CB collection, 7 days a week, can be confirmed in settings with different collection organizations, particularly so at sites where delayed clamping at times greater than 60–120 s are used.

In spite of all the above difficulties and limitations, it would be highly undesirable to significantly reduce the number of UCB solidarity donations and the availability of public cord blood banking programs, as many hemopoietic transplant candidates do not currently find a compatible donor within their family or in the family-unrelated donor registries, particularly so in ethnic minorities [19]. Sustainability of the public programs and responsible use of the available resources could be improved by reorganization and consolidation of the processing and storage facilities, which have already been undertaken in some countries [11]. Moreover, new products obtained from UCB units donated for solidarity purposes but not qualified for hemopoietic transplant due to low cell count could contribute not only to offer new therapeutic options to patients but also to improve cost recovery by the public banks. In this regard, annual reports from a large public cord blood bank during 2012–2016 show that 87% of 11,120 UCB solidary collections were not processed for storage into the hemopoietic inventory due to volume below 60 mL or total nucleated cell count below 1.5×10^9 (L. Lecchi, written communication, 23 February 2017). Overall, other recent reports from public banks adopting high total nucleated cell thresholds show that more than 80% of UCB units are not banked for hemopoietic transplant purposes. Such high proportions of routinely and aseptically collected units of biological material donated by carefully qualified donors screened for transmissible diseases, whose data are registered in fully traceable electronic data sets, offer the opportunity for novel products development.

This chapter describes some innovative products developed from UCB and discusses the perspectives of their use in the laboratory and for therapeutic applications.

7.2 Cell and Tissue Culture Media

Platelets contain a large number of biological molecules—including coagulation factors and adhesion molecules, enzymes, chemokines, cytokines, and growth factors that collectively constitute the so-called platelet granule cargo—which play important roles in physiological and pathological processes related to hemostasis, immunity, angiogenesis, and tissue repair (Table 7.1). Due to the above biological functions and the relative easiness of procurement, expired platelet concentrates originally collected from adult volunteer blood donors for transfusion purposes have been used since the early 1980s to prepare several homemade and commercial additives to supplement minimal essential media for in vitro cell and tissue culture [20, 21].

More recently, increased interest in their use was generated by concern caused by the traditional use of fetal bovine serum as a supplement in media for good manufacturing practice (GMP) culture of advanced therapy medicinal products (ATMPs), which carries the risk of xenoimmunization and transmission of nonhuman pathogenic agents to ATMP recipients.

Table 7.1 A partial list of cell growth factors present in human platelets [20, 21]

Molecule	Acronym
Brain-derived neurotrophic factor	BDNF
Basic fibroblast growth factor	bFGF
Bone morphogenetic protein	BMP
Connective tissue growth factor	CTGF
Epidermal growth factor	EGF
Hepatocyte growth factor	HGF
Insulin-like growth factor-1	IGF-1
Matrix metalloproteinase	MMP
Platelet-derived growth factor	PDGF
Platelet factor 4	PF4
Stromal cell-derived factor-1	SDF-1
Transforming growth factor-beta	TGF-beta
Tissue inhibitor of metalloproteinases	TIMPs
Vascular endothelial growth factor	VEGF

Recent reviews describe in detail the processing technologies and the in vitro biological efficacy of media supplements so far obtained from expired platelet concentrates prepared from blood of adult donors [20, 21]. New studies aimed at a comparative analysis of adult versus UCB platelets identified interesting quantitative and qualitative variations of their proteomes [22], suggesting that culture media supplements obtained from the two sources could support alternative differentiation pathways during cell culture. Further studies are needed to define appropriate standards for platelet-derived supplements from adult blood and UCB. Moreover, programs should be developed to harmonize national and international regulatory pathways for their procurement and use in different jurisdictions.

7.3 Platelet Gel

The active role played by platelets in tissue regeneration prompted the use of platelets from blood obtained from both autologous and allogeneic adult donors for the preparation of a blood component named "platelet gel." Gelification occurs after addition of calcium gluconate and/or batroxobin to platelets obtained by differential centrifugation of whole blood and suspended in plasma at a concentration 4–5 times higher than physiologic values. Both intact platelets and a platelet lysate obtained by one or more freezing and thawing procedures can be used for the preparation of the gel.

In 2008, a patented procedure was developed for the preparation of a platelet lysate (PL) and platelet gel (PG) from UCB units not suitable for hemopoietic transplantation due to low cell count [23] (US patent no. 8501170: Platelet fraction deriving from placental blood). UCBPL is obtained by freeze/thaw of a UCB platelet concentrate obtained by differential centrifugation. UCBPG is prepared from UCBPL by addition of calcium gluconate and/or batroxobin.

A national exercise for the standardization of UCBPG was performed in Italy in 2013–2014 with the coordination of the Italian National Blood Centre. The study involved 14 public UCB banks which produced 1080 CBPG units with mean (SD) volume of 11.4 (4.4) mL and platelet concentration equal to 1003 (229) per μL, corresponding to 47.7 (17.8) % mean platelet recovery [24]. In 2015 the Blood and Tissue Bank in Barcelona (Spain) joined the Italian program and started local studies on UCBPG.

Small successful preclinical studies in an animal model of pleural lesion [25] and clinical studies on the repair of skin ulcers in recessive dystrophic epidermolysis bullosa [26, 27] suggest that UCBPG could represent a valid tool for the treatment of different surgical lesions and skin ulcers.

7.4 Plasma and Serum

UCB plasma (UCBP) contains significant amounts of cytokines and growth factors with pro- and anti-effects on inflammation, angiogenesis, cell proliferation, and tissue remodeling. Similar to plasma from adult blood, it has been documented that UCBP contains high levels of vascular endothelial growth factor (VEGF) and platelet-derived growth factor (PDGF)-BB, substantial amounts of fibroblast growth factor (FGF), hepatocyte growth factor (HGF), and transforming growth factor (TGF)-beta1 and minimal amounts of PDGF-AB [22].

However, UCBP has some particularities. For instance, levels of VEGF and PDGF-BB were demonstrated to be higher if compared to those measured in adult peripheral blood [28]. Interestingly, UCBP contains detectable levels of soluble NKG2D ligands such as sMICA/B and sULBP1, which act as immunosuppressive molecules preventing NK and CD8+ T cell reactivity [29]. These proteins, which are not found in plasma from healthy adult subjects, are related to the unique immunological state derived from fetomaternal tolerance. Altogether, these evidences support the use of UCBP in different regenerative therapeutic applications including immune modulation.

In this sense, uses in ophthalmology have taken the lead. Based on similarities in the biochemical composition of natural tears and human plasma, the latter has been used to obtain ophthalmic preparations suitable for the treatment of corneal lesions including the dry eye syndrome (DES) [30]. Knowledge of the unique composition of UCBP and the very high safety profile of newborns as donors may overcome possible complications associated with the use of patient's autologous serum, as the latter could contain undesirable inflammatory molecules involved in ocular damage. The allogeneic approach could additionally facilitate the off-the-shelf access to the therapy to many patients in need.

Clinical studies on the use of allogeneic eye drops obtained from umbilical cord blood (UCBED) have been carried out in Italy [28, 31–33]. Initially, serum was used rather than plasma to avoid possible interference of the anticoagulant. Investigators in Bologna evaluated the efficacy of UCBED derived from cord blood serum pools in different conditions like severe DEs, ophthalmic chronic graft versus host disease, and Sjögren syndrome. These authors reported significant improvements both in corneal healing and pain during 1 month of therapy. Other international studies support the use of cord blood for the preparation of biological ophthalmic products [34, 35].

Based on the above experiences, investigators from Barcelona started a study to test a plasma-derived product in the treatment of corneal ulcers. To make manufacturing of the eye drops affordable within the current cord blood bank setting, this group developed a clinical-grade product obtained from standardized frozen platelet-rich plasma obtained to manufacture UCBPG. The latter contains an average ten billion platelets in 10 mL of UCBP. A platelet lysate generated after thawing of the platelet-rich plasma is diluted 1:2 with saline solution, immediately centrifuged to remove residual platelets and stroma and cryopreserved until use in special containers that facilitate the eye application. This group recently started a randomized clinical trial to evaluate the safety and efficacy of this product in neurotrophic keratitis (clinical trial no. EudraCT 2016-001841-23).

7.5 Red Blood Cells

Investigators of a public UCB bank in Rome, Italy, developed studies to define the feasibility and safety of a program of allogeneic transfusion of red blood cells from UCB in premature newborns [36]. This program was prompted by some evidence suggesting that the transfusion of red blood cells from adult donors to premature babies could have a causative role in the development of retinopathy of prematurity (ROP), necrotizing enterocolitis, and bronchopulmonary dysplasia. These risks have been associated with increased oxygen delivery by adult versus fetal hemoglobin.

During about 10 months, these researchers selected 63 infants with gestational age <30 weeks and/or birth weight <1500 g. About a third of these infants (n = 20) required transfusion of red blood cells and were included in a program involving the transfusion of red blood cells from adult blood or UCB, based on the availability and the compatibility of such units. Globally, 9 patients in the experimental arm received 23 and 4 units obtained from UCB and adult blood, respectively. In parallel, 11 patients in the control group received 24 units obtained from adult blood. The authors found no adverse events with either transfusion strategy. Similar incidences of ROP, necrotizing enterocolitis, bronchopulmonary dysplasia, sepsis, and intraventricular hemorrhage were found in the two groups during a median follow-up interval of 57 days. Moreover, a careful hemovigilance program showed the absence of febrile, allergic and hemolytic reactions, and respiratory complications in the newborns transfused with UCB. From an operational point of view, the authors reported that they could cover the patient's needs with UCB in 45% and 78% of the initial and successive transfusion requests, with greater difficulties for Rh (D) negative patients, respectively. The limited statistical power of this study does not allow drawing firm conclusions on the efficacy of red blood cells from UCB in preventing or reducing the frequency of ROP, necrotizing enterocolitis, and bronchopulmonary dysplasia in premature newborns. However, the absence of complications and adverse events related to the transfusion of UCB red blood cells to this fragile category of transfusion recipients is an encouraging element to promote additional, multicenter studies. Performance of a sufficiently powered clinical trial will probably require collaborative programs between UCB banks to ensure sufficiently large inventories of UCB red blood cell units to fully match the patients' transfusion requirements.

Another program of allogeneic transfusion of red blood cells from UCB in pediatric patients was developed in the Kilifi District, Kenya, as a complement to the transfusion of units of red blood cells by local adult donors, not always available or often unsuitable for clinical use. During 12 months, 74 red blood cell units from UCB were transfused to 55 pediatric patients with severe anemia, without adverse events related to this type of blood component [37].

7.6 Ethical, Legal, and Social Issues

Similar to many other medical activities, UCB collection, processing, storage, and use require careful consideration of a number of ethical, legal, and social issues (ELSI). ELSI related to the solidary UCB donation to public banks and therapeutic use in hemopoietic stem cell transplant programs versus those related to private programs of UCB conservation for autologous or family use have been analyzed and thoroughly discussed in several publications [1, 2]. The current perspectives of developing novel reagents and therapeutics from biological materials contained in UCB clearly require additional discussion of items such as ownership of UCB and its products, maternal and paternal informed consent for the novel uses, for-profit versus not-for-profit processing, and distribution by biotechnology manufacturing industry [38–46]. Moreover, pathways and procedures should be harmonized at the international level by national competent authorities to regulate the procurement of UCB units not suitable for the hemopoietic transplant inventory by industry interested in the development and commercial distribution of novel reagents and therapeutics.

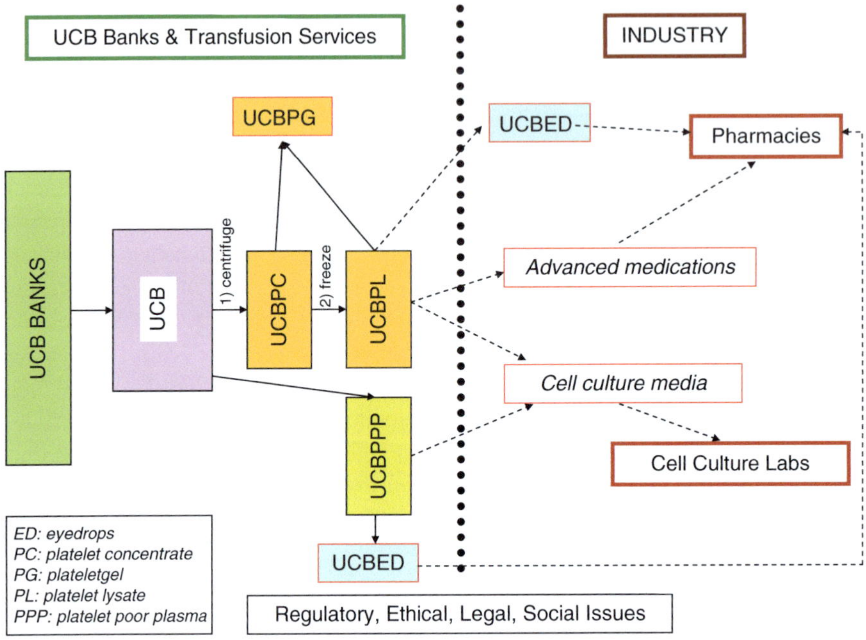

Fig. 7.1 A collaborative framework between UCB banks and industry for the development of novel reagents and therapeutics from UCB

7.7 Conclusion

UCB is a human biological material with unique properties that make it attractive for use in diverse regenerative medicine applications. Until now, UCB banks were primarily focused on building large inventories of well-characterized units based on their content in hematopoietic stem and progenitor cells as an alternative source to adult donors for allogeneic bone marrow transplantation. Nowadays, new uses of UCB-derived components and GMP reagents are foreseen. Especially UCBPG and UCBED show encouraging results in wound healing and in ophthalmic disorders. From these initial experiences, public UCB bank platforms could be broadly developed into programs of "multicomponent" UCB banking through cooperation with biotechnology industries (Fig. 7.1). This will facilitate the use of large amounts of currently discarded UCB units and be the basis of large-scale manufacturing of these new reagents and therapeutics. This transformation could use the regulatory framework of blood derivatives obtained from plasma donated by adult donors as a suitable reference model for the relationship between blood banks and pharma industries. Careful review of ethical and regulatory issues will be necessary to fully exploit the novel uses of UCB voluntary donated by non-remunerated parents for public use.

Acknowledgments *Disclosures*: Paolo Rebulla, a co-inventor of patented procedures for the preparation of blood components from cord blood, is founding member and equity holder of Episkey srl. No disclosures were declared by Sergio Querol and Alejandro Madrigal.

References

1. Petrini C. Umbilical cord blood banking: from personal donation to international public registries to global bioeconomy. J Blood Med. 2014;5:87–97.
2. Ballen KK, Verter F, Kurtzberg J. Umbilical cord blood donation: public or private? Bone Marrow Transplant. 2015;50:1271–8.
3. Ballen KK, Gluckman E, Broxmeyer HE. Umbilical cord blood transplantation: the first 25 years and beyond. Blood. 2013;122:491–8.
4. NetCord-FACT International Standards for Cord Blood Collection, Banking, and Release for

Administration. 6th ed. http://www.factweb.org/forms/store/CommercePlusFormPublic/search?action=Publications.

5. Rocha V, Broxmeyer HE. New approaches for improving engraftment after cord blood transplantation. Biol Blood Marrow Transplant. 2010;16(1 Suppl):S126–32.
6. Aljitawi OS, Paul S, Ganguly A, et al. Erythropoietin modulation is associated with improved homing and engraftment after umbilical cord blood transplantation. Blood. 2016;128:3000–10.
7. Versluis J, Labopin M, Ruggeri A, et al. On behalf of the Acute Leukemia Working Party of the European Society for Blood and Marrow Transplantation (EBMT). Alternative donors for allogeneic hematopoietic stem cell transplantation in poor-risk AML in CR1. Blood Adv. 2017;1:477–85.
8. Dahlberg A, Milano F. Cord blood transplantation: rewind to fast forward. Bone Marrow Transplant. 2017;52(6):799–802. https://doi.org/10.1038/bmt.2016.336.
9. Fuchs EJ. Related haploidentical donors are a better choice than matched unrelated donors: point. Blood Adv. 2017;1:397–400.
10. Shaw BE. Related haploidentical donors are a better choice than matched unrelated donors: counterpoint. Blood Adv. 2017;1:401–6.
11. Magalon J, Maiers M, Kurtzberg J, et al. Banking or bankrupting: strategies for sustaining the economic future of public cord blood banks. PLoS One. 2015;10(12):e0143440.
12. Mercer JS, Erickson-Owens DA. Rethinking placental transfusion and cord clamping issues. J Perinat Neonatal Nurs. 2012;26:202–17.
13. World Health Organization. Optimal timing of cord clamping for the prevention of iron deficiency anaemia in infants. Geneva (CH): WHO; 2015. http://www.who.int/elena/titles/cord_clamping/en.
14. Committee on Obstetric Practice. Committee Opinion No. 684: delayed umbilical cord clamping after birth. Obstet Gynecol. 2017;129:e5–e10.
15. Al-Qahtani R, Al-Hedythi S, Arab S, Aljuhani A, Jawdat D. Factor predicting total nucleated cell counts in cord blood units. Transfusion. 2016;56:2352–4.
16. Wu S, Xie G, Wu J, Chen J, Lu Y, Li Y, Tang X, Liao C. Influence of maternal, infant, and collection characteristics on high-quality cord blood units in Guangzhou Cord Blood Bank. Transfusion. 2015;55:2158–67.
17. Frändberg S, Waldner B, Konar J, Rydberg L, Fasth A, Holgersson J. High quality cord blood banking is feasible with delayed clamping practices. The eight-year experience and current status of the national Swedish Cord Blood Bank. Cell Tissue Bank. 2016;17:439–48.
18. Allan DS, Scrivens N, Lawless T, et al. Delayed clamping of the umbilical cord after delivery and implications for public cord blood banking. Transfusion. 2016;56:662–5.
19. Majhail NS, Nayyar S, Santibanez MEB, Murphy EA, Denzen EM. Racial disparities in hematopoietic cell transplantation in the United States. Bone Marrow Transplant. 2012;47:1385–90.
20. Astori G, Amati E, Bambi F, et al. Platelet lysate as a substitute for animal serum for the ex-vivo expansion of mesenchymal stem/stromal cells: present and future. Stem Cell Res Ther. 2016;7:93.
21. Burnouf T, Strunk D, Koh MB, Schallmoser K. Human platelet lysate: replacing fetal bovine serum as a gold standard for human cell propagation? Biomaterials. 2016;76:371–87.
22. Longo V, Rebulla P, Pupella S, Zolla L, Rinalducci S. Proteomic characterization of platelet gel releasate from adult peripheral and cord blood. Proteomics Clin Appl. 2016;10:870–82.
23. Parazzi V, Lazzari L, Rebulla P. Platelet gel from cord blood: a novel tool for tissue engineering. Platelets. 2010;21:549–54.
24. Rebulla P, Pupella S, Santodirocco M, et al. Multicentre standardisation of a clinical grade procedure for the preparation of allogeneic platelet concentrates from umbilical cord blood. Blood Transfus. 2016;14:73–9.
25. Rosso L, Parazzi V, Damarco F, et al. Pleural tissue repair with cord blood platelet gel. Blood Transfus. 2014;12(Suppl 1):s235–42.
26. Tadini G, Pezzani L, Ghirardello S, Rebulla P, Esposito S, Mosca F. Cord blood platelet gel treatment of dystrophic recessive epidermolysis bullosa. BMJ Case Rep. 2015;2015:bcr2014207364.
27. Tadini G, Guez S, Pezzani L, et al. Preliminary evaluation of cord blood platelet gel for the treatment of skin lesions in children with dystrophic epidermolysis bullosa. Blood Transfus. 2015;13:153–8.
28. Versura P, Buzzi M, Giannaccare G, et al. Cord blood serum-based eye drops: the impact of donor haematological and obstetric factors on the variability of epidermal growth factor levels. Blood Transfus. 2014;12(Suppl 1):s44–50.
29. Cox ST, Laza-Briviesca R, Pearson H, et al. Umbilical cord blood plasma contains soluble NKG2D ligands that mediate loss of natural killer cell function and cytotoxicity. Eur J Immunol. 2015;45:2324–34.
30. van der Meer PF, Seghatchian J, Marks DC. Quality standards, safety and efficacy of blood-derived serum eye drops: a review. Transfus Apher Sci. 2016;54:164–7.
31. Versura P, Profazio V, Buzzi M, et al. Efficacy of standardized and quality-controlled cord blood serum eye drop therapy in the healing of severe corneal epithelial damage in dry eye. Cornea. 2013;32:412–8.
32. Giannaccare G, Fresina M, Vagge A, Versura P. Synergistic effect of regenerating agent plus cord blood serum eye drops for the treatment of resistant neurotrophic keratitis: a case report and a hypothesis for pathophysiologic mechanism. Int Med Case Rep J. 2015;8:277–81.
33. Versura P, Buzzi M, Giannaccare G, et al. Targeting growth factor supply in keratopathy treatment: comparison between maternal peripheral blood and cord blood as sources for the preparation of topical eye drops. Blood Transfus. 2016;14:145–51.

34. Sharma N, Goel M, Velpandian T, Titiyal JS, Tandon R, Vajpayee RB. Evaluation of umbilical cord serum therapy in acute ocular chemical burns. Invest Ophthalmol Vis Sci. 2011;52:1087–92.
35. Oh HJ, Jang JY, Li Z, Park SH, Yoon KC. Effects of umbilical cord serum eye drops in a mouse model of ocular chemical burn. Curr Eye Res. 2012;37:1084–90.
36. Bianchi M, Giannantonio C, Spartano S, et al. Allogeneic umbilical cord blood red cell concentrates: an innovative blood product for transfusion therapy of preterm infants. Neonatology. 2015;107:81–6.
37. Hassall OW, Thitiri J, Fegan G, et al. Safety and efficacy of allogeneic umbilical cord red blood cell transfusion for children with severe anaemia in a Kenyan hospital: an open-label single-arm trial. Lancet Haematol. 2015;2:e101–7.
38. Stewart C, Kerridge I. Umbilical cord blood banking and the next generation of human tissue regulation: an agenda for research. J Law Med. 2012;19:423–9.
39. Stewart CL, Aparicio LC, Kerridge IH. Ethical and legal issues raised by cord blood banking—the challenges of the new bioeconomy. Med J Aust. 2013;199:290–2.
40. Goold I. Why does it matter how we regulate the use of human body parts? J Med Ethics. 2014;40:3–9.
41. King NM, Perrin J. Ethical issues in stem cell research and therapy. Stem Cell Res Ther. 2014;5:85.
42. Sundell IB, Setzer TJ. Age, sex, and religious beliefs impact the attitude towards cord blood banking. J Stem Cells. 2015;10:33–41.
43. Armson BA, Allan DS, Casper RF, Cord Blood Banking Working Group. Umbilical cord blood: counselling, collection, and banking. J Obstet Gynaecol Can. 2015;37:832–46.
44. Brown N, Williams R. Cord blood banking—bio-objects on the borderlands between community and immunity. Life Sci Soc Policy. 2015;11:11.
45. Roura S, Pujal JM, Gálvez-Montón C, Bayes-Genis A. The role and potential of umbilical cord blood in an era of new therapies: a review. Stem Cell Res Ther. 2015;6:123.
46. Guilcher GM, Fernandez CV, Joffe S. Are hybrid umbilical cord blood banks really the best of both worlds? J Med Ethics. 2015;41:272–5.

8 The Characteristics and Therapeutic Application of Perinatal Mesenchymal Stem Cell-Derived Exosomes

Fengxia Ma

Abstract

Exosomes are secretory vesicles with diameters range from 40 to 100 nm, which are considered to play an important role in intercellular communication. Exosomes contain proteins, lipids, microRNAs, mRNAs, and other noncoding RNAs. It is demonstrated that exosomes derived from diverse cell types have different therapeutic potentials. Recently, perinatal mesenchymal stem cells (MSCs) have attracted much attention due to their unique characteristics. Therefore, perinatal MSC-derived exosomes may provide novel therapeutic potentials. In this chapter, we will summarize the characteristics of exosomes. Moreover, we will focus on recent reports on perinatal MSC-derived exosomes and consider their therapeutic potentials. Finally, we will introduce some strategies to modify exosome for tracking their fate in vivo and enhancing their therapeutic efficiency.

F. Ma (✉)
State Key Laboratory of Experimental Hematology, Institute of Hematology and Hospital of Blood Disease, Chinese Academy of Medical Sciences, Tianjin, China
e-mail: mafengxia@ihcams.ac.cn

8.1 Introduction

The secretion of vesicles by cells is a common and evolutionary conserved process because both prokaryotic and eukaryotic cells secrete vesicles [1]. Vesicles were originally described during reticulocytes maturation nearly 30 years ago [2, 3]. It was thought that vesicles were involved in discarding unnecessary membrane proteins from cells. Since then, vesicles have been purified from almost all cell types, such as immune cells [4], tumor cells [5], epithelial cells [6], neuronal cells [7], fibroblasts [8], stem cells [9], etc. Furthermore, vesicles have been isolated from most body fluids such as the blood [10], urine [11], breast milk [12], cerebrospinal fluid [13], bronchoalveolar lavage fluid [14], and synovial fluid [15]. Up to now, several types of vesicles have been defined based on their cellular origin and/or biological characteristics or biogenesis. Three major classes of vesicles are exosomes, microvesicles, and apoptotic bodies [1]. Here, we focus on exosomes because they are the most well-known class of vesicles [1]. They are homogenous, and their diameters range from 40 to 100 nm. The membranes of exosomes are phospholipid bilayers. Exosomes have specific protein markers such as tetraspanins (CD63, CD9, CD81, CD82), heat shock proteins (HSP70, HSP90), Alix and tumor susceptibility gene 101 (Tsg101), etc. Exosomes can deliver their contents to recipient cells. Therefore, exosomes function as a new form

Z. C. Han et al. (eds.), *Perinatal Stem Cells*, https://doi.org/10.1007/978-981-13-2703-2_8

of intercellular communication between both neighboring and distant cells [1].

8.2 The Biogenesis of Exosomes

Exosomes are formed by invaginations of the plasma membrane [16, 17]. The first invaginations of plasma membrane form endosomes, and then parts of endosome membrane invaginate and selectively incorporate molecules in the cytoplasm and form intraluminal vesicles (ILVs). Once the endosome contains ILVs, it is called multivesicular body (MVB), which fuse with the plasma membrane and lead to exosomes release [16, 17].

The sorting of protein complexes into MVB is a highly regulated process. It is indicated that the endosomal sorting complex required for transport (ESCRT) plays a role in the ILV formation and protein sorting [18, 19]. Furthermore, ESCRT-independent protein sorting also exists [1].

Recent studies suggested a possible correlation between AGO2 and exosomal miRNA sorting [20]. Knockout of AGO2 could decrease the types or abundance of miRNAs in HEK293T-derived exosomes. It has also shown a relationship between miRNA-induced silencing complex (miRISC) and exosomal miRNA sorting [21]. Thus, specific sequences present in certain miRNAs may guide their sorting. Furthermore, some proteins also regulate sorting of miRNAs into exosome in a miRNA sequence-independent manner.

In fact, lipid and other RNA sorting also occurs during exosome biogenesis [22, 23]. Protein, lipid, and RNA sorting leads to a specific content of exosomes.

8.3 The Content of Exosomes

Exosomes have ubiquitous proteins and parental cell-specific proteins [1, 24]. The ubiquitous proteins include tetraspanins (CD9, CD63, CD81, CD82), cell structure and motility protein (actins, myosin), heat shock proteins (HSP70, HSP90), antigen-presentation protein (MHC-I, MHC-II), cell adhesion molecules (integrins, MFGE8), metabolic enzymes (fatty acid synthase, peroxidases), and signaling proteins (kinases, G proteins). In addition, exosomes have specific proteins from parental cells, like CD11b, CD18, and Mac1 in neutrophil-derived exosomes [25].

Valadi firstly demonstrated that exosomes could contain mRNA and microRNA [26]. Exosomal miRNAs can be transferred to recipient cells and be able to silence target genes [27, 28]. In addition to the classic role of targeting miRNA, exosomal miR-21 and miR-29a act as ligands binding to toll-like receptors (TLRs) and activating immune cells [29]. Apart from mRNAs and miRNAs, other RNAs are also observed within exosomes, such as Y-RNAs, tRNAs, snRNA, and long noncoding RNAs [30–32]. It was shown that double-stranded DNA existed inside exosomes, and they are attractive, potential biomarker candidates in the early detection of cancers and the monitoring of treatment response [33].

The content of exosomes varies according to cell conditions. Heat shock, hypoxia, oxidative stress, and infections lead to remarkable changes in the content of exosomes [34–36]. Exosome content is sensitive to external stimuli, therefore, characterizing the molecular machinery sorting specific RNAs, and proteins toward exosome are essential [37–39].

8.4 The Interaction of Exosomes and Recipient Cells

After uptaken by recipient cells, exosomes are able to activate downstream signaling pathways in recipient cells [40–42]. Exosome proteins such as MHC I and II [43] and tetraspanins [44] are active by triggering integrins and calcium signaling [42], mitogen-activated protein kinase (MAPK) [41], or natural killer group 2D (NKG2D) signaling [45]. ICAM-1 (intercellular adhesion molecule 1) and LFA-1 (leukocyte function-associated antigen-1) play a role in the capture of exosomes by recipient cells [46, 47]. Moreover, exosomes also have phospholipases leading to the formation of bioactive lipid mediators which can interact with G-protein-coupled receptors and nuclear receptors in recipient cells [48].

Based on the interaction of exosomes and recipient cells, exosomes behave as a signaling vehicle. For example, activating cell surface receptors, merging their contents into the recipient cell, and delivering their content into recipient cells.

8.5 Isolation and Identification of Exosomes

The most extensively used isolation method for exosomes is ultracentrifugation. Cells debris and larger particles are removed by filter and/or centrifugation from 200 × *g* to 100,000 × *g*. Then exosomes are isolated by ultracentrifugation at ≥100,000 × *g* for at least 2 h. This method results in highly purified exosomes; however, specialized equipment and much time are needed. Moreover, this is a non-specific purification method, and exosomes are easy to be contaminated with protein aggregates and other particles [49].

Immunoaffinity is also used for exosome isolation [50]. An antibody conjugated to magnetic beads and against exosome-specific antigen such as CD63 is used to purify exosomes from conditioned medium or other biological fluids. However, most exosome-specific antigens are not absolutely specific to exosomes, which result in contamination of other protein and particles.

Some alternatives are tried. Cheruvanky et al. isolate exosomes using ultrafiltration that are rapid and do not need special equipment [51]. Polyethylene glycol (PEG) is a condensation polymer of ethylene oxide and water, which has been used in the precipitation of proteins [52]. It also could be used to precipitate exosomes [53]. Then exosomes can be isolated using low-speed centrifugation. Recently, corporations have developed many kinds of exosome isolation kits and exosome precipitation solutions, which provide convenient and efficient methods to isolate exosomes.

To identify exosomes, common methods are electron microscopy and nanoparticle tracking analysis to know the morphology and size distribution of exosomes. Flow cytometric analysis and Western blotting are used to identify protein markers of exosomes such as CD63, CD8, and CD9. Normally, these methods are usually used in combination for exosome identification.

8.6 Drug Delivery of Exosomes

Recently, exosomes are considered as the most promising candidate for a natural drug delivery vehicle because of their features [54, 55]. Firstly, protein, DNA, and RNA could be loaded into exosomes. Secondly, exosomes have a longer circulating half-life. Thirdly, exosomes could deliver their contents into recipient cells [56]. Fourthly, exosomes possess homing capacity [57]. Furthermore, evidence suggests that the homing capacity of exosomes depends on their cell source. Finally, exosomes are easy to membrane modifications [58].

Drugs could be loaded into exosomes either in vivo or in vitro. In vitro drug loading means drugs loaded into purified exosomes. A transient physical (e.g., electroporation) or chemical disruption (e.g., lipofection) of the exosome membrane could load small molecules such as siRNA. Incorporating drugs into exosomes during their biogenesis is for drugs that cannot be loaded into purified exosomes such as cytosolic and transmembrane proteins. Furthermore, genetical engineering could be used to load some drugs. Therefore, exosome display technology is developed [59, 60], which is a new method to control exosome protein content. Exosome display of soluble antigens and membrane proteins is achieved upon fusion of proteins with the C1C2 domain of lactadherin (also named MFGE8), localizing in exosome membrane [61]. The exosome display technology opens many possibilities to design novel therapies and generate new diagnostic tools.

8.7 Perinatal Mesenchymal Stem Cells Are an Ideal Source of Exosomes

Mesenchymal stem cells (MSCs) are promising stem cells for tissue regeneration and cell therapy because of their ability of self-renewal, multilineage differentiation, immunomodulation, and anti-inflammation. Therefore, they have been widely used in treatment of cardiovascular, neurodegenerative, tissue injury, and autoimmune diseases according to their biological properties [62, 63]. MSCs can be isolated from nearly all tissues, such as bone marrow, adipose, perinatal tissues (umbilical cord, amniotic membrane, placenta). Compared with BM-MSCs, perinatal MSCs have several advantages such as abundant supply, painless collection procedure, less chance of microbial contamination, better expandable ability, low immunogenicity, and strong immunosuppressive potential [64, 65]. Therefore, perinatal MSCs can be an ideal source of exosomes.

8.8 Biological Roles and Therapeutic Application of Exosomes Derived from Perinatal Mesenchymal Stem Cells

Exosomes can mimic the function of their parent cells. The therapeutic application of Exo is even more promising than cells. Compared with their parent cells, exosomes have no risk of aneuploidy, lower possibility of immune rejection, and lung barrier following in vivo allogeneic administration. Therefore, exosomes derived from perinatal MSCs may be highly effective therapeutic agent in regulating inflammation and immune response, promoting angiogenesis, participating in tissue repair.

8.8.1 Inflammation and Immune Response

Many studies suggest that MSCs can produce the immunomodulatory cytokines and regulate both the innate and the adaptive immune systems [66]. Upon stimulated by pro-inflammatory cytokines, MSCs secrete immunosuppressive cytokines including transforming growth factor (TGF)-β, prostaglandin E2, interleukin 10 (IL-10), etc. [67]. It was shown that bone marrow MSC-derived exosomes suppressed pro-inflammatory factor TNF-α and IL-1β secretion but increased anti-inflammatory factor TGF-β secretion when they were cultured with peripheral blood mononuclear cells (PBMCs). In addition, exosome may lead to the conversion of Th1 into Th2 cells and the reduction of Th17 differentiation. However, the level of Treg cells was increased. These results suggested that MSC-derived exosome has the immunomodulatory potentials. The immunoregulation of MSCs depended on indoleamine 2,3-dioxygenase (IDO). No significant changes of IDO happened in PBMCs cocultured with MSC-derived exosome. It was hypothesized that the immune-modulating activities and mechanisms of exosome and MSCs might differ [68].

Similar to bone marrow MSC-derived exosomes, human umbilical cord MSC (UC-MSC)-derived exosomes decreased $CD4^+$ and $CD8^+$ cells proliferation and increased the proportion of $CD4^+CD25^+FoxP3^+$ Treg cells [69]. Furthermore, human UC-MSC-derived exosomes reduced inflammation after burn [70]. UC-MSC exosome successfully reversed the augment of tumor necrosis factor α (TNF-α), interleukin-1β (IL-1β) levels, and the reduction of IL-10 levels. It was reported that LPS-preconditioned UC-MSC modified macrophage polarization for resolution of chronic inflammation, and exosomes played an important roles in that [71].

8.8.2 Angiogenesis

Angiogenesis is new capillary formation from existing blood vessels, which is involved in many cytokines such as fibroblast growth factor (FGF), vascular endothelial growth factor (VEGF), and angiogenin [72]. Proteomic array analysis showed that exosomes derived from MSCs contain growth factors such as VEGF, TGFβ, HGF, and IL-8 [73]. It was shown that uptake of bone marrow MSC exosomes by human umbilical vein endothelial cells gave rise to the increase of tube formation. It was found that bone marrow MSC exosomes could activate several signaling pathways (Akt, ERK, and STAT3) and induce the production of a number of growth factors such as insulin-like growth factor-1 (IGF1), hepatocyte growth factor (HGF), nerve growth factor (NGF), and stromal-derived growth factor-1 (SDF1) [74].

Similar to bone marrow MSC-derived exosomes, it was demonstrated that exosomes from UC-MSC enhanced angiogenesis in the repair of skin burn injury [75]. Wnt/β-catenin activation was critical in UC-MSC-exo-induced angiogenesis. Another research group also found that exosomes come from Akt-modified UC-MSCs (Akt-Exo) which were effective in promoting angiogenesis and improving cardiac function [76]. Akt-Exo significantly promoted endothelial cell proliferation and migration in vitro and blood vessel formation in vivo.

8.8.3 Tissue Repair

8.8.3.1 Renal Injury

The therapeutic effect of MSC-derived microvesicles on tissue repair was first investigated in renal injury. It was demonstrated that BM-MSC-derived MVs protected toxic kidney injury, ischemia reperfusion-induced kidney injury, because proliferation and resistance to apoptosis in tubular epithelial cells were observed after cocultured with MVs [77]. Similar results were shown for UC-MSC-derived exosomes [78]. The same research group showed that after a single administration of human Wharton-Jelly MSC-derived microvesicles (hWJMSC-MVs) in acute kidney injury (AKI) rats, renal cell apoptosis and inflammation were alleviated, and proliferation was enhanced [79]. MVs could suppress the expression of chemoattractant factor for macrophages and decrease the number of macrophages in the kidney. Lately, renal function was improved, and renal fibrosis was abrogated. They further demonstrated that hWJMSC-MVs might protect the kidney by suppressing the oxidative stress in the early stage of kidney injury [80].

8.8.3.2 Cardiac Injury

MSCs transplantation has been reported to reduce infarct size but improve left ventricular ejection, vascular density, and myocardial perfusion in most animal models of acute myocardial injury (AMI) [81]. In a clinical trial, it was shown that infusion of allogeneic MSCs to patients with AMI could improve pulmonary function and left ventricular function [82]. However, after transplantation, most MSCs were blocked in the lungs, and very few MSC would engraft, differentiate, and replace damaged cardiac tissues. Therefore, it was supposed that the therapeutic effect of MSC transplantation was a result of paracrine effects [81]. It was shown that bone marrow MSC-derived exosomes improved heart function after ischemic injury by stimulating neovascularization and restraining the inflammation response [83].

UC-MSC-derived exosomes (UC-MSC-exo) also have a protective effect on AMI. Administration of UC-MSC-exo significantly resulted in the improvement of cardiac function and the reduction of cardiac fibrosis [84]. Moreover, UC-MSC-exo not only inhibited myocardial cell apoptosis but promoted the tube formation and migration of human umbilical vein endothelial cells.

Similar to MSC, cytokines and growth factors may be responsible for the injury protection of exosome. At the same time, miRNAs transferred from MSCs to damaged cardiac cells may have some effects.

8.8.3.3 Hepatic Injury

Mesenchymal stem cell (MSC) therapy is considered as an effective alternative treatment for hepatic diseases [85]. MSCs can differentiate into hepatocytes. Furthermore, MSCs transplantation resulted in the reduction of hepatocyte apoptosis and liver fibrosis but the promotion of hepatocyte regeneration and function by secreting trophic factors and suppressing inflammatory responses. Despite these advantages, MSCs also have some disadvantages such as promoting fibrosis and tumor cell growth [86, 87].

MSC-derived exosomes may have similar effects on hepatic diseases but avoid the disadvantages of MSCs. It was reported that exosomes derived from human UC-MSCs relieved liver fibrosis by inhibiting epithelial-to-mesenchymal transition (EMT) and protecting hepatocytes [88]. It was found that transplantation of exosomes derived from UC-MSC (UC-MSC-exo) mitigated hepatic inflammation and collagen deposition in fibrotic liver. In addition to transplantation of exosomes, there is a potential approach for the use of exosomes in the therapy for hepatic injury including engineered exosomes containing specific biological or small molecule drugs and delivering them to the target cells.

8.9 Engineered Exosomes

As mentioned above, exosomes have therapeutic potentials on many kinds of diseases and can be an ideal vehicle for drug delivery. However, when injected systemically in animal models, exosomes accumulated mainly in the liver, kidney, and spleen, leading to few homing of exosomes in the target tissues or organs [89, 90]. This biodistribution profile and off-target effects protect exosomes from clinically effective therapeutics.

As their parent cells, exosomes can be engineered to display homing peptides or ligands on their surface, which will significantly decrease the off-target effect and increase therapeutic efficacy. Displaying ligands/homing peptides on the surface of exosomes can be achieved by fusing them with the transmembrane proteins that are rich in exosomes [60]. For example, a neuron-specific RVG encoding sequence was fused with a exosomal membrane protein LAMP2B. By overexpressing this fusion gene in dendritic cells, Woods et al. demonstrated that engineered dendritic cells secreted exosomes with RVG peptide on their membranes [58]. After loaded with BACE1 siRNA, the therapeutic potential of exosome-mediated siRNA delivery was demonstrated by the strong mRNA (60%) and protein (62%) knockdown of BACE1, a therapeutic target in Alzheimer's disease. Similarly, another research group tried to engineer mouse immature dendritic cells (imDCs) to express Lamp2b and iRGD peptide fusion gene. Then Dox was loaded into iRGD-engineered exosomes. iRGD-engineered exosomes showed highly homing and Dox delivery to breast cancer cells, leading to inhibition of tumor growth without obvious toxicity [91].

To study the in vivo fate of exosomes, Takahashi Y et al. designed a fusion protein consisting of *Gaussia* luciferase (Gluc) and C1C2 domain of murine lactadherin. Exosomes derived from murine melanoma cells expressing this fusion protein had strong luciferase activity [89]. Gluc has a short half-life in vivo (20 min), allowing for real-time testing without undesirable excessive signal accumulation. Different from Takahashi's work, my research group are trying to use human UC-MSC to express humanized Gluc and lactadherin infusion protein to produce exosomes labeled with GLuc.

To enhance the therapeutic potential of exosomes, Xu MF and his team isolated exosomes from mesenchymal stem cells (MSCs) overexpressing GATA-4 (MSC^{GATA-4}) [92]. Exosomes from MSC^{GATA-4} resulted in increased survival and reduced apoptosis of cardiomyocytes. Direct intramyocardial transplantation of exosomes significantly restored cardiac function and reduced infarct size. Several anti-apoptotic miRs were highly expressed in Exo^{GATA-4}. The enhanced protective effects observed in CM were diminished by inhibition of miR-19a. It was concluded that ExoGATA-4 transferred miRs into damaged cardiomyocytes, triggering cell survival.

8.10 Concluding Remarks

In this chapter, I have summarized the characteristics of exosomes and focused on recent reports on the therapeutic potentials of perinatal MSC-derived exosomes. Perinatal MSCs are ideal parent cells of exosomes because of their high secretory and expandable ability and low immunogenicity. In addition, the production and function of exosomes can be improved by preconditioning or genetic modification of the parent cells. I hypothesize that exosomes derived from perinatal MSCs could be used as therapeutic agent and drug or gene delivery vehicles which can cross blood-brain barrier.

References

1. EL Andaloussi S, Mäger I, Breakefield XO, et al. Extracellular vesicles: biology and emerging therapeutic opportunities. Nat Rev Drug Discov. 2013;12:347–57.
2. Pan BT, Johnstone RM. Fate of the transferrin receptor during maturation of sheep reticulocytes

in vitro: selective externalization of the receptor. Cell. 1983;33:967–78.

3. Pan BT, Teng K, Wu C, et al. Electron microscopic evidence for externalization of the transferrin receptor in vesicular form in sheep reticulocytes. J Cell Biol. 1985;101:942–8.
4. Nazimek K, Ptak W, Nowak B, et al. Macrophages play an essential role in antigen-specific immune suppression mediated by T CD8+ cell-derived exosomes. Immunology. 2015;146:23–32.
5. Chowdhury R, Webber JP, Gurney M, et al. Cancer exosomes trigger mesenchymal stem cell differentiation into pro-angiogenic and pro-invasive myofibroblasts. Oncotarget. 2015;6:715–31.
6. Moon HG, Cao Y, Yang J, et al. Lung epithelial cell-derived extracellular vesicles activate macrophage-mediated inflammatory responses via ROCK1 pathway. Cell Death Dis. 2015;6:e2016.
7. Budnik V, Ruiz-Cañada C, Wendler F. Extracellular vesicles round off communication in the nervous system. Nat Rev Neurosci. 2016;17:160–72.
8. Baroni S, Romero-Cordoba S, Plantamura I, et al. Exosome-mediated delivery of miR-9 induces cancer-associated fibroblast-like properties in human breast fibroblasts. Cell Death Dis. 2016;7:e2312.
9. Tan CY, Lai RC, Wong W, et al. Mesenchymal stem cell-derived exosomes promote hepatic regeneration in drug-induced liver injury models. Stem Cell Res Ther. 2014;5:76.
10. Hannafon BN, Trigoso YD, Calloway CL, et al. Plasma exosome microRNAs are indicative of breast cancer. Breast Cancer Res. 2016;18:90.
11. Tataruch-Weinert D, Musante L, Kretz O, et al. Urinary extracellular vesicles for RNA extraction: optimization of a protocol devoid of prokaryote contamination. J Extracell Vesicles. 2016;5:30281.
12. Qin W, Tsukasaki Y, Dasgupta S, et al. Exosomes in human breast milk promote EMT. Clin Cancer Res. 2016;22:4517–24.
13. Gui Y, Liu H, Zhang L, et al. Altered microRNA profiles in cerebrospinal fluid exosome in Parkinson disease and Alzheimer disease. Oncotarget. 2015;6:37043–53.
14. Gregson AL, Hoji A, Injean P, et al. Altered exosomal RNA profiles in bronchoalveolar lavage from lung transplants with acute rejection. Am J Respir Crit Care Med. 2015;192:1490–503.
15. Kim SJ, Chen Z, Essani AB, et al. Identification of a novel Toll-like receptor 7 endogenous ligand in rheumatoid arthritis synovial fluid that can provoke arthritic joint inflammation. Arthritis Rheumatol. 2016;68:1099–110.
16. Kowal J, Tkach M, Théry C. Biogenesis and secretion of exosomes. Curr Opin Cell Biol. 2014;29:116–25.
17. Dreyer F, Baur A. Biogenesis and functions of exosomes and extracellular vesicles. Methods Mol Biol. 2016;1448:201–16.
18. Katzmann DJ, Odorizzi G, Emr SD. Receptor downregulation and multivesicular-body sorting. Nat Rev Mol Cell Biol. 2002;3:893–905.
19. Raiborg C, Rusten TE, Stenmark H. Protein sorting into multivesicular endosomes. Curr Opin Cell Biol. 2003;15:446–55.
20. Guduric-Fuchs J, O'Connor A, Camp B, et al. Selective extracellular vesicle-mediated export of an overlapping set of microRNAs from multiple cell types. BMC Genomics. 2012;13:357.
21. Gibbings DJ, Ciaudo C, Erhardt M, et al. Multivesicular bodies associate with components of miRNA effector complexes and modulate miRNA activity. Nat Cell Biol. 2009;11:1143–9.
22. Chevallier J, Chamoun Z, Jiang G, et al. Lysobisphosphatidic acid controls endosomal cholesterol levels. J Biol Chem. 2008;283:27871–80.
23. Shurtleff MJ, Temoche-Diaz MM, Karfilis KV, et al. Y-box protein 1 is required to sort microRNAs into exosomes in cells and in a cell-free reaction. Elife. 2016;5. pii: e19276.
24. Xiao Z, Blonder J, Zhou M, et al. Proteomic analysis of extracellular matrix and vesicles. J Proteomics. 2009;72:34–45.
25. Timár CI, Lorincz AM, Csépányi-Kömi R, et al. Antibacterial effect of microvesicles released from human neutrophilic granulocytes. Blood. 2013;121(3):510–8.
26. Valadi H, Ekstrom K, Bossios A, et al. Exosome mediated transfer of mRNAs and microRNAs is a novel mechanism of genetic exchange between cells. Nat Cell Biol. 2007;9:654–9.
27. Montecalvo A, Larregina AT, Shufesky WJ, et al. Mechanism of transfer of functional microRNAs between mouse dendritic cells via exosomes. Blood. 2012;119:756–66.
28. Ismail N, Wang Y, Dakhlallah D, et al. Macrophage microvesicles induce macrophage differentiation and miR-223 transfer. Blood. 2013;121:984–95.
29. Fabbri M, Paone A, Calore F, et al. MicroRNAs bind to Toll-like receptors to induce prometastatic inflammatory response. Proc Natl Acad Sci U S A. 2012;109:E2110–6.
30. Bellingham SA, Coleman BM, Hill AF. Small RNA deep sequencing reveals a distinct miRNA signature released in exosomes from prion-infected neuronal cells. Nucleic Acids Res. 2012;40:10937–49.
31. Huang X, Yuan T, Tschannen M, et al. Characterization of human plasma-derived exosomal RNAs by deep sequencing. BMC Genomics. 2013;14:319.
32. Gezer U, Ozgur E, Cetinkaya M, et al. Long non-coding RNAs with low expression levels in cells are enriched in secreted exosomes. Cell Biol Int. 2014;38:1076–9.
33. Thakur BK, Zhang H, Becker A, et al. Double-stranded DNA in exosomes: a novel biomarker in cancer detection. Cell Res. 2014;24:766–9.
34. Zhan R, Leng X, Liu X, et al. Heat shock protein 70 is secreted from endothelial cells by a non-classical pathway involving exosomes. Biochem Biophys Res Commun. 2009;387:229–33.
35. De Jong OG, Verhaar MC, Chen Y, et al. Cellular stress conditions are reflected in the protein and

RNA content of endothelial cell-derived exosomes. J Extracell Vesicles. 2012;16:1.
36. Kalamvoki M, Du T, Roizman B. Cells infected with herpes simplex virus 1 export to uninfected cells exosomes containing STING, viral mRNAs, and microRNAs. Proc Natl Acad Sci U S A. 2014;111:E4991–6.
37. Kore RA, Abraham EC. Inflammatory cytokines, interleukin-1 and tumor necrosis factor-alpha upregulated in glioblastoma multiforme, raise the levels of CRYAB in exosomes secreted by U373 glioma cells. Biochem Biophys Res Commun. 2014;453:326–31.
38. Kato T, Miyaki S, Ishitobi H, et al. Exosomes from IL-1β stimulated synovial fibroblasts induce osteoarthritic changes in articular chondrocytes. Arthritis Res Ther. 2014;16:R163.
39. Ekström EJ, Bergenfelz C, von Bülow V, et al. WNT5A induces release of exosomes containing pro-angiogenic and immunosuppressive factors from malignant melanoma cells. Mol Cancer. 2014;13:88.
40. Nazarenko I, Rana S, Baumann A, et al. Cell surface tetraspanin Tspan8 contributes to molecular pathways of exosome-induced endothelial cell activation. Cancer Res. 2010;70:1668–78.
41. Clayton A, Turkes A, Dewitt S, et al. Adhesion and signaling by B cell-derived exosomes: the role of integrins. FASEB J. 2004;18:977–9.
42. Calzolari A, Raggi C, Deaglio S, et al. TfR2 localizes in lipid raft domains and is released in exosomes to activate signal transduction along the MAPK pathway. J Cell Sci. 2006;119:4486–98.
43. Keller S, Sanderson MP, Stoeck A, et al. Exosomes: from biogenesis and secretion to biological function. Immunol Lett. 2006;107:102–8.
44. Perez-Hernandez D, Gutierrez-Vazquez C, Jorge I, et al. The intracellular interactome of tetraspanin-enriched microdomains reveals their function as sorting machineries toward exosomes. J Biol Chem. 2013;288:11649–61.
45. Viaud S, Terme M, Flament C, et al. Dendritic cell-derived exosomes promote natural killer cell activation and proliferation: a role for NKG2D ligands and IL-15Ralpha. PLoS One. 2009;4:e4942.
46. Nolte-'t Hoen EN, Buschow SI, Anderton SM, et al. Activated T cells recruit exosomes secreted by dendritic cells via LFA-1. Blood. 2009;113:1977–81.
47. Segura E, Guerin C, Hogg N, et al. $CD8^+$ dendritic cells use LFA-1 to capture MHC-peptide complexes from exosomes in vivo. J Immunol. 2007;179:1489–96.
48. Subra C, Grand D, Laulagnier K, et al. Exosomes account for vesicle-mediated transcellular transport of activatable phospholipases and prostaglandins. J Lipid Res. 2010;51:2105–20.
49. Xu R, Greening DW, Zhu HJ, et al. Extracellular vesicle isolation and characterization: toward clinical application. J Clin Invest. 2016;126:1152–62.
50. Clayton A, Court J, Navabi H, et al. Analysis of antigen presenting cell derived exosomes, based on immuno-magnetic isolation and flow cytometry. J Immunol Methods. 2001;247:163–74.
51. Cheruvanky A, Zhou H, Pisitkun T, et al. Rapid isolation of urinary exosomal biomarkers using a nanomembrane ultrafiltration concentrator. Am J Physiol Renal Physiol. 2007;292:F1657–61.
52. Ingham KC. Precipitation of proteins with polyethylene glycol. Methods Enzymol. 1990;182:301–6.
53. Weng Y, Sui Z, Shan Y, et al. Effective isolation of exosomes with polyethylene glycol from cell culture supernatant for in-depth proteome profiling. Analyst. 2016;141(15):4640–6.
54. Ha D, Yang N, Nadithe V. Exosomes as therapeutic drug carriers and delivery vehicles across biological membranes: current perspectives and future challenges. Acta Pharm Sin. 2016;6:287–96.
55. Lai RC, Yeo RW, Tan KH, et al. Exosomes for drug delivery—a novel application for the mesenchymal stem cell. Biotechnol Adv. 2013;31:543–51.
56. Tomasoni S, Longaretti L, Rota C, et al. Transfer of growth factor receptor mRNA via exosomes unravels the regenerative effect of mesenchymal stem cells. Stem Cells Dev. 2013;22:772–80.
57. Hood JL, San RS, Wickline SA. Exosomes released by melanoma cells prepare sentinel lymph nodes for tumor metastasis. Cancer Res. 2011;71:3792–801.
58. Alvarez-Erviti L, Seow Y, Yin H, et al. Delivery of siRNA to the mouse brain by systemic injection of targeted exosomes. Nat Biotechnol. 2011;29:341–5.
59. Fang Y, Wu N, Gan X, et al. Higher-order oligomerization targets plasma membrane proteins and HIV gag to exosomes. PLoS Biol. 2007;5:e158.
60. Delcayre A, Estelles A, Sperinde J, et al. Exosome display technology: applications to the development of new diagnostics and therapeutics. Blood Cells Mol Dis. 2005;35:158–68.
61. Véron P, Segura E, Sugano G, et al. Accumulation of MFG-E8/lactadherin on exosomes from immature dendritic cells. Blood Cells Mol Dis. 2005;35:81–8.
62. Baksh D, Song L, Tuan RS. Adult mesenchymal stem cells: characterization, differentiation, and application in cell and gene therapy. J Cell Mol Med. 2004;8:301–16.
63. Kode JA, Mukherjee S, Joglekar MV, et al. Mesenchymal stem cells: immunobiology and role in immunomodulation and tissue regeneration. Cytotherapy. 2009;11:377–91.
64. Baksh D, Yao R, Tuan RS. Comparison of proliferative and multilineage differentiation potential of human mesenchymal stem cells derived from umbilical cord and bone marrow. Stem Cells. 2007;25:1384–92.
65. Yang ZX, Han ZB, Ji YR, et al. CD106 identifies a subpopulation of mesenchymal stem cells with unique immunomodulatory properties. PLoS One. 2013;8:e59354.
66. Gao F, Chiu SM, Motan DA, et al. Mesenchymal stem cells and immunomodulation: current status and future prospects. Cell Death Dis. 2016;7:e2062.
67. English K. Mechanisms of mesenchymal stromal cell immunomodulation. Immunol Cell Biol. 2013;91:19–26.

68. Chen W, Huang Y, Han J, et al. Immunomodulatory effects of mesenchymal stromal cells-derived exosome. Immunol Res. 2016;64:831–40.
69. Liu M, Wang J, Liu M, et al. Study of immunomodulatory function of exosomes derived from human umbilical cord mesenchymal stem cells. Zhonghua Yi Xue Za Zhi. 2015;95:2630–3.
70. Li X, Liu L, Yang J, et al. Exosome derived from human umbilical cord mesenchymal stem cell mediates MiR-181c attenuating burn-induced excessive inflammation. EBioMedicine. 2016;8:72–82.
71. Ti D, Hao H, Tong C, et al. LPS-preconditioned mesenchymal stromal cells modify macrophage polarization for resolution of chronic inflammation via exosome-shuttled let-7b. J Transl Med. 2015;13:308.
72. Adams RH, Alitalo K. Molecular regulation of angiogenesis and lymphangiogenesis. Nat Rev Mol Cell Biol. 2007;8:464–78.
73. Anderson JD, Johansson HJ, Graham CS, et al. Comprehensive proteomic analysis of mesenchymal stem cell exosomes reveals modulation of angiogenesis via nuclear factor-kappaB signaling. Stem Cells. 2016;34:601–13.
74. Shabbir A, Cox A, Rodriguez-Menocal L, et al. Mesenchymal stem cell exosomes induce proliferation and migration of normal and chronic wound fibroblasts, and enhance angiogenesis in vitro. Stem Cells Dev. 2015;24:1635–47.
75. Zhang B, Wu X, Zhang X, et al. Human umbilical cord mesenchymal stem cell exosomes enhance angiogenesis through the Wnt4/β-catenin pathway. Stem Cells Transl Med. 2015;4:513–22.
76. Ma J, Zhao Y, Sun L, et al. Exosomes derived from Akt-modified human umbilical cord mesenchymal stem cells improve cardiac regeneration and promote angiogenesis via activating platelet-derived growth factor D. Stem Cells Transl Med. 2016;6:51–9.
77. Bruno S, Grange C, Deregibus MC, et al. Mesenchymal stem cell-derived microvesicles protect against acute tubular injury. J Am Soc Nephrol. 2009;20:1053–67.
78. Zhou Y, Xu H, Xu W, et al. Exosomes released by human umbilical cord mesenchymal stem cells protect against cisplatin-induced renal oxidative stress and apoptosis in vivo and in vitro. Stem Cell Res Ther. 2013;4:34.
79. Zou X, Zhang G, Cheng Z, et al. Microvesicles derived from human Wharton's Jelly mesenchymal stromal cells ameliorate renal ischemia-reperfusion injury in rats by suppressing CX3CL1. Stem Cell Res Ther. 2014;5:40.
80. Zhang G, Zou X, Miao S, et al. The anti-oxidative role of micro-vesicles derived from human Wharton-Jelly mesenchymal stromal cells through NOX2/gp91(phox) suppression in alleviating renal ischemia-reperfusion injury in rats. PLoS One. 2014;9:e92129.
81. Ranganath SH, Levy O, Inamdar MS, et al. Harnessing the mesenchymal stem cell secretome for the treatment of cardiovascular disease. Cell Stem Cell. 2012;10:244–58.
82. Hare JM, Traverse JH, Henry TD, et al. A randomized, double-blind, placebo-controlled, dose-escalation study of intravenous adult human mesenchymal stem cells (prochymal) after acute myocardial infarction. J Am Coll Cardiol. 2009;54:2277–86.
83. Teng X, Chen L, Chen W, et al. Mesenchymal stem cell-derived exosomes improve the microenvironment of infarcted myocardium contributing to angiogenesis and anti-inflammation. Cell Physiol Biochem. 2015;37:2415–24.
84. Zhao Y, Sun X, Cao W, et al. Exosomes derived from human umbilical cord mesenchymal stem cells relieve acute myocardial ischemic injury. Stem Cells Int. 2015;2015:761643.
85. Liu WH, Song FQ, Ren LN, et al. The multiple functional roles of mesenchymal stem cells in participating in treating liver diseases. J Cell Mol Med. 2015;19:511–20.
86. Li C, Kong Y, Wang H, et al. Homing of bone marrow mesenchymal stem cells mediated by sphingosine 1-phosphate contributes to liver fibrosis. J Hepatol. 2009;50:1174–83.
87. Ma F, Chen D, Chen F, et al. Human umbilical cord mesenchymal stem cells promote breast cancer metastasis by interleukin-8- and interleukin-6-dependent induction of CD44(+)/CD24(-) cells. Cell Transplant. 2015;24:2585–99.
88. Li T, Yan Y, Wang B, et al. Exosomes derived from human umbilical cord mesenchymal stem cells alleviate liver fibrosis. Stem Cells Dev. 2013;22:845–54.
89. Takahashi Y, Nishikawa M, Shinotsuka H, et al. Visualization and in vivo tracking of the exosomes of murine melanoma B16-BL6 cells in mice after intravenous injection. J Biotechnol. 2013;165:77–84.
90. Lai CP, Mardini O, Ericsson M, et al. Dynamic biodistribution of extracellular vesicles in vivo using a multimodal imaging reporter. ACS Nano. 2014;8:483–94.
91. Tian Y, Li S, Song J, et al. A doxorubicin delivery platform using engineered natural membrane vesicle exosomes for targeted tumor therapy. Biomaterials. 2014;35:2383–90.
92. Yu B, Kim HW, Gong M, et al. Exosomes secreted from GATA-4 overexpressing mesenchymal stem cells serve as a reservoir of anti-apoptotic microRNAs for cardioprotection. Int J Cardiol. 2015;182:349–60.

9 Therapeutic Application of Perinatal Mesenchymal Stem Cells in Diabetes Mellitus

Y. Cheng, J. Shen, and H. J. Hao

Abstract

Diabetes mellitus (DM) is a major global problem, and its complications seriously threaten the human health. The failure of pancreatic islet β-cells and insulin resistance are the main mechanism of the occurrence and the development of DM, and this can be induced by immune damage and inflammatory response. The current antidiabetic drugs, which do not focus on the link of the onset of diabetes, can only relieve or alleviate the symptoms, delay the progression of the disease, but cannot cure the disease. A developing field in diabetes therapy is the application of some stem populations, in particular those defined as "perinatal," derived from fetus-associated tissues usually discarded at birth. Umbilical cord-derived mesenchymal stem cells (UC-MSCs) are multipotent, self-renewing adult stem cells showing immunoregulatory properties and are capable to secrete an array of soluble cytokines and growth factors. Accumulating evidence showed that UC-MSCs can transdifferentiate into insulin-producing cells (IPCs) in vitro. Animal experiments and small sample clinical trials have validated the effectiveness and safety of UC-MSC therapy for diabetes. This chapter reviews the progress that has been made in UC-MSC treatment for diabetes, discusses the underlying mechanisms, and also raises the remaining hurdles.

Y. Cheng
Department of Endocrinology, Chinese PLA General Hospital, Beijing, China

Department of Molecular Biology, Institute of Basic Medicine, School of Life Science, Chinese PLA General Hospital, Beijing, China

J. Shen
Department of Endocrinology, Chinese PLA 309 Hospital, Beijing, China

H. J. Hao (✉)
Department of Molecular Biology, Institute of Basic Medicine, School of Life Science, Chinese PLA General Hospital, Beijing, China

9.1 Introduction

The pancreas is an essential organ for proper nutrient metabolism. The mature pancreas contains exocrine and endocrine parts. The exocrine compartment, which accounts for greater than 90% of pancreatic tissue, consists of acinar and ductal cells. Acinar cells synthesize digestive enzymes that catalyze the breakdown of proteins, carbohydrates, and lipids. The ductal cells line the channels that transport these secretions to the gastrointestinal tract. The endocrine compartment, which accounts for 1–2% of the total organ volume, consists of about one million structured endocrine glands called the islet of Langerhans. Multiple cell types reside in the islet, including insulin-producing β-cells, glucagon-producing α-cells, somatostatin-producing δ-cells, and pancreatic polypeptide-producing PP cells.

Z. C. Han et al. (eds.), *Perinatal Stem Cells*, https://doi.org/10.1007/978-981-13-2703-2_9

Functioning largely independent from exocrine cells, the endocrine compartment is responsible for glucose homeostasis by releasing hormones into the blood stream, and insulin released from β-cells is the main hypoglycemizing hormone that regulates blood glucose concentration.

Diabetes mellitus (DM) now is one of the leading health problems of the world. Its prevalence is expected to rise inexorably worldwide over the coming decades; a research published in 2010 revealed its prevalence in China as high as 9.7%,and the World Health Organization (WHO) indicated that the number of diabetic patients would reach 300 million by the year 2025 [1]. DM is classified into two related but distinct diseases. Type 1 diabetes (T1D) is caused by absolute insulin deficiency due to autoimmune destruction of insulin-secreting pancreatic β-cells, while type 2 diabetes (T2D), which is commonly associated with obesity, results from inadequate β-cell mass and function that can no longer compensate for insulin resistance of peripheral tissues. The continuing hyperglycemia results in both microvascular and macrovascular complications with high morbidity and mortality, such as atherosclerosis, kidney failure, blindness, and coronary artery disease. Currently available therapeutic regimens for T1D and T2D involve regular insulin injections and oral medication with sulfonylurea. However, although the current treatments can ameliorate glycemic abnormality in diabetic patients, they fail to control the blood sugar precisely and cannot prevent complications associated with diabetes. Pancreas transplantation and islet-cell transplantation are relatively curative. The success of the Edmonton protocol for pancreatic islet transplantation has sparked new interests in the treatment for DM. However, the requirement for immunosuppression and the scarcity of donors largely hinders the clinical application of this strategy. Therefore, the search for alternative approaches is of paramount clinical interest [2, 3].

In the last decade, stem cell biology has become an interesting topic in diabetes therapy, with abundant researches focusing on differentiation of insulin-producing cells (IPCs) from embryonic stem cells (ESCs) due to its virtually endless proliferative capacity and potential to differentiate into most, if not all, somatic cell types [4–9]. 2006 sees a breakthrough in this field by Baetge and his colleagues, developing a stepwise protocol to direct differentiation of ESCs into IPCs via mimicking the major stages of β-cell organogenesis during development [10]. In the following years, the stepwise protocol got widely recognized, and researchers dedicated to optimize the protocol to improve the differentiation efficiency and to induce IPC responsiveness to glucose stimuli close to the manner of adult human islets [11–17]. In 2014, two successive reports recorded the cellular level evidence of glucose responsiveness in ESC-derived IPCs [18, 19]. However, the responses were far smaller and slower than what the researchers observe in batches of healthy human islets. In addition, in 2015, Russ et al. reported that these IPCs showed limited glucose responsiveness after transplantation [20]. Thus, there is still a long way to go to direct ESCs to fully functional β-cells that can cure diabetes after transplantation.

Another type of stem cell that draws great attention for diabetes therapy is mesenchymal stem cells (MSCs). MSCs are multipotent adult stem cells that have been noted for the capacity to differentiate into cells of various lineages, including the bone, cartilage, and adipose tissues. Importantly, MSCs also show immunoregulatory properties and are demonstrated to secrete an array of soluble cytokines and growth factors. These properties rendered MSCs promising cell sources for diabetes therapy [21–25]. Unlike ESCs that are still under preclinical investigation, a series of small sample clinical trials have already confirmed the efficacy and safety for MSCs in treating T1D and T2D [26–28]. MSCs are suggested to be present virtually in any vascularized tissue throughout the whole body. As in diabetes therapy, the application of MSCs derived from bone marrow, adipose tissue, and perinatal tissues has been most extensively investigated. Bone marrow-derived mesenchymal stem cells (BM-MSCs) and adipose-derived stromal cells (ADSCs) allow autologous cell transplantation with no limitations due to host immune reactions. However, isolation of BM-MSCs inevitably requires a highly invasive procedure and harvests limited amount of cells. In comparison, adipose tissue is easily available and relatively abundant, being a more attractive cell source for therapeutic purposes. However, the number and function of autologous ADSCs are dependent on the donor's age and medical condition. In particular, recent data indicated that ADSCs from diabetic donors exhibit reduced proliferative potential and increased apoptosis and secrete less effective growth factors and cytokines. Therefore, tissue-engineering methods must be improved to ensure the success of regenerative medicine when autologous ADSCs are used. Perinatal tissue-derived MSCs also have gained great attention in recent years. These tissues usually are discarded as medical waste, and cells can be isolated easily with simple and reproducible lab techniques with few ethical or safety issues. Especially, perinatal tissues contain the youngest MSCs featuring the highest potential and low immunogenicity, rendering these cells promising for the vast field of regenerative medicine [29, 30]. In the present chapter, we will discuss the contribution of the umbilical cord-derived MSCs (UC-MSCs) in the treatment of diabetes and also raise the remaining hurdles.

9.2 Pancreas Development

Pancreas development is the result of a carefully orchestrated series of embryologic events. The recent decades have seen a huge expansion in our knowledge about endocrine pancreas development. Numerous lineage-restricted transcription factors and signaling pathways that are important for pancreas development have been identified, and the specific roles have been investigated by genetic and cell biological methods [31, 32]. If the readers are interested in the detailed description of the molecular regulation of pancreas development, some excellent reviews in the field can be referred to [33–37]. In this chapter, we will only provide some basic notions, which might help to comprehend the rationale behind in vitro differentiation experiments for UC-MSCs and the choice of markers to monitor the success of these attempts. Our knowledge of pancreas development mainly derives from rodent models, which means we must be careful when the results were translated into the differentiation strategies since the well-defined events in rodent pancreas development may be not exactly recapitulated for the human organ.

During organogenesis, pancreas development first becomes morphologically evident with the emergence of a dorsal and ventral epithelial bud protruding the primitive gut epithelium. These two pancreatic buds subsequently elongate and eventually fuse to a single organ. At the same time, branching morphogenesis occur when cell proliferation leads to a rapid grow of the pancreatic buds, which change their shape and begin to form branched tubular structures. This early phase of pancreatic development has been referred to as the primary transition. Next, the pancreatic epithelium further expands and branches into a more complex yet highly ordered tubular network. Simultaneously, the undifferentiated pancreatic epithelium differentiated into endocrine, acinar, and ductal cells. This later phase was referred to as the secondary transition. In the end of the secondary transition, the pancreas became a highly organized organ, with clusters of acinar cells coalesced around the ends of the ductal network and endocrine cell aggregates scattered throughout the organ.

Once the ventral and dorsal domains are specified, a very important transcription factor Pdx1 become specifically expressed in the pancreatic regions and is a marker of common pancreatic progenitors. Loss of Pdx1 function results in an early block in pancreatic outgrowth and differentiation in both mice and humans [38–40]. In addition to an early role in pancreatic bud outgrowth, studies using tetracycline-inducible Pdx1 inactivation demonstrated that pdx1 is also specifically required for subsequent differentiation of endocrine and exocrine cells [41]. Pdx1 is specially expressed in β-cells in adult pancreas and is essential for maintaining mature β-cell function. Loss of Pdx1 protein between 3 and 5 weeks after birth resulted in a dramatic decrease in insulin, Nkx6.1, and Glut2 expression in β-cells and overt diabetes in 3–5-month-old mice [42]. During the secondary transition, a subset of cells in the pancreatic progenitor pools initiate the expression of the transcription factor Ngn3, which marks the onset of endocrine cell differentiation. Precise control of Ngn3 induction is the key factor balancing the endocrine versus ductal cell fate decision [43, 44]. Ngn3 positive cells then generate all the five different endocrine cell types. During this process, the transcription factors Pax4, Pdx1, Nkx2.2, and Nkx6.1 act as critical β-cell determinants [45–48], while Arx expression induced α-cell identity [49–51]. Forced expression of Pax4, Pdx1, or Nkx6.1 in endocrine precursors favors the β-cell fate [52, 53]. Conversely, forced expression of Arx leads to excess α-cells and reduced β-cells [54].

9.3 Biological Characteristics of UC-MSCs

9.3.1 The History of UC-MSCs

BM-MSCs are the historically accepted "gold standard" of MSCs, as they are the first identified MSCs in the 1970s when Friedenstein et al. demonstrated that the fibroblast-like cell population after long-term human bone marrow culture are a rare, heterogeneous, stromal population of multipotent non-hematopoietic progenitor cells that are capable to differentiate into multiple mesenchymal lineages, including the bone, fat, and cartilage. But currently, it is well accepted that MSCs can be extracted virtually from any vascularized tissue throughout the whole body [55].

The human umbilical cord, which develops from the yolk sac and allantois, arises at day 13 of embryonic development and connects the developing embryo or fetus and the placenta. The umbilical cord contains two umbilical arteries and one umbilical vein, both embedded by an outer covering of one or more layers of amniotic epithelium with an interior mass of gelatinous substance named Wharton's jelly after Thomas Wharton (1614–1673), an English physician and anatomist. The umbilical cord tissue before was considered as medical waste with no scientific value. This point of view was completely revised in 1991, when McElreavey et al. for the first time reported isolation of fibroblast-like cells from Wharton's jelly [56]. In 2004, these fibroblast-like cells were proved to be MSCs as they met the minimal criteria provided by the International Society for Cellular Therapy to define MSCs [57]: (a) plastic adherent ability; (b) negative for hematopoietic and endothelial markers such as CD11b, CD14, CD31, CD34, and CD45 and positive for CD90, CD105, CD44, CD73, CD9, and very low levels of CD80; and (c) capacity to differentiate into adipocytes, osteoblasts, and chondrocytes.

9.3.2 The Phenotype of UC-MSCs

To date, a lot of efforts have been put into the exploration of the expression profile of surface markers and pluripotency markers of UC-MSCs [58, 59]. Overall, MSCs from UC fit with the minimal criteria for MSCs. Specially, UC-MSCs have also been demonstrated to express markers of ES cells, such as Oct-4, SSEA4, nucleostemin, SOX-2, and Nanog.

9.3.3 Proliferative Potential of UC-MSCs

One major property of MSCs is the self-renewal ability. Several studies aimed to compare the proliferative potential of MSCs derived from different tissues, and it is well demonstrated that UC-MSCs display the highest proliferative rate [60]. Based on the cell doubling time calculation, the cell doubling time of the UC-MSCs was approximately 40 h, while it was approximately 45 h for ADSCs and 70 h for BM-MSCs. The CFU-F assay based on Castro-Malaspina's method showed that the mean of CFU-F (colony-forming unit-fibroblast) colonies per 1×10^6 nucleated cells was significantly higher in UC-MSCs (800, range 300–2000) than in BM-MSCs (36, range 16–64). CFU-F frequency determined by limiting dilution assay further revealed a higher frequency of CFU-F in UC-MSCs (1:1609 ± 0.18) than in BM-MSCs (1:35700 ± 0.01)[61]. According to another report, typical CFU-F efficiency (the ratio of number of cells forming colonies under clonal conditions and number of cells seeded directly after isolation) for BM-MSCs ranged from 0.001 to 0.01%, while in cell isolates from five different donors of UC, 0.2–1.8% CFU-Fs developed in cultures[62].

9.3.4 The Secretion Pattern of UC-MSCs

MSCs secrete a variety of bioactive compounds that mediate their therapeutic activity in diseases. The secretion pattern of UC-MSCs differs significantly from MSCs from BM-MSCs and ADSCs. Several groups of researchers collectively reported the almost complete absence of the well-known proangiogenic factor vascular endothelial growth factor (VEGF) in the culture medium of UC-MSCs with the concentration thousands of times lower than BM-MSCs and ADSCs [63, 64]. MCP-1 secretion was demonstrated to be significantly higher for WJ-MSC than MSCs from other sources. Considering anti-inflammatory cytokines, the data are controversial. One research showed a significant higher concentration of IL-1Ra for WJ-MSCs [63], but another study claimed that the concentration of IL-1Ra for WJ-MSCs was significantly lower [62]. In addition, UC-MSCs secrete significantly higher amounts of several important cytokines and hematopoietic growth factors, including G-CSF, GM-CSF, IL-6, and IL-8, compared to BM-MSCs and ADSCs [65, 66]. It has also been reported that UC-MSCs exhibited increased secretion of neurotrophic factors nerve growth factor (NGF), and the authors believed that UC-MSCs might be more favorable for diseases of the nervous system [67].

9.3.5 Multi-lineage Differentiation Potential of UC-MSCs

Another major property of MSCs is the multilineage differentiation potential. A large body of literature tested the differentiation capacity of UC-MSCs in vitro, and the results indicated that in a response to inductive cues, UC-MSCs were able to differentiate into chondrocytes, adipocytes, osteoblasts, odontoblast-like cells, dermal fibroblasts, smooth muscle cells, skeletal muscle cells, cardiomyocytes, hepatocyte-like cells, sweat gland cells, endothelial cells, neuroglia cells (oligodendrocytes), dopaminergic neurons, and importantly, insulin-producing cells (IPCs).

Comparison of the differentiation potential of UC-MSCs and MSCs from other sources (bone marrow and adipose tissue) is the subject of numerous studies. For adipogenesis, the lipids were observed 3 days after induction for ADSCs, while the time period was 6 days for BM-MSCs and 8 days for UC-MSCs. The ratio of differentiated adipocytes from the total cells for ADSCs was also the highest: it was 52 ± 3.2% for the ADSCs, 45 ± 1.5% for the UC-MSCs, and 39 ± 1% for the BM-MSCs. We can see that although the speed of adipocyte differentiation for the UC-MSCs was slower than that of the BM-MSCs, the final induction ratio for the UC-MSC was higher than that of the BM-MSCs. As regards osteogenesis, yellow substances accumulated in the MSCs from all tissue sources, particularly in the UC-MSCs, after 3 weeks of induction, and gradually, these sediments increased in size to form round or polygonal nodules. At the end of observation, the average numbers of bone nodules from 1 well were 19 ± 1.8 for the UC-MSCs (0.01 ± 0.001%), 11 ± 1.7 for the ADSCs (0.006 ± 0.0009%), and 7.5 ± 1.3 for the BM-MSCs (0.004 ± 0.0007%) [65]. For neuronal differentiation, when UC-MSCs and BM-MSCs were exposed to the combination of sonic hedgehog (SHH), fibroblast growth factor 8 (FGF8), along with basic fibroblast growth factor (bFGF), the efficiencies of transdifferentiation from both sources of MSCs to the dopaminergic cell type were similar [68]. In another study, MSCs isolated from adipose, bone marrow, and Wharton's jelly could all be induced to form primary neurospheres expressing neural progenitor markers using epidermal growth factor (EGF) and bFGF. ADSC-derived progenitors failed to self-renew, while primary neurospheres derived from the other two sources were able to do so. When primary neurospheres from three sources were differentiated to neuronal lineage on fibronectin-coated dishes, the efficiency varied a lot. Progenitors derived from UC-MSCs could be induced into neural-like cells with higher efficiency compared to BM-MSCs and ADSCs [69]. Following endothelial differentiation, UC-MSCs had higher expression of the endothelial-specific markers than BM-MSCs. Both UC-MSCs and BM-MSCs after endothelial differentiation were able to form the capillary network, but the product differentiated from UC-MSCs had significantly higher total tubule length, diameter, and area than that differentiated from BM-MSCs [70]. When pancreatic differentiation potential was compared between UC-MSCs and BM-MSCs, the data suggested that differentiated UC-MSCs expressed significantly higher level of the pancreatic-specific transcription factor PDX1 in both mRNA and protein levels than differentiated BM-MSCs, and the percentage of differentiated UC-MSCs expressing pancreatic-specific marker C-peptide was 72% higher than differentiated BM-MSCs[71]. Overall, the differentiation potential of UC-MSCs is at least comparable with MSCs from adipose and BM.

9.3.6 Immunoregulatory Property of UC-MSCs

Apart from the multi-lineage potential, UC-MSCs possess various immunoregulatory properties, rendering them as appealing candidates for cell-based therapies. UC-MSCs expressed very low level of HLA class I and lack the expression of HLA-DR. Notably, compared to BM-MSCs obtained from aged donors, HLA-I expression was much lower in UC-MSCs, while the intracellular concentrations of the immunosuppressive molecule HLA-G are much higher. In addition, UC-MSCs do not express the co-stimulatory molecules CD40, CD80, and CD86 and do not elicit proliferative responses of allogeneic T cells in vitro. These findings collectively suggest that UC-MSCs may be characterized by inherently low immunogenicity. However, another study reported that UC-MSCs may not be as immune-privileged as thought, as they can trigger weak allogeneic immune activation in vitro and will eventually undergo rejection following xenogeneic transplantation.

Much attention has been paid on the immunosuppressive potential of UC-MSCs on T lymphocytes. Many reports consistently showed that UC-MSCs are able to inhibit T-cell proliferation regardless of the stimulus used to activate T cells and augment the generation of regulatory T cells (Tregs) [72–76]. Several studies suggested that the immunomodulatory effect of MSCs on T cells might be mediated, at least in part, by soluble factors. A study co-cultured UC-MSCs with human peripheral blood mononuclear cells (hPBMCs) and identified prostaglandin E2 (PGE2) as the major factor by blocking experiments, as inhibition of PGE2 synthesis almost completely inhibited the immunosuppressive effects, whereas neutralization of transforming growth factor-β1 (TGF-β1), HGF, and indoleamine 2,3-dioxygenase (IDO), and nitric oxide (NO) activities had little effects [77]. However, later data showed that abundant IFN-γ released by T lymphocyte could induce the expression of IDO in UC-MSCs. UC-MSCs increased apoptosis of T lymphocyte as the expression of caspase3 was significantly increased, while Bcl2 and CDK4mRNA expression decreased dramatically. Addition of the IDO inhibitor restored T-lymphocyte proliferation, reduced apoptosis, and induced resumption of the cell cycle. In addition, the changes in caspase3, CDK4, and Bcl2 expression were reversed by the inhibitor. These findings demonstrate that UC-MSCs induce T-lymphocyte apoptosis and cell cycle arrest partly by expressing abundant IDO [78]. Additional reports also confirmed the role of IDO in mediating T-cell immunosuppressive activity of UC-MSCs [79–81]. In addition, one of the reports also showed that UC-MSCs constitutively express a negative regulator of T-cell activation, B7-H1, and its expression is increased after IFN-γ treatment. Addition of a B7-H1-specific blocking antibody abrogated the T-cell immunosuppressive activity [80]. Moreover, two of the reports recognized the importance of cell-cell contact in the interaction between UC-MSCs and T lymphocyte, as blocking cell-cell contact using a Transwell chamber largely attenuated the immunosuppression mediated by UC-MSCs. Interestingly, one paper identifies monocytes as an essential intermediary through which UC-MSCs exert their suppressive effects on T-cell proliferation, as the removal of monocytes but not B cells from hPBMCs reduced the immunosuppressive effects of UC-MSCs on T-cell proliferation[82].

The immunomodulatory effects of UC-MSCs on B lymphocytes have been poorly investigated and undoubtedly need further elucidation, as the information will provide crucial clues for the introduction of UC-MSCs into the clinic. Only two researches till today discussed this issue, and surprisingly, the results are controversial. One report stated that UC-MSCs markedly suppressed B-cell differentiation as shown by the decreased number of CD138+ cells and reduced levels of IgM and IgG production in co-culture [83], while another report showed that treatment with human UC-MSC resulted in an increase of proliferation, differentiation of B cells into plasma cells, and production of antibodies in vitro [84].

Natural killer (NK) cells represent the major lymphocyte population in peripheral blood and display graft-versus-leukemia (GvL) effects after hematopoietic stem cell transplantation (HSCT) [85, 86]. Two groups reported that UC-MSCs could potently suppress NK cell cytotoxicity via soluble factors. One group identified prostaglandin E2 (PGE2) that played crucial roles in suppression of NK cells [87]. Another study showed that UC-MSCs secreted considerable amounts of activin-A, which could suppress IFN-γ production by NK cells by a loss of function experiment via neutralization of activin-A [88]. The authors also found that inhibition of PGE2 production could also significantly restore IFN-γ production.

After injury, resident tissue macrophages are supplemented by an active recruitment of blood monocytes, which then differentiate into macrophages and dendritic cells. Macrophages are abundant during all stages of tissue repair and have an important influence on the progress and resolution of tissue damage [89–92]. Macrophages have been categorized conventionally into pro-inflammatory M1 and tissue-repairing M2 phenotypes. Using a co-culture system, UC-MSCs were confirmed to induce the acquisition of M2-macrophage features in terms of morphology, surface markers, migratory properties, and antigen presentation capacity during monocyte differentiation [93]. Importantly, in vivo data showed that UC-MSCs played a central role in macrophage polarization in various disease models, which could account for the therapeutic effect of UC-MSCs. For example, in an experimental acute myocardial infarction model, UC-MSC infusion reduced the overall macrophage/monocyte levels, including pro-inflammatory M1 macrophages, while the proportion of alternatively activated M2-type macrophages was significantly increased in the circulation and heart. Therefore, expression of pro-inflammatory factors such as IL-1β and IL-6 decreased, and anti-inflammatory factors like IL-10 significantly increased in the infarct area. Fewer apoptotic cardiomyocytes without changes in angiogenesis were also detected in the infarct area [94]. After renal ischemia-reperfusion injury, UC-MSC infusion resulted in significant improvements in renal function, with a lower tubular injury score together with more proliferative and fewer apoptotic tubular cells in kidney tissue. Moreover, UC-MSCs reduced the infiltration of macrophages into renal interstitium, while the proportion of anti-inflammatory M2 macrophages was markedly increased. Macrophage depletion during the repair phase eliminated the protective role of UC-MSCs [95]. Macrophages from systemic lupus erythematosus (SLE) patients showed lower expression of M2 macrophage marker CD206 and reduced phagocytic activity, which were corrected by co-culture with UC-MSCs in vitro. UC-MSC transplantation to SLE patients improved the proportion of CD206+ macrophages and their phagocytic activity [96]. Very inspiringly, our published data showed that UC-MSC infusion produced significant antidiabetic effects and promoted insulin sensitivity in an experimental T2D rat model via directing macrophages in adipose tissue into an alternatively activated phenotype (M2, anti-inflammatory). Furthermore, we concluded that IL-6 secreted by UC-MSCs elicited the M2 polarization by in vitro knockdown of IL-6 and transplanting IL-6 knockdown UC-MSCs into T2D rats [97].

9.4 Differentiation of UC-MSCs into IPCs

WJ-MSCs gained much attention in the regenerative medicine due to their ability to differentiation into several mature cell types. A series of recent reports demonstrated that WJ-MSCs could be successfully induced into IPCs. In addition, in vivo experiments also showed that transplantation of IPCs derived from WJ-MSCs could ameliorate hyperglycemia in experimental diabetic animals.

The first attempt to differentiate WJ-MSCs into IPCs was performed by Chao and coworkers through a stepwise culture protocol [98]. Interestingly, a key step in this protocol was to induce the expression of nestin, a neural progenitor marker, in WJ-MSCs using the neuronal conditioned medium (NCM) that was made from rat brain. The author claimed the similarities between islet cells and neurons. In some invertebrate species, such as Drosophila, brain neurons were the main source of circulating insulin. It has also been shown that multipotent precursor cells from the adult mouse pancreas can form clonal colonies, which express both neural and pancreatic precursor markers, generating distinct populations of neurons and glial cells, pancreatic endocrine cells, and pancreatic exocrine and stellate cells. Moreover, a study conducted by Hori suggested that brain-derived human neural progenitor cells formed clusters of glucose-responsive IPCs when exposed to a series of signals that regulate in vivo pancreatic islet development. Therefore, it was not surprised that when the nestin positive cells derived from WJ-MSCs were placed in a medium containing nicotinamide, and B27, which are known factors to promote β-cell differentiation and maturation, islet-like clusters became evident. These clusters showed mRNA expression of insulin and other pancreatic β-cell-related genes, such as Pdx1, Hlxb9, Nkx2.2, Nkx6.1, and Glut-2. Importantly, these clusters exhibited glucose responsiveness similar to pancreatic β-cells. When islet-like cell clusters were transplanted into the liver of STZ-induced diabetic rats via laparotomy, hyperglycemia was normalized 3 days afterward, and stable blood glucose levels were maintained for 9 weeks until sacrifice. Glucose clearance was also normalized determined by IPGT test.

Moshrefi and coworkers in 2015 attempted to assess the effect of different concentrations of NCM on nestin and PDX1 expression of the differentiated UC-MSCs [99]. The proportion of nestin-positive cells was 17%, 31%, and 85%, after induction in a medium containing 20%, 40%, and 60% NCM concentration, respectively. Then, these cells were cultured in medium supplemented with nicotinamide and insulin-transferrin-selenium (ITS) for 1 week to induce Pdx1 expression. As assessed by RT-PCR, the optical density of Pdx1 expression was 0.44, 0.59, and 0.83 in 20%, 40%, and 60% NCM concentrations, respectively. As a result, the author indicated that 60% NCM was favorable for generating IPCs from WJ-MSCs.

There are also many other researches that did not utilize NCM as an inductive cue. For example, Wu and colleagues carried out a comparative study to compare the differentiation potential of WJ-MSCs and BM-MSCs toward IPCs using a medium containing nicotinamide, activin A, HGF, exendin-4, and pentagastrin without a step toward nestin-positive cells [71]. By this simple method, both UC-MSCs and BM-MSCs formed islet-like cell clusters as soon as 3 days after culture. The clusters formed by UC-MSCs were larger than those formed by BM-MSCs. Notably, the higher expression levels of PDX1 were assessed in differentiated UC-MSCs with respect to differentiated BM-MSCs. Likewise, secretion of insulin and C-peptide was comparably higher in the differentiated WJ-MSCs. The authors did not evaluate glucose responsiveness of the islet-like clusters in vitro or in vivo. So whether this simple method could generate functional islet clusters as the stepwise strategy did was unknown.

Next, Wang and coworkers tried a new stepwise differentiation protocol without the use of NCM [100]. Cells were cultured for 7 days in CMRL1066 medium containing β-nerve growth factor, activin-A, nicotinamide, and epidermal growth factor (EGF). Then, the culture medium was changed to DMEM/F12 for the next 3 days. Finally, nicotinamide, insulin-transferrin-selenium, and basic fibroblastic growth factor (bFGF) were added, and incubation was continued for the next 7 days. At the end of differentiation, mRNA levels for genes related to β-cell development and function such as MafA, Pax4, Nkx2.2, Glut2, and insulin were significantly increased in differentiated cells. Secretory vesicles showing insulin and C-peptide were observed in differentiated cells by immunogold electron microscopy. The product also responded very well to glucose stimulation in vitro. Furthermore, after transplantation, IPCs were located in the liver and were able to restore physiological glycemia in NOD mice, which develops a spontaneous form of autoimmune diabetes that closely resembles the human disease.

This protocol was then well accepted and modified in several reports. A report slightly modified this protocol as following: cells were cultured in DMEM/F12 medium containing nicotinamide, activin-A and EGF for 1 week, and then the culture medium was changed to DMEM/F12 for another week followed by an addition of nicotinamide, bFGF, and ITS for another 2 weeks [101]. Similar results were obtained in this study. Particularly, differentiated cells were transplanted in rats that received 90% pancreatectomy and showed elevated blood glucose levels of 2 days after surgery. The values of the sham group kept elevating throughout the experimental period. However, in the cell transplantation group, blood sugar levels significantly decreased 2 weeks after transplantation and gradually reversed to a normal level at the end of observation. The function of the transplanted cells was also tested by IPGT test at the end of the experiment, and the glucose clearance was comparable with the normal control.

Tsai and coworkers reported another three-stage strategy that was modified from previously established protocol by Wang [102]. At the first stage, cells were counted and resuspended in medium containing DMEM/F12 with sodium butyrate, 2-mercaptoethanol, ITS, transferrin, and activin A. Stage I lasted for 2 days. On the third day, the culture medium was changed to a medium containing DMEM/F12 with ITS-X and taurine. On the fifth day, the cell culture was replaced by a medium containing DMEM/F12 with ITS, taurine, glucagon-like peptide-1 (GLP-1), nicotinamide, and nonessential amino acids. Aggregated cell clusters formed after three-stage protocol. The author detected expression of Pax1, Pax4, and insulin at the mRNA level in the clusters. The differentiated cells secreted around 50 pmol/L of C-peptide at a low glucose concentration (5.5 mM) and approximately five times as much at a high glucose concentration. Interestingly, the differentiated IPCs were transplanted into STZ-induced T1D rats through the portal vein instead of the kidney capsule, which is more usually taken as a transplantation spot. By 6 weeks posttransplantation, the function of transplanted IPCs was evidenced by the appearance of both human C-peptide and human cell nuclei in the lobules of the rat liver. Also, the blood sugar levels were reduced significantly 4 weeks after transplantation. These findings suggest that the IPCs differentiated from human UC-MSCs were able to function as islet-like structures following transplantation.

It seemed that protocols devised by Chao, Wu, or Wang all generate functional IPCs. Therefore a study conducted by Seyedi compared these three commonly reported protocols with a new modified differentiation protocol employed by the author [103]. The details of each protocol can be referred to in the article. Briefly, protocol I replicated the protocol proposed by Chao, protocol II was more like the protocol used by Wu, and protocol III followed the method proposed by Wang. The modified protocol named protocol IV employed by the author also used the NCM to induce vimentin expression in UC-MSCs, and the author determined the appropriate time of NCM exposure to be 5 days. Different from the protocol proposed by Chao, the subsequently used medium was supplemented with more growth factors including nicotinamide, exendin-4, pentagastrin, HGF, and B27. Protocols I, II, and IV induced UC-MSCs into islet-like clusters, but in protocol III, although the morphology of the induced cells changed into round, no obvious cluster was observed. The cells in protocol

IV formed more apparent clusters than the other protocols. Insulin and dithizone staining showed that protocol IV produced more and larger islets than the other protocols. All protocols produced glucose-responsiveness IPCs, but insulin secretion after high glucose stimulation in the new modified protocol IV was significantly higher than the other groups. The authors concluded that appropriate period of nestin preselection stage followed by induction with adequate inducing agents will enhance UC-MSCs to differentiate into more mature IPCs in a greater number.

Other researches intending to enhance the differentiation process from UC-MSCs to IPCs also emerged. One study explored the effect of the culture system on differentiation [104]. As confirmed by gene expression and protein synthesis, UC-MSCs successfully differentiated into IPCs in both two-dimensional (2D) and three-dimensional (3D) environment. Both groups secreted insulin and responded to the glucose challenge test. However, cells cultured in the 3D environment secreted insulin at a level significantly higher (about 40-fold) than 2D group. Exposure of induced cells to low- and high-glucose environments resulted in twofold insulin secretion in the monolayer group and a 5.5-fold insulin secretion in the hanging drops group. This study indicated that 3D-culture might have a beneficial effect on the maturation of IPCs. Another study determined laminin 411 as a potent inducer of UC-MSC differentiation into IPCs [105]. Laminin is a heterotrimer glycoprotein. At least 19 laminin isoforms have been identified. Laminin is a key component of the basement membrane and is involved in the structural scaffold, cell proliferation, and differentiation. Previous studies showed that in the presence of laminin-1, the 13.5-day mouse fetus differentiation resulted in a dramatic increase in the number of β-cells by nearly 60-fold. In addition, laminin combined with fibronectin promoted expression of islet cell markers in hepatic oval cells. In this study, laminin 411 markedly induced the expression of markers for pancreatic precursor cells, efficiently induced IPC differentiation from UC-MSCs, and upregulated the insulin expression at both mRNA and protein levels.

Pdx1, as we mentioned in the paragraph of pancreas development, is one of the most important transcription factors in pancreatic differentiation. Accumulating evidence suggested that via ectopic expression of Pdx1 in vitro or in vivo, liver cells could be reprogrammed toward β-like cells [106–109]. Attempts have also been made to induce UC-MSCs into IPCs by Pdx1 gene manipulation. A research differentiated UC-MSCs with in vitro-synthesized Pdx1 messenger mRNA into islet-like cell clusters. hUC-MSCs were confirmed by both biomarker detection and functional differentiation. The upregulated expression of Pdx1 protein can be detected soon after transfection and remains detectable for 36 h. Morphology analysis and dithizone staining confirmed the induction of islet-like structures. Results obtained from RT-PCR revealed the expression of some key pancreatic transcription factors, such as Pdx1, NeuroD, NKX6.1, Glut-2, and insulin, in islet-like cell clusters. Differentiated cells expressed both insulin and C-peptide at protein level. Glucose stimulation test also validated the function of induced islet-like cell clusters [110]. Another study infected UC-MSCs with recombinant adenovirus carrying the Pdx1 gene and treated UC-MSCs with several inducing factors for differentiation into islet β-like cells. UC-MSCs were aggregated and presented like islet-like bodies. Insulin and dithizone staining of these cells were positive. In addition, the expression of β-cell-related genes was confirmed [111].

9.5 Clinical Application of UC-MSCs for DM

Several small sample clinical trials in succession investigated the effect of UC-MSCs on T2D. A study tested the effect of UC-MSCs on six patients with different histories of T2D [112]. Approximately 10^6 UC-MSCs/kg body was administered intravenously twice at a 2-week interval. Of the 6 patients, 3 became insulin-free for varying lengths of time between 25 and 43 months, and the additional 3 patients showed reduced requirements for insulin dose. In the insulin-free group, the C-peptide production significantly increased after the transplantation, while there were no statistically significant changes observed in the insulin-dependent group.

In another study, 22 patients with a fasting blood glucose level ≥7.0 mmol/L and HbA1c ≥7% received UC-MSC infusions through 1 intravenous injection and 1 intrapancreatic endovascular injection [113]. The glycated hemoglobin levels showed a progressive decline after transplantation, with a maximum decrease 3 months after treatment (baseline, 8.20 ± 1.69%; 1 month, 7.08 ± 1.09%; and 3 months, 6.89 ± 0.90%), maintained stable for the next 6 months, and slightly rebounded to 7.0 ± 0.60% at the 12-month follow-up. The fasting blood glucose levels also showed a downward trend in 6 months after intervention and slightly elevated at the 12-month follow-up. The β-cell function also peaked at 6-month follow-up. Seven out of 17 patients who received insulin therapy before transplantation became insulin-free for a mean time of 9 months. Other nine patients reduced insulin requirement by various degrees.

Very recently, a published study evaluated the long-term efficacy and safety of infusion of UC-MSCs on T2D [114]. Sixty-one patients with a clinical and laboratory diagnosis of T2DM according to the criteria outlined by the American Diabetes Association were enrolled and randomly divided into two groups. Patients in group I received UC-MSCs intravenously twice with a 4-week interval, and patients in group II were treated with normal saline as control. At the initiation of therapy, all patients had been treated with diet, exercise, oral hypoglycemic agents, and insulin injections at stable doses for at least 2 months. Following UC-MSC infusion, a gradual reduction in fasting blood glucose (FBG) and postprandial blood glucose (PBG) was detected during the follow-up. FBG and PBG levels were at the lowest at the 3-month follow-up or the 6-month follow-up post-therapy and remained stable till the 24th month, after which, FBG and PBG moderately increased during the remaining follow-up time. A similar dynamic change of HbA1c was also detected after transplantation. Insulin withdrawal was observed in 10 patients out of 31 patients that received UC-MSC injection. These patients remained insulin-free for 12.5 ± 6.8 months. In total, 58.1% (18/31) of patients exhibited a ≥50% reduction in insulin requirement; in 5 of the remaining 13 patients, daily insulin dosage was reduced by 15–50%, whereas the insulin dosage requirements of 8 patients were maintained or reduced by <15%.

However, there is 1 study which showed that UC-MSC infusion only significantly reduced the FBG and PBG in 18 T2D patients. There was a gradual fall in the exogenous insulin requirements, but the decrease did not reach statistic difference [115].

To date, two reports have tested the effect of UC-MSCs on T1D. Hu and coworkers performed a study in which 29 patients with newly onset T1D were enrolled and divided into 2 groups by randomized blocks; 15 patients participated in the WJ-MSC treatment group (group I), while the other 14 patients participated in control group (group II) [116]. UC-MSCs were administrated twice with a 4-week interval by intravenous delivery. FBG levels didn't show significant difference between the two groups during the 24-month follow-up. But PBG levels in group I reached the lowest level at the end of first year after therapy, then began to rise lightly, but still remained a better control than that in group II patients. The mean value of HbA1c showed a gradual decrease and reached the lowest level at the 6th month of follow-up (baseline 6.8 ± 0.57%, 6 months 5.5 ± 0.67%) after cell transplantation and fluctuated slightly over the following 18 months. Consistently, there was a progressive increase of mean fasting C-peptide levels after treatment, and the level peaked at the end of first year of follow-up and then had a little decrease. At the end of the analysis period, 3 out of 15 patients discontinued insulin injection. Eight of the remaining 12 patients reduced their daily insulin dosage by more than 50% of the baseline. One patient reduced the daily insulin dosage by 15–50%, and the remaining two patients maintained the insulin dosage.

In a more recent study, 42 patients with T1D were randomized to either control group or stem cell transplantation group (SCT group) in which UC-MSCs plus autologous bone marrow mononuclear cells (BM-MNCs) were infused through selective pancreatic artery [117]. After 1 year, improvement of metabolic measures was observed in patients receiving stem cell therapy. The primary end point, the C-peptide area under the curve (AUCC-Pep) during a 3-h oral glucose tolerance test, increased 105.7% from basal (6.6 ± 6.1 to 13.6 ± 8.1 pmol/mL/180 min), with 15 of 21 patients showing increased levels at 1 year in the SCT group. HbA1c levels in the transplantation group were significantly lowered compared to the baseline. Progressive and significant reductions in insulin dose requirements after transplantation were observed at indicated time points after transplantation. The insulin requirement was reduced 29.2% in the transplantation group (0.9 ± 0.2 to 0.6 ± 0.2 IU/day/kg).

Although inspiring results have been obtained, several issues still need to be addressed. First, all the five clinical trials mentioned above are phase I or phase I/II study. The largest number of patients enrolled is 61. In addition, only two of the five studies are randomized controlled study. Therefore, the antidiabetic effect of UC-MSCs must be tested in larger randomized controlled study.

Second, in some of the clinical studies, UC-MSCs are administrated intravenously. But in others, superior pancreaticoduodenal artery or splenic artery delivery was chosen as the administration route. Selective pancreatic artery infusion possibly could enhance the number of UC-MSCs homing into the pancreas [118], but whether the amount of homing MSCs is parallel with the effect on diabetes control is unknown. Unless solid evidence is provided, superior pancreaticoduodenal artery delivery is not a wise choice for UC-MSC transplantation as it is a highly invasive procedure.

Third, the clinical dose of UC-MSCs in diabetes ranges from 0.9×10^6 to 3×10^6 cells/kg body weight of the recipient, and UC-MSCs were usually administrated twice at a 2-week interval. In patients with steroid-refractory acute graft-versus-host disease, high dose (8×10^6 cells/kg) and low dose (2×10^6 cells/kg) did not reveal significant differences in response rate or relapse of the primary disease [119], while repeated infusion of MSCs at certain intervals seems to affect the results [120–122]. Moreover, compared with one single infusion that only transiently lowered blood glucose level, multiple intravenous infusions of BM-MSCs completely reversed hyperglycemia in experimental type 2 diabetic rats [123]. So we suspected that a strategy of multiple infusions with shorter interval in the beginning of the stem cell therapy might exert better effects for UC-MSCs in diabetes therapy. In addition, 2 of the 5 clinical trials mentioned that the maximum antidiabetic effect was reached around 3–6 months after infusion, but there was a moderate elevation of FBG, PBG, and HbA1c levels 12 months after treatment. If an additional infusion was performed at or before this time point, metabolic measures of the patients might maintain at the same level or possibly further improved. We think this point of view deserves deeper investigation.

Efforts in enhancing the therapeutic effect of UC-MSCs on DM are necessary. Sitagliptin is a dipeptidyl-peptidase IV (DPP4) inhibitor that prevents the inactivation of GLP-1 and glucose-dependent insulinotropic polypeptide (GIP), leading to increased insulin release and decreased glucagon levels. The clinical trials revealed that sitagliptin is effective in reducing HbA1c values, fasting plasma glucose (FPG) levels, and 2-h PBG levels in patients with type 2 diabetes. Combined therapy of UC-MSCs and sitagliptin might be beneficial. Hu and colleagues verified this hypothesis in a T2D rat model [124]. After therapy, FBG in UC-MSC group, sitagliptin group, and UC-MSCs plus sitagliptin group began to decrease. Two weeks later, FBG reached the lowest level in all three groups, but FBG in UC-MSCs and sitagliptin group showed a fluctuation after that while FPG in UC-MSCs plus sitagliptin group maintained at a low level, although fasting C-peptide and the number of β-cells in UC-MSC group and UC-MSCs plus sitagliptin group showed no significant difference at the end of the experiment. Another way to enhance the effect might be preconditioning UC-MSCs. It is noteworthy that the secretome of MSCs does not maintain unchangeable but relies on the environmental stimuli [125]. For example, under hypoxia, a large number of genes involved in glycolysis and metabolism are unregulated [126]. Therefore, adding appropriate stimuli to precondition UC-MSCs in order to enhance the expression of identified effective growth factors might promote the therapeutic effect of UC-MSC transplantation. Our unpublished data suggested that high glucose stimulation for 24 h before transplantation effectively promoted the expression of FGF21 in UC-MSCs and transplantation with preconditioned UC-MSCs in T2D rats showed improved islet function and insulin sensitivity as compared with standard MSC infusion.

Next, two reports studied the biological effects of gestational DM on the cellular properties of UC-MSCs [127, 128]. Both reports demonstrated that UC-MSCs derived from gestational DM patients showed decreased cell growth, reduced cell viability, and earlier cellular senescence compared with UC-MSCs collected from healthy donors. Moreover, these cells displayed significantly lower osteogenic and adipogenic differentiation potentials. We postulated that not only DM but also other major pregnancy complications might also influence the property of UC-MSCs. These findings underscore the importance of the critical assessment of the biological impact of maternal-fetal conditions in applications of UC-MSCs in DM.

Last but not least, almost in each trial mentioned above, there were patients who had no response to UC-MSC treatment, while others responded very well. It is crucial to analyze what makes the huge differences between patients and, if possible, find out a marker to predict the therapeutic effects in each patient to avoid invalid injection.

9.6 Conclusion

Due to mixed phenotype and immaturity of the products, there is a long way to go for transdifferentiation from UC-MSCs into IPCs to be applicable in clinic. Direct transplantation of UC-MSCs is a safe procedure and has been intensely investigated in clinical studies. Based on the results of small sample clinical data, transplantation of UC-MSCs benefits patients with T1D and T2D. The therapeutic effects varied among studies. Optimal source of UC-MSCs, doses and dose frequency, and transplantation route should be determined by larger randomized trials in the future. However, the mechanism by which UC-MSCs exert antidiabetic effect remains largely unknown and urgently needs further investigation. The immunomodulatory effects of UC-MSCs might play key roles in UC-MSC-based diabetes therapy. In addition, appropriate adjustment to the in vitro culture system might upregulate the expression of some genes that favored diabetes therapy, which will greatly enhance the effects of UC-MSC infusion.

Acknowledgment The authors' work was supported by the 863 Projects of the Ministry of Science and Technology of PR China (No. 2013AA020105 and 2012AA020502).

Disclosure: None of the authors have any conflicts of interest relating to this work.

References

1. Shen J, Cheng Y, Han Q, et al. Generating insulin-producing cells for diabetic therapy: existing strategies and new development. Ageing Res Rev. 2013;12:469–78.
2. Borowiak M, Melton DA. How to make beta cells? Curr Opin Cell Biol. 2009;21:727–32.
3. Zhou Q, Melton DA. Pathways to new beta cells. Cold Spring Harb Symp Quant Biol. 2008;73:175–81.
4. Guo T, Hebrok M. Stem cells to pancreatic beta-cells: new sources for diabetes cell therapy. Endocr Rev. 2009;30:214–27.
5. Hori Y. Insulin-producing cells derived from stem/progenitor cells: therapeutic implications for diabetes mellitus. Med Mol Morphol. 2009;42:195–200.
6. Johnson JD. The quest to make fully functional human pancreatic beta cells from embryonic stem cells: climbing a mountain in the clouds. Diabetologia. 2016;59:2047–57.
7. Mayhew CN, Wells JM. Converting human pluripotent stem cells into beta-cells: recent advances and future challenges. Curr Opin Organ Transplant. 2010;15:54–60.
8. Mfopou JK, Chen B, Sui L, et al. Recent advances and prospects in the differentiation of pancreatic cells from human embryonic stem cells. Diabetes. 2010;59:2094–101.
9. Zhang D, Jiang W, Shi Y, et al. Generation of pancreatic islet cells from human embryonic stem cells. Sci China C Life Sci. 2009;52:615–21.
10. D'amour KA, Bang AG, Eliazer S, et al. Production of pancreatic hormone-expressing endocrine cells

from human embryonic stem cells. Nat Biotechnol. 2006;24:1392–401.
11. Cai J, Yu C, Liu Y, et al. Generation of homogeneous PDX1(+) pancreatic progenitors from human ES cell-derived endoderm cells. J Mol Cell Biol. 2010;2:50–60.
12. Champeris Tsaniras S, Jones PM. Generating pancreatic beta-cells from embryonic stem cells by manipulating signaling pathways. J Endocrinol. 2010;206:13–26.
13. Mfopou JK, Chen B, Mateizel I, et al. Noggin, retinoids, and fibroblast growth factor regulate hepatic or pancreatic fate of human embryonic stem cells. Gastroenterology. 2010;138:2233–45, 2245 e2231-2214.
14. Phillips BW, Hentze H, Rust WL, et al. Directed differentiation of human embryonic stem cells into the pancreatic endocrine lineage. Stem Cells Dev. 2007;16:561–78.
15. Thatava T, Nelson TJ, Edukulla R, et al. Indolactam V/GLP-1-mediated differentiation of human iPS cells into glucose-responsive insulin-secreting progeny. Gene Ther. 2011;18:283–93.
16. Van Hoof D, D'amour KA, German MS. Derivation of insulin-producing cells from human embryonic stem cells. Stem Cell Res. 2009;3:73–87.
17. Zhang D, Jiang W, Liu M, et al. Highly efficient differentiation of human ES cells and iPS cells into mature pancreatic insulin-producing cells. Cell Res. 2009;19:429–38.
18. Pagliuca FW, Millman JR, Gürtler M, et al. Generation of functional human pancreatic β cells in vitro. Cell. 2014;159:428–39.
19. Rezania A, Bruin JE, Arora P, et al. Reversal of diabetes with insulin-producing cells derived in vitro from human pluripotent stem cells. Nat Biotechnol. 2014;32:1121–33.
20. Russ HA, Parent AV, Ringler JJ, et al. Controlled induction of human pancreatic progenitors produces functional beta-like cells in vitro. EMBO J. 2015;34:1759–72.
21. Dave S. Mesenchymal stem cells derived in vitro transdifferentiated insulin-producing cells: a new approach to treat type 1 diabetes. Adv Biomed Res. 2014;3:266.
22. Dominguez-Bendala J, Lanzoni G, Inverardi L, et al. Concise review: mesenchymal stem cells for diabetes. Stem Cells Transl Med. 2012;1:59–63.
23. Hashemian SJ, Kouhnavard M, Nasli-Esfahani E. Mesenchymal stem cells: rising concerns over their application in treatment of type one diabetes mellitus. J Diabetes Res. 2015;2015:675103.
24. Katuchova J, Harvanova D, Spakova T, et al. Mesenchymal stem cells in the treatment of type 1 diabetes mellitus. Endocr Pathol. 2015;26:95–103.
25. Klinker MW, Wei CH. Mesenchymal stem cells in the treatment of inflammatory and autoimmune diseases in experimental animal models. World J Stem Cells. 2015;7:556–67.
26. Bhansali A, Asokumar P, Walia R, et al. Efficacy and safety of autologous bone marrow-derived stem cell transplantation in patients with type 2 diabetes mellitus: a randomized placebo-controlled study. Cell Transplant. 2014;23:1075–85.
27. Bhansali A, Upreti V, Khandelwal N, et al. Efficacy of autologous bone marrow-derived stem cell transplantation in patients with type 2 diabetes mellitus. Stem Cells Dev. 2009;18:1407–16.
28. Carlsson PO, Schwarcz E, Korsgren O, et al. Preserved beta-cell function in type 1 diabetes by mesenchymal stromal cells. Diabetes. 2015;64:587–92.
29. Arutyunyan I, Elchaninov A, Makarov A, et al. Umbilical cord as prospective source for mesenchymal stem cell-based therapy. Stem Cells Int. 2016;2016:6901286.
30. Watson N, Divers R, Kedar R, et al. Discarded Wharton jelly of the human umbilical cord: a viable source for mesenchymal stromal cells. Cytotherapy. 2015;17:18–24.
31. Afelik S, Jensen J. Notch signaling in the pancreas: patterning and cell fate specification. Wiley Interdiscip Rev Dev Biol. 2013;2:531–44.
32. Serup P. Signaling pathways regulating murine pancreatic development. Semin Cell Dev Biol. 2012;23:663–72.
33. Edlund H. Pancreatic organogenesis--developmental mechanisms and implications for therapy. Nat Rev Genet. 2002;3:524–32.
34. Gittes GK. Developmental biology of the pancreas: a comprehensive review. Dev Biol. 2009;326:4–35.
35. Puri S, Hebrok M. Cellular plasticity within the pancreas--lessons learned from development. Dev Cell. 2010;18:342–56.
36. Shih HP, Wang A, Sander M. Pancreas organogenesis: from lineage determination to morphogenesis. Annu Rev Cell Dev Biol. 2013;29:81–105.
37. Stanger BZ, Hebrok M. Control of cell identity in pancreas development and regeneration. Gastroenterology. 2013;144:1170–9.
38. Jonsson J, Carlsson L, Edlund T, et al. Insulin-promoter-factor 1 is required for pancreas development in mice. Nature. 1994;371:606–9.
39. Offield MF, Jetton TL, Labosky PA, et al. PDX-1 is required for pancreatic outgrowth and differentiation of the rostral duodenum. Development. 1996;122:983–95.
40. Stoffers DA, Ferrer J, Clarke WL, et al. Early-onset type-II diabetes mellitus (MODY4) linked to IPF1. Nat Genet. 1997;17:138–9.
41. Holland AM, Hale MA, Kagami H, et al. Experimental control of pancreatic development and maintenance. Proc Natl Acad Sci U S A. 2002;99:12236–41.
42. Gannon M, Ables ET, Crawford L, et al. pdx-1 function is specifically required in embryonic beta cells to generate appropriate numbers of endocrine cell types and maintain glucose homeostasis. Dev Biol. 2008;314:406–17.

43. Johansson KA, Dursun U, Jordan N, et al. Temporal control of neurogenin3 activity in pancreas progenitors reveals competence windows for the generation of different endocrine cell types. Dev Cell. 2007;12:457–65.
44. Wang S, Jensen JN, Seymour PA, et al. Sustained Neurog3 expression in hormone-expressing islet cells is required for endocrine maturation and function. Proc Natl Acad Sci U S A. 2009;106:9715–20.
45. Collombat P, Mansouri A, Hecksher-Sorensen J, et al. Opposing actions of Arx and Pax4 in endocrine pancreas development. Genes Dev. 2003;17:2591–603.
46. Henseleit KD, Nelson SB, Kuhlbrodt K, et al. NKX6 transcription factor activity is required for alpha- and beta-cell development in the pancreas. Development. 2005;132:3139–49.
47. Papizan JB, Singer RA, Tschen SI, et al. Nkx2.2 repressor complex regulates islet beta-cell specification and prevents beta-to-alpha-cell reprogramming. Genes Dev. 2011;25:2291–305.
48. Sosa-Pineda B, Chowdhury K, Torres M, et al. The Pax4 gene is essential for differentiation of insulin-producing beta cells in the mammalian pancreas. Nature. 1997;386:399–402.
49. Collombat P, Hecksher-Sorensen J, Broccoli V, et al. The simultaneous loss of Arx and Pax4 genes promotes a somatostatin-producing cell fate specification at the expense of the alpha- and beta-cell lineages in the mouse endocrine pancreas. Development. 2005;132:2969–80.
50. Dhawan S, Georgia S, Tschen SI, et al. Pancreatic beta cell identity is maintained by DNA methylation-mediated repression of Arx. Dev Cell. 2011;20:419–29.
51. Yang YP, Thorel F, Boyer DF, et al. Context-specific alpha-to-beta-cell reprogramming by forced Pdx1 expression. Genes Dev. 2011;25:1680–5.
52. Collombat P, Xu X, Ravassard P, et al. The ectopic expression of Pax4 in the mouse pancreas converts progenitor cells into alpha and subsequently beta cells. Cell. 2009;138:449–62.
53. Schaffer AE, Taylor BL, Benthuysen JR, et al. Nkx6.1 controls a gene regulatory network required for establishing and maintaining pancreatic Beta cell identity. PLoS Genet. 2013;9:e1003274.
54. Collombat P, Hecksher-Sorensen J, Krull J, et al. Embryonic endocrine pancreas and mature beta cells acquire alpha and PP cell phenotypes upon Arx misexpression. J Clin Invest. 2007;117:961–70.
55. Si YL, Zhao YL, Hao HJ, et al. MSCs: biological characteristics, clinical applications and their outstanding concerns. Ageing Res Rev. 2011;10:93–103.
56. Mcelreavey KD, Irvine AI, Ennis KT, et al. Isolation, culture and characterisation of fibroblast-like cells derived from the Wharton's jelly portion of human umbilical cord. Biochem Soc Trans. 1991;19:29S.
57. Wang HS, Hung SC, Peng ST, et al. Mesenchymal stem cells in the Wharton's jelly of the human umbilical cord. Stem Cells. 2004;22:1330–7.
58. Lee M, Jeong SY, Ha J, et al. Low immunogenicity of allogeneic human umbilical cord blood-derived mesenchymal stem cells in vitro and in vivo. Biochem Biophys Res Commun. 2014;446:983–9.
59. Weiss ML, Anderson C, Medicetty S, et al. Immune properties of human umbilical cord Wharton's jelly-derived cells. Stem Cells. 2008;26:2865–74.
60. Jin HJ, Bae YK, Kim M, et al. Comparative analysis of human mesenchymal stem cells from bone marrow, adipose tissue, and umbilical cord blood as sources of cell therapy. Int J Mol Sci. 2013;14:17986–8001.
61. Lu LL, Liu YJ, Yang SG, et al. Isolation and characterization of human umbilical cord mesenchymal stem cells with hematopoiesis-supportive function and other potentials. Haematologica. 2006;91:1017–26.
62. Wegmeyer H, Broske AM, Leddin M, et al. Mesenchymal stromal cell characteristics vary depending on their origin. Stem Cells Dev. 2013;22:2606–18.
63. Amable PR, Teixeira MV, Carias RB, et al. Protein synthesis and secretion in human mesenchymal cells derived from bone marrow, adipose tissue and Wharton's jelly. Stem Cell Res Ther. 2014;5:53.
64. Kuchroo P, Dave V, Vijayan A, et al. Paracrine factors secreted by umbilical cord-derived mesenchymal stem cells induce angiogenesis in vitro by a VEGF-independent pathway. Stem Cells Dev. 2015;24:437–50.
65. Li X, Bai J, Ji X, et al. Comprehensive characterization of four different populations of human mesenchymal stem cells as regards their immune properties, proliferation and differentiation. Int J Mol Med. 2014;34:695–704.
66. Pires AO, Mendes-Pinheiro B, Teixeira FG, et al. Unveiling the differences of secretome of human bone marrow mesenchymal stem cells, adipose tissue-derived stem cells, and human umbilical cord perivascular cells: a proteomic analysis. Stem Cells Dev. 2016;25:1073–83.
67. Du WJ, Chi Y, Yang ZX, et al. Heterogeneity of proangiogenic features in mesenchymal stem cells derived from bone marrow, adipose tissue, umbilical cord, and placenta. Stem Cell Res Ther. 2016;7:163.
68. Datta I, Mishra S, Mohanty L, et al. Neuronal plasticity of human Wharton's jelly mesenchymal stromal cells to the dopaminergic cell type compared with human bone marrow mesenchymal stromal cells. Cytotherapy. 2011;13:918–32.
69. Balasubramanian S, Thej C, Venugopal P, et al. Higher propensity of Wharton's jelly derived mesenchymal stromal cells towards neuronal lineage in comparison to those derived from adipose and bone marrow. Cell Biol Int. 2013;37:507–15.
70. Chen MY, Lie PC, Li ZL, et al. Endothelial differentiation of Wharton's jelly-derived mesenchymal stem cells in comparison with bone marrow-derived mesenchymal stem cells. Exp Hematol. 2009;37:629–40.

71. Wu LF, Wang NN, Liu YS, et al. Differentiation of Wharton's jelly primitive stromal cells into insulin-producing cells in comparison with bone marrow mesenchymal stem cells. Tissue Eng Part A. 2009;15:2865–73.
72. Manochantr S, U-pratya Y, Kheolamai P, et al. Immunosuppressive properties of mesenchymal stromal cells derived from amnion, placenta, Wharton's jelly and umbilical cord. Intern Med J. 2013;43:430–9.
73. Najar M, Raicevic G, Boufker HI, et al. Adipose-tissue-derived and Wharton's jelly-derived mesenchymal stromal cells suppress lymphocyte responses by secreting leukemia inhibitory factor. Tissue Eng Part A. 2010;16:3537–46.
74. Vellasamy S, Sandrasaigaran P, Vidyadaran S, et al. Mesenchymal stem cells of human placenta and umbilical cord suppress T-cell proliferation at G0 phase of cell cycle. Cell Biol Int. 2013;37:250–6.
75. Vellasamy S, Tong CK, Azhar NA, et al. Human mesenchymal stromal cells modulate T-cell immune response via transcriptomic regulation. Cytotherapy. 2016;18:1270–83.
76. Yang H, Sun J, Li Y, et al. Human umbilical cord-derived mesenchymal stem cells suppress proliferation of PHA-activated lymphocytes in vitro by inducing CD4(+)CD25(high)CD45RA(+) regulatory T cell production and modulating cytokine secretion. Cell Immunol. 2016;302:26–31.
77. Chen K, Wang D, Du WT, et al. Human umbilical cord mesenchymal stem cells hUC-MSCs exert immunosuppressive activities through a PGE2-dependent mechanism. Clin Immunol. 2010;135:448–58.
78. Li X, Xu Z, Bai J, et al. Umbilical cord tissue-derived mesenchymal stem cells induce T lymphocyte apoptosis and cell cycle arrest by expression of indoleamine 2, 3-dioxygenase. Stem Cells Int. 2016;2016:7495135.
79. He H, Nagamura-Inoue T, Takahashi A, et al. Immunosuppressive properties of Wharton's jelly-derived mesenchymal stromal cells in vitro. Int J Hematol. 2015;102:368–78.
80. Tipnis S, Viswanathan C, Majumdar AS. Immunosuppressive properties of human umbilical cord-derived mesenchymal stem cells: role of B7-H1 and IDO. Immunol Cell Biol. 2010;88:795–806.
81. Yoo KH, Jang IK, Lee MW, et al. Comparison of immunomodulatory properties of mesenchymal stem cells derived from adult human tissues. Cell Immunol. 2009;259:150–6.
82. Cutler AJ, Limbani V, Girdlestone J, et al. Umbilical cord-derived mesenchymal stromal cells modulate monocyte function to suppress T cell proliferation. J Immunol. 2010;185:6617–23.
83. Che N, Li X, Zhou S, et al. Umbilical cord mesenchymal stem cells suppress B-cell proliferation and differentiation. Cell Immunol. 2012;274:46–53.
84. Ji YR, Yang ZX, Han ZB, et al. Mesenchymal stem cells support proliferation and terminal differentiation of B cells. Cell Physiol Biochem. 2012;30:1526–37.
85. Ruggeri L, Mancusi A, Burchielli E, et al. NK cell alloreactivity and allogeneic hematopoietic stem cell transplantation. Blood Cells Mol Dis. 2008;40:84–90.
86. Vivier E, Raulet DH, Moretta A, et al. Innate or adaptive immunity? The example of natural killer cells. Science. 2011;331:44–9.
87. Chatterjee D, Marquardt N, Tufa DM, et al. Role of gamma-secretase in human umbilical-cord derived mesenchymal stem cell mediated suppression of NK cell cytotoxicity. Cell Commun Signal. 2014;12:63.
88. Chatterjee D, Marquardt N, Tufa DM, et al. Human umbilical cord-derived mesenchymal stem cells utilize activin-A to suppress interferon-gamma production by natural killer cells. Front Immunol. 2014;5:662.
89. Chazaud B. Macrophages: supportive cells for tissue repair and regeneration. Immunobiology. 2014;219:172–8.
90. Forbes SJ, Rosenthal N. Preparing the ground for tissue regeneration: from mechanism to therapy. Nat Med. 2014;20:857–69.
91. Novak ML, Koh TJ. Phenotypic transitions of macrophages orchestrate tissue repair. Am J Pathol. 2013;183:1352–63.
92. Wynn TA, Vannella KM. Macrophages in tissue repair, regeneration, and fibrosis. Immunity. 2016;44:450–62.
93. Selleri S, Bifsha P, Civini S, et al. Human mesenchymal stromal cell-secreted lactate induces M2-macrophage differentiation by metabolic reprogramming. Oncotarget. 2016;7:30193–210.
94. Dayan V, Yannarelli G, Billia F, et al. Mesenchymal stromal cells mediate a switch to alternatively activated monocytes/macrophages after acute myocardial infarction. Basic Res Cardiol. 2011;106:1299–310.
95. Li W, Zhang Q, Wang M, et al. Macrophages are involved in the protective role of human umbilical cord-derived stromal cells in renal ischemia-reperfusion injury. Stem Cell Res. 2013;10:405–16.
96. Deng W, Chen W, Zhang Z, et al. Mesenchymal stem cells promote CD206 expression and phagocytic activity of macrophages through IL-6 in systemic lupus erythematosus. Clin Immunol. 2015;161:209–16.
97. Xie Z, Hao H, Tong C, et al. Human umbilical cord-derived mesenchymal stem cells elicit macrophages into an anti-inflammatory phenotype to alleviate insulin resistance in type 2 diabetic rats. Stem Cells. 2016;34:627–39.
98. Chao KC, Chao KF, Fu YS, et al. Islet-like clusters derived from mesenchymal stem cells in Wharton's

Jelly of the human umbilical cord for transplantation to control type 1 diabetes. PLoS One. 2008;3:e1451.

99. Moshrefi M, Yari N, Nabipour F, et al. Transplantation of differentiated umbilical cord mesenchymal cells under kidney capsule for control of type I diabetes in rat. Tissue Cell. 2015;47:395–405.
100. Wang HS, Shyu JF, Shen WS, et al. Transplantation of insulin-producing cells derived from umbilical cord stromal mesenchymal stem cells to treat NOD mice. Cell Transplant. 2011;20:455–66.
101. Yu YB, Bian JM, Gu DH. Transplantation of insulin-producing cells to treat diabetic rats after 90% pancreatectomy. World J Gastroenterol. 2015;21:6582–90.
102. Tsai PJ, Wang HS, Shyr YM, et al. Transplantation of insulin-producing cells from umbilical cord mesenchymal stem cells for the treatment of streptozotocin-induced diabetic rats. J Biomed Sci. 2012;19:47.
103. Seyedi F, Farsinejad A, Moshrefi M, et al. In vitro evaluation of different protocols for the induction of mesenchymal stem cells to insulin-producing cells. In Vitro Cell Dev Biol Anim. 2015;51:866–78.
104. Seyedi F, Farsinejad A, Nematollahi-Mahani SA, et al. Suspension culture alters insulin secretion in induced human umbilical cord matrix-derived mesenchymal cells. Cell J. 2016;18:52–61.
105. Qu H, Liu X, Ni Y, et al. Laminin 411 acts as a potent inducer of umbilical cord mesenchymal stem cell differentiation into insulin-producing cells. J Transl Med. 2014;12:135.
106. Aviv V, Meivar-Levy I, Rachmut IH, et al. Exendin-4 promotes liver cell proliferation and enhances the PDX-1-induced liver to pancreas transdifferentiation process. J Biol Chem. 2009;284:33509–20.
107. Ferber S, Halkin A, Cohen H, et al. Pancreatic and duodenal homeobox gene 1 induces expression of insulin genes in liver and ameliorates streptozotocin-induced hyperglycemia. Nat Med. 2000;6:568–72.
108. Jin CX, Li WL, Xu F, et al. Conversion of immortal liver progenitor cells into pancreatic endocrine progenitor cells by persistent expression of Pdx-1. J Cell Biochem. 2008;104:224–36.
109. Shternhall-Ron K, Quintana FJ, Perl S, et al. Ectopic PDX-1 expression in liver ameliorates type 1 diabetes. J Autoimmun. 2007;28:134–42.
110. Wang XL, Hu P, Guo XR, et al. Reprogramming human umbilical cord mesenchymal stromal cells to islet-like cells with the use of in vitro-synthesized pancreatic-duodenal homebox 1 messenger RNA. Cytotherapy. 2014;16:1519–27.
111. He D, Wang J, Gao Y, et al. Differentiation of PDX1 gene-modified human umbilical cord mesenchymal stem cells into insulin-producing cells in vitro. Int J Mol Med. 2011;28:1019–24.
112. Guan LX, Guan H, Li HB, et al. Therapeutic efficacy of umbilical cord-derived mesenchymal stem cells in patients with type 2 diabetes. Exp Ther Med. 2015;9:1623–30.
113. Liu X, Zheng P, Wang X, et al. A preliminary evaluation of efficacy and safety of Wharton's jelly mesenchymal stem cell transplantation in patients with type 2 diabetes mellitus. Stem Cell Res Ther. 2014;5:57.
114. Hu J, Wang Y, Gong H, et al. Long term effect and safety of Wharton's jelly-derived mesenchymal stem cells on type 2 diabetes. Exp Ther Med. 2016;12:1857–66.
115. Kong D, Zhuang X, Wang D, et al. Umbilical cord mesenchymal stem cell transfusion ameliorated hyperglycemia in patients with type 2 diabetes mellitus. Clin Lab. 2014;60:1969–76.
116. Hu J, Yu X, Wang Z, et al. Long term effects of the implantation of Wharton's jelly-derived mesenchymal stem cells from the umbilical cord for newly-onset type 1 diabetes mellitus. Endocr J. 2013;60:347–57.
117. Cai J, Wu Z, Xu X, et al. Umbilical cord mesenchymal stromal cell with autologous bone marrow cell transplantation in established type 1 diabetes: a pilot randomized controlled open-label clinical study to assess safety and impact on insulin secretion. Diabetes Care. 2016;39:149–57.
118. Yaochite JN, Caliari-Oliveira C, De Souza LE, et al. Therapeutic efficacy and biodistribution of allogeneic mesenchymal stem cells delivered by intrasplenic and intrapancreatic routes in streptozotocin-induced diabetic mice. Stem Cell Res Ther. 2015;6:31.
119. Kebriaei P, Isola L, Bahceci E, et al. Adult human mesenchymal stem cells added to corticosteroid therapy for the treatment of acute graft-versus-host disease. Biol Blood Marrow Transplant. 2009;15:804–11.
120. Herrmann R, Sturm M, Shaw K, et al. Mesenchymal stromal cell therapy for steroid-refractory acute and chronic graft versus host disease: a phase 1 study. Int J Hematol. 2012;95:182–8.
121. Prasad VK, Lucas KG, Kleiner GI, et al. Efficacy and safety of ex vivo cultured adult human mesenchymal stem cells (Prochymal) in pediatric patients with severe refractory acute graft-versus-host disease in a compassionate use study. Biol Blood Marrow Transplant. 2011;17:534–41.
122. Zhou H, Guo M, Bian C, et al. Efficacy of bone marrow-derived mesenchymal stem cells in the treatment of sclerodermatous chronic graft-versus-host disease: clinical report. Biol Blood Marrow Transplant. 2010;16:403–12.
123. Hao H, Liu J, Shen J, et al. Multiple intravenous infusions of bone marrow mesenchymal stem cells reverse hyperglycemia in experimental type 2 diabetes rats. Biochem Biophys Res Commun. 2013;436:418–23.

124. Hu J, Wang F, Sun R, et al. Effect of combined therapy of human Wharton's jelly-derived mesenchymal stem cells from umbilical cord with sitagliptin in type 2 diabetic rats. Endocrine. 2014;45:279–87.
125. Caplan AI, Dennis JE. Mesenchymal stem cells as trophic mediators. J Cell Biochem. 2006; 98:1076–84.
126. Ohnishi S, Yasuda T, Kitamura S, et al. Effect of hypoxia on gene expression of bone marrow-derived mesenchymal stem cells and mononuclear cells. Stem Cells. 2007;25:1166–77.
127. Kim J, Piao Y, Pak YK, et al. Umbilical cord mesenchymal stromal cells affected by gestational diabetes mellitus display premature aging and mitochondrial dysfunction. Stem Cells Dev. 2015;24:575–86.
128. Wajid N, Naseem R, Anwar SS, et al. The effect of gestational diabetes on proliferation capacity and viability of human umbilical cord-derived stromal cells. Cell Tissue Bank. 2015;16:389–97.

10 Perspective of Therapeutic Angiogenesis Using Circulating Endothelial Progenitors from Umbilical Cord Blood

S. Ferratge, J. Boyer, N. Arouch, F. Chevalier, and G. Uzan

Abstract

Circulating endothelial progenitor cells give rise in culture to endothelial colony-forming cells (ECFCs). These cells retain stem cells features and display all characteristics of endothelial cells. Cord blood-derived ECFCs (CB-ECFCs) exhibit a high clonogenic and proliferative capacities and exhibit vascular repair potential beneficial for the treatment of ischemic diseases. However, the link between the functional properties of CB-ECFCs and their immaturity is still poorly investigated. We showed that these cells have a high clonogenic potential and are capable to generate efficiently induced pluripotent stem cells. Moreover, we analyzed the expression of a selected panel of genes involved in embryonic stem cell functions. We defined a new and unique stem cell transcriptional signature for CB-ECFCs for a better characterization and stratification according to their stem cell features. Our results show that CB-ECFCs are immature cells harboring specific functions such as formation of colonies and proliferation and formation of vascular structures in vitro and in vivo. We then improved the yield of CB-ECFC production for obtaining more functional cells in a shorter culture time. We used glycosaminoglycans (GAG) as constituents of the extracellular matrix which potentiate heparin-binding growth factor activities. GAG mimetics were designed, with the capacity of optimizing the yield of ECFC production, by increasing the number of colonies and improving adhesion, proliferation, migration, and self-renewal. GAG mimetics have thus great interest for vascular regeneration in combination with ECFC.

Keywords

Vascular repair · Circulating endothelial progenitors · Plasticity · Stem cells

10.1 Introduction

Circulating endothelial progenitor cells were discovered in adult peripheral blood in 1997 [1]. These cells, named EPCs, are potentially able to participate to vascular healing by integrating into damaged vessels, where they differentiate into functional endothelial cells [2]. It is thus admitted that the physiological role of EPCs is the maintenance of the vascular integrity [3–5]. Different groups working in the field of EPCs have identified in culture two distinct cell populations, which differ in the delay of appearance of first colonies. The specificity of two populations was described by Yoder [6]. Both populations secrete pro-angiogenic factors that promote neo-angiogenesis.

S. Ferratge · J. Boyer · N. Arouch · F. Chevalier
G. Uzan (✉)
Inserm U1197, Hôpital Paul Brousse,
Villejuif Cedex, France
e-mail: Georges.uzan@inserm.fr

Z. C. Han et al. (eds.), *Perinatal Stem Cells*, https://doi.org/10.1007/978-981-13-2703-2_10

On the one hand, colony-forming unit-endothelial cells (CFU-ECs) are derived from monocyte/macrophage progenitor cells, and they are not able to participate to vessel formation in vivo. On the other hand, ECFCs express specific endothelial markers (CD144, KDR, wWF, CD31, and CD146), but neither CD45, a general hematopoietic marker, nor CD14, a leukocyte marker, is expressed by these cells. ECFCs are thus genuine endothelial cells that also display high proliferative and clonogenic potentials [6–9]. Interestingly, when transplanted in vivo in different mouse models, ECFCs incorporate into pre-existing vascular networks and form stable vessels [6, 10, 11].

Umbilical cord blood contains high amount of different types of stem cells and also constitutes a valuable source of EPCs. These cord blood EPCs give rise to a high number of highly clonogenic ECFCs (CB-ECFCs) that can be extensively expanded in culture. CB-ECFCs produce much more endothelial cells than adult peripheral blood-derived ECFCs (PB-ECFCs) [12]. Moreover, since CB-ECFCs are still immature, they have not yet acquired specialized functions. In this context, we have shown that when CB-ECFCs are exposed to specific instructive stimuli, they can acquire in culture different specific specialized phenotypes such as that of blood brain barrier microvascular endothelial cells or arterial endothelial cells [13].

For using EPCs as a cell therapy product, we still need to define more precisely their capacities, and it is important to provide information about their stem cell features. This question was first addressed by analyzing the expression of the stem cell marker CD133 by these cells [14, 15]. However, this definition was challenged by different studies [16, 17].

Since EPCs are very rare cells, we have used CB-ECFC population as a working model, because these cells are homogeneous and well-characterized and they provide enough material to work on. First we focused on the capacity of CB-ECFC to generate secondary colonies. Then, we analyzed if immature CB-ECFCs were able to generate induced pluripotent stem cells (iPSCs). Many somatic cells are able to generate more or less efficiently iPSCs [18, 19]. Moreover, the kinetics and efficacy of the reprogramming process depend on the cell type and on their initial immaturity [18]. Indeed, immature somatic cells display higher reprogramming capacity than more mature cells.

We found that CB-ECFCs generated iPSCs with a much higher efficacy than adult mature endothelial cells (human aortic endothelial cells, HAECs) and fibroblasts. iPSC colonies also appeared much earlier with CB-ECFCs than with control endothelial cells. Moreover, ECFC-derived iPSCs differentiated into the three germ layers and produced functional endothelial cells. The yield and kinetics of this production are similar to that obtained with hESCs. Then, to confirm that CB-ECFC displays stem cell features, we analyzed the expression of a panel of stem cell markers expressed by hESCs. We found that CB-ECFCs expressed some of these stem cell genes. These genes were almost undetectable in HAECs, suggesting that CB-ECFC has a stem cell transcriptional signature that is not detected in mature endothelial cells.

In summary, we showed that CB-ECFCs display stem cell properties including better reprograming capacity. Then, iPSCs derived from CB-ECFC provide a valuable source of pluripotent stem cells available, for example, for drug screening. Finally, the transcriptional stem cell signature found on CB-ECFC can be used to stratify them according to their immature status.

The culture time of cell therapy products is an important parameter to take into account, in order to limit the emergence of genetic abnormalities that may appear when cells are cultured during long periods. This culture time should be reduced as much as possible. Concerning CB-ECFC production, two parameters are important: first, the number of colonies obtained and second, the number of endothelial cells yielded from these colonies. By adding glycosaminoglycan (GAG) polysaccharidic compounds, we have shown that we could improve these two parameters.

Proteoglycans associate sulfated GAG molecules to a core protein. These polysaccharidic structures interact with proteins to strengthen their architecture. They also stabilize the extracellular matrix (ECM). Moreover, by interacting with heparin-binding domain of many growth factors (HBGF), GAG are involved in the regulation of a wide range of biological processes [20]. These biological effects of GAG are mediated by their specific structural organization, which modulate HBGF activities [21]. Structural and functional analogs of natural GAG have been designed, which are chemically defined and well-characterized GAG [22]. These GAG mimetics were validated in different animal models. These studies demonstrated that these compounds greatly increased the efficacy of wound healing. Several properties of GAG mimetics can explain these biological effects. First, they resist to glycanase degradation which improves their availability for HBGF functional improvement [23]. Indeed, GAG mimetics efficiently bind to VEGF and bFGF, improving the interaction with their specific receptors. This results in improving the biological activity of endothelial cells and consequently in vivo angiogenesis [24]. Moreover, GAG mimetics have been shown to increase the mobilization of bone marrow hematopoietic stem cells and mesenchymal stromal cells [25, 26].

Since GAG mimetics improve the functions of broad panel of cells including stem cells, we selected a lead GAG mimetic and tested its ability to potentiate CB-ECFC functional properties. In the context of further therapeutic applications, this lead GAG mimetic may improve CB-ECFC proliferation capacity, thus limiting the culture time, and may repair the damaged matrix in the targeted diseased area.

Taken together, our results show that cord blood ECFCs are immature cells, able to be expanded in culture, and able to be differentiated into different specialized endothelial cells. Production of ECFCs could be improved by matrix components like GAG.

10.2 Results and Discussion

10.2.1 Functional Features Assessing CB-ECFC Immaturity

ECFCs are the EPC outgrowth characterized in culture by their clonogenic, proliferative, and pro-angiogenic properties. The degree of immaturity of CB-ECFCs may have a profound impact on their functional properties. Supporting this hypothesis, we found that CB-ECFCs are more clonogenic and more proliferative than PB-ECFCs. They display a higher telomerase activity [12], and when injected in animal models of ischemia, they form functional and long-lasting vessels. AB-ECFCs are less efficient and form vessels which rapidly regress [10]. However, the degree of EPC immaturity is still poorly documented. EPC stemness/immaturity was mainly evaluated by measuring the level of expression of CD133, a stem cell marker. However, the accuracy of this marker for defining EPC stemness is now challenged [16]. CD34, another marker of hematopoietic stem and progenitor cells [27], is also expressed in most mature vascular structures, and it is not a relevant marker of immature endothelial cells. For this reason, defining CB-ECFC stemness with new markers is the central question addressed by this study.

10.2.1.1 Clonogenicity of CB-ECFCs

The capacity of a cell population to give rise to colonies in culture has been widely used for defining its degree of stemness. In culture, CB-ECFCs can give rise to primary, secondary, and even tertiary colonies, underlining their immaturity. In order to provide quantitative data about this property, we used a method similar to that used for quantifying the number of colonies yielded by hematopoietic stem cells. It is based on limiting dilution analysis and a Poisson's representation. We found that among six cells seeded from the primary colonies, one was able to generate a secondary colony. This indicates that primary colonies of CB-ECFCs contain subpopulations with high clonogenic potential, which probably display self-renewal capacities.

10.2.1.2 Generation of iPSCs from CB-ECFCs

Going forward for assessing ECFC immaturity, we tested their capacity of generating iPSCs. Many somatic cells were analyzed according to their capacity to be reprogrammed into iPSCs. They yielded iPSCs colonies with variable kinetics and efficiencies. It is logical to assume that the reprogramming efficacy could be influenced by the differentiation state of the initial somatic cells. This has been demonstrated in 2009 by Eminli. He showed that hematopoietic progenitors generated iPSCs more frequently (7–28%) than more mature blood cells (0.02–0.6%) [18]. We measured the efficacy of iPSC generation as an indicator for defining the degree of CB-ECFC stemness. Using the Yamanaka's standard protocol, we compared CB-ECFCs and mature endothelial cells (HAECs) for their capacity of being reprogrammed into iPSCs. CB-ECFCs were in average 80 times more efficient higher than HAECs for generating iPS colonies, and these colonies appeared 2 times faster. The two control fibroblast samples yielded results similar to those obtained with HAECs. This further confirms that ECFCs are immature cells.

10.2.1.3 Expression of Stem Cell Genes by CB-EFCs

We then hypothesized that ECFCs express a set of genes involved in stem cell biology that could explain the high reprogramming efficiency displayed by these cells. We thus compared CB-ECFCs, HAECs, and adult fibroblasts according to the level of expression of stem cell genes, using hESCs as a positive control of stemness. We first analyzed the transcriptional activity of 92 genes selected according to work of 17 international teams. This work was performed of a panel of 59 hESC lines [28]. Among the 92 genes, some are considered as early differentiation markers, and the others are expressed in non-differentiated hESCs. ECFCs and HAECs highly expressed endothelial markers (PECAM-1, CD144, and CD34), while hESCs and fibroblast did not express these endothelial markers. Unlike HAEC, CB-ECFCs expressed the sialoprotein podocalyxin a marker of hematopoietic and hemangioblastic cells. *ISL1*, an early marker of the mesoderm, was also significantly more expressed in CB-ECFCs than in HAECs and fibroblasts. We also revealed in CB-ECFCs the overexpression of seven genes related to hESC stemness *DNMT3B*, *GBX2*, *GDF3*, *GRB7*, *NANOG*, *SOX*, and *TDGF1*. These genes, which are not expressed in HAECs and fibroblasts, define the molecular signature of hESC lines [29]. Unexpectedly, ECFCs, HAECs, and fibroblasts expressed lowly but significantly *OCT3/4*. This transcription factor plays an important role in hESCs, since it modulates the self-renewal and pluripotency capacity of these cells [30]. Consistent with our results, another study has reported that *OCT3/4* is also expressed in peripheral blood mononuclear cells [31]. We then confirmed by RT-qPCRT this pattern of stem cell genes. CB-ECFCs thus combine a strong endothelial signature and expression of *DNMT3B*, *GDF3*, *GRB7*, *NANOG*, and *SOX 2*, all related to stemness. Expression of *ISL1* was also confirmed. *GDF3* and *DNMT3B* displayed the highest transcription levels observed in ECFCs.

GDF3 and *ISL1* were expressed at similar levels in CB-ECFCs and hESCs. CB-ECFCs express high levels of GDF3, a member of the TGFβ superfamily [30]. This is surprising, because this gene is expressed in hESCs, and its expression decreases while these cells enter into differentiation [32]. By inhibiting the BMP pathway, GDF3 maintains hESCs undifferentiated [33]. In culture, multipotent progenitors expressing the early mesodermal differentiation marker *ISL1* can either differentiate into endothelial cells, cardiomyocytes, and smooth muscle cells. It has been also reported that in endothelial cells, *ISL1* expression enhances their proliferative, migratory, and adhesion capacities. It also enhances vascularization in mice [34]. *ISL1* expression in ECFCs could thus, at the same time, maintain the stemness of these cells, while it also induces their endothelial properties. In future studies, we will investigate the potential correlation between *ISL1* expression and main ECFC functions. One of the most expressed genes in ECFCs, *DNMT3B*, is a DNA methyltransferase which is not directly involved in self-renewal but in the control of mesodermal or hematopoietic differentiation [35]. *DNMT3B* interferes with *OCT3/4* and *NANOG* transcription by methylating their promoters during embryonic cell differentiation [36]. *DNMT3B* also participates in biological functions including angiogenesis and regulation of cell mobility [37]. *NANOG* and *SOX2* are key regulating factors of hESCs, which, together with *OCT3/4*, are among the master transcriptional factors. In hESCs, they exert together combined actions inside feedback loops required for maintaining self-renewal and pluripotency [38]. Furthermore, *NANOG* may be also involved in endothelial cell physiology since this transcription factor is expressed at low levels in HUVECs and mature endothelial cells. In HUVECs, *VEGFR-2* gene transcription is regulated by the *NANOG* and may thus play a central role in endothelial cell functions [39].

In conclusion, all the genes we have identified are involved in stem or progenitor cell functions, and some of them also play a role in mature endothelial cell functions. These different genes could interact with each other in networks. Finally, our study identifies a unique transcriptional signature specific to immature CB-ECFCs. This signature may participate to the status of these progenitors, displaying at the same time both stem/progenitor and endothelial features.

In order to better characterize EPC, it is important to evaluate precisely their degree of immaturity. We have provided specific tools for measuring CB-ECFC stemness. We have provided evidences showing that these cells display a high clonogenic potential and they can efficiently give rise to iPSCs. Since EPCs can be isolated from a simple blood taking or from cord blood, they represent an easy and valuable source of pluripotent stem cells available for research, cell therapy, and pharmaceutical studies. Consistent with the stemness of these cells is the unique stem cell transcriptional signature of CB-ECFCs that we have defined, showing that these cells are genuine endothelial cells but with stem cell features that distinguish them from mature adult endothelial cells. This stem cell signature relies on the expression of *DNMT3B*, *GDF3*, *GRB7*, *NANOG*, and *SOX2* genes. The precise and specific role of these genes in CB-ECFC functions remains however to be investigated.

10.2.2 GAG Mimetics Improve CB-ECFC Efficacy

Several strategies have been developed to improve EPC proliferation and functional capacities. Most of them rely on combination of sets of growth factors added in the culture media, including VEGF, FGF-2, SCF, TPO, GCSF, IL-3, and IL-6 [40–42]. However, even with improved growth factor combination, active serum-free media compatible with the clinical use of these cells are not yet available.

The majority of these cytokines and growth contain a heparin-binding domain. We thus assumed that GAG, the polysaccharide components of the ECM, may improve the bioactivity of the cytokines and growth factors in ECFCs. These active growth factors could be added to the culture medium or secreted by the cells. However, natural GAG are not well structurally defined and may include contaminants hampering their clinical use. We thus developed an alternative strategy based on GAG mimetic products. Preliminary studies have selected a lead compound named [OTR4131]. It is synthesized using robust and standardized methods [22] and is compatible with clinical use.

10.2.2.1 [OTR4131] Increases the Number of Colonies Obtained from Cord Blood EPCs

Patients suffering from vascular diseases are often aged, they have less EPCs, and their endothelial progeny has poor functional capacities. ECFCs produced from these autologous EPCs are not the ideal cell therapy product. Alternatively, ECFC derived from allogenic cord blood EPCs could be efficiently expanded and display high angiogenic capacities. It would be convenient to use existing cord blood banks as a source of immune-compatible source of EPCs. However, the existing freezing conditions of CB blood used for banking are not well adapted to EPC and yield lower number of ECFC [43]. These freezing conditions must be improved and adapted to ECFC production. Today, if we use cord blood or adult blood for producing ECFCs dedicated to clinical use, we have to improve the yield of colony formation and the proliferation capacity of these cells for reaching the cellular therapeutic dose in the shorter culture time.

We show that addition of [OTR4131] GAG mimetic during the first day of CB mononuclear cell seeding results in a twofold increase of the number of ECFC colonies. One possible mechanism of action of [OTR4131] at this crucial initial step is the improvement of adhesion properties of EPC. Indeed, we show that when ECFCs are detached from the culture plate with trypsin, [OTR4131] then improves significantly their adhesion. Furthermore, addition of [OTR4131] to the culture medium along the first 7 days of culture results in a threefold increase of the number of colonies. This strongly suggests that the GAG mimetic acts not only at the early step of colony formation but also on later, likely on cell proliferation and self-renewal of ECFC. This is supported by our results showing that [OTR4131] also improve ECFC proliferation rate and formation of secondary colonies. Growing evidences show that GAG are involved in stem cell behavior. EXT1 gene codes for the enzyme involved in the elongation of the heparan sulfate chains. Inhibition of the expression of this gene by siRNA knockdown in mouse embryonic stem (ES) cell triggers the differentiation of 50% of the ES colonies, even if LIF and serum are maintained in the culture medium. A critical decrease of the proliferation rate of these ES cells is also observed. Moreover, chondroitin sulfate, another GAG species, is also involved in stem cell function. It interferes with the signaling pathways required for self-renewal, proliferation, and cell lineage commitment of neural stem cells. These studies demonstrate heparan sulfate and chondroitin sulfate GAG are important regulators of stem cell self-renewal and lineage commitment capacities [44]. In this context, we clearly demonstrate that [OTR4131], by acting on ECFC clonogenicity and proliferation, is able to regulate stem cell properties as it was demonstrated for natural GAG.

10.2.2.2 [OTR4131] Improves ECFC Proliferation

By interacting with HBGF, GAG are implicated in endothelial cell proliferation. Indeed, heparin potentiates the proliferation of endothelial cells induced by both FGF-1 and VEGF165 [45]. Binding of heparan sulfate to FGF-2 and VEGF165 increases the affinity to their respective receptors [46]. We demonstrate that [OTR4131] potentiates ECFC proliferation induced by VEGF and FGF-2 added to the medium supplemented with low serum concentration (0.5%). The action of [OTR4131] is much higher on FGF-2 than on VEGF. This may result from different affinities toward each growth factor. Several studies analyzing the influence of GAG structure on their specific binding to HBGF support this hypothesis [45, 47]. Our data are consistent with previous results obtained with [OTR4120], another GAG mimetic. [OTR4120] binds to VEGF165 and potentiates the affinity to its receptors. This leads to an improved proliferation and migration of endothelial cells induced by VEGF [24].

10.2.2.3 [OTR4131] Promotes ECFC Migration of ECFC

We show that [OTR4131] increases the VEGF-dependent migration of CB-ECFC. Even in the absence of VEGF, [OTR4131] also potentiates migration, probably because this molecule interacts with different chemo-attractant molecules present in the medium or secreted by these cells. Consistently, we have already shown that [OTR4131] also improves the migration of rat MSCs. Moreover, [OTR4131] mobilizes bone marrow hematopoietic stem cells according to a mechanism involving interaction with SDF-1 [25]. Another study demonstrates that heparan sulfate carrying acetyl groups, like [OTR4131], bind with increased affinity to chemokines such as CCR2 [48]. This cytokine is involved in EPC homing toward a neovascularization site [49]. Finally, the migratory effect of [OTR4131] opens clinical perspectives, because if injected to an injured tissue, it could attract transplanted cell therapy product and endogenous stem cells into the injury site.

10.2.2.4 Role of [OTR4131] in Angiogenesis

The [OTR4131] mimetic has no significant effect on ECFC angiogenic potential, as shown by in vitro Matrigel™ tube formation assay and in vivo Matrigel™ plug assay. However, another GAG mimetic with a structure similar to that of [OTR4131] is able to stimulate angiogenesis in vivo in chicken model of embryonic chorioallantoic membrane development [24] or in model of myocardial ischemia developed in pigs [50]. It is likely that these GAG mimetics act on angiogenesis by repairing the damaged extracellular matrix during tissue remodeling process occurring in embryonic development or in the healing process following injury. GAG mimetics may also improve the efficacy of HBGF present in the injured tissues. This situation is not reproduced in Matrigel™ experiments, because there is no tissue remodeling. Within this matrix, the GAG mimetics have no degraded endogenous GAG to replace, and thus their effect cannot be evidenced in this model. Moreover, HBGF such as VEGF and FGF-2 are present at high concentrations in Matrigel™, and this may hide the effects of their interactions with GAG mimetics. Indeed, during the culture experiments, in order to reveal effects of the GAG mimetics on ECFC, we had to adapt culture conditions by using low serum concentration and growth factors at minimal concentrations. Conversely, in the in vivo Matrigel™ plug assay, it is difficult to modulate the experimental conditions.

We have demonstrated that [OTR4131] is able to improve ECFC colony formation frequency. This is a critical step, required to improve ECFC amplification. This GAG mimetic also promotes cell self-renewal, adhesion, and migration. In addition, in low serum culture condition, [OTR4131] potentiates the proliferative effect of FGF-2 and VEGF, two important angiogenic growth factors.

In a perspective of using CB- or PB-ECFCs as cell therapy products for ischemic diseases, [OTR4131] could be a promising tool for improving the yield of ECFC production from these blood sources and for promoting their angiogenic capacities. [OTR4131] can also be an important component of clinical grade serum-free medium. Moreover, our work opens perspectives for therapeutic matrix repair strategy, based on GAG mimetics. These molecules could be combined to innovative tissue engineering approaches. GAG mimetics will potentiate treatments with exogenous growth factor and improve the effects of cell therapy products.

10.3 Conclusion

Cord blood EPCs give rise to active ECFCs that still retain stem cell features, as shown by different convergent evidences. The stem cell genes identified can be useful for predicting the stemness of CB-ECFCs, thus their further clinical value for the treatment of ischemic diseases.

Moreover, our recent data show these cells could be cultured during more than 3 months and they continue to proliferate and do not display any sign of replicative senescence.

The proliferation capacity of these cells can be even improved with matrix components such as GAG. Specific GAG mimetics could be used for targeting specific CB-ECFC functions, improving their clinical interest for the treatment of ischemic diseases.

References

1. Asahara T, Murohara T, Sullivan A, Silver M, van der Zee R, Li T, Witzenbichler B, Schatteman G, Isner JM. Isolation of putative progenitor endothelial cells for angiogenesis. Science. 1997;275:964–7.
2. Masuda H, Asahara T. Post-natal endothelial progenitor cells for neovascularization in tissue regeneration. Cardiovasc Res. 2003;58:390–8.
3. Iwakura A, Luedemann C, Shastry S, Hanley A, Kearney M, Aikawa R, Isner JM, Asahara T, Losordo DW. Estrogen-mediated, endothelial nitric oxide synthase-dependent mobilization of bone marrow-derived endothelial progenitor cells contributes to reendothelialization after arterial injury. Circulation. 2003;108:3115–21.
4. Takahashi T, Kalka C, Masuda H, Chen D, Silver M, Kearney M, Magner M, Isner JM, Asahara T. Ischemia- and cytokine-induced mobilization of bone marrow-derived endothelial progenitor cells for neovascularization. Nat Med. 1999;5:434–8.
5. Walter DH, Rittig K, Bahlmann FH, Kirchmair R, Silver M, Murayama T, Nishimura H, Losordo DW, Asahara T, Isner JM. Statin therapy accelerates reendothelialization: a novel effect involving mobilization and incorporation of bone marrow-derived endothelial progenitor cells. Circulation. 2002;105:3017–24.
6. Yoder MC, Mead LE, Prater D, Krier TR, Mroueh KN, Li F, Krasich R, Temm CJ, Prchal JT, Ingram DA. Redefining endothelial progenitor cells via clonal analysis and hematopoietic stem/progenitor cell principals. Blood. 2007;109:1801–9.
7. Barclay GR, Tura O, Samuel K, Hadoke PW, Mills NL, Newby DE, Turner ML. Systematic assessment

in an animal model of the angiogenic potential of different human cell sources for therapeutic revascularization. Stem Cell Res Ther. 2012;3:23.
8. Bompais H, Chagraoui J, Canron X, Crisan M, Liu XH, Anjo A, Tolla-Le Port C, Leboeuf M, Charbord P, Bikfalvi A, Uzan G. Human endothelial cells derived from circulating progenitors display specific functional properties compared with mature vessel wall endothelial cells. Blood. 2004;103:2577–84.
9. Prater DN, Case J, Ingram DA, Yoder MC. Working hypothesis to redefine endothelial progenitor cells. Leukemia. 2007;21:1141–9.
10. Au P, Daheron LM, Duda DG, Cohen KS, Tyrrell JA, Lanning RM, Fukumura D, Scadden DT, Jain RK. Differential *in vivo* potential of endothelial progenitor cells from human umbilical cord blood and adult peripheral blood to form functional long-lasting vessels. Blood. 2008;111:1302–5.
11. Sieveking DP, Buckle A, Celermajer DS, Ng MK. Strikingly different angiogenic properties of endothelial progenitor cell subpopulations: insights from a novel human angiogenesis assay. J Am Coll Cardiol. 2008;51:660–8.
12. Ingram DA, Mead LE, Tanaka H, Meade V, Fenoglio A, Mortell K, Pollok K, Ferkowicz MJ, Gilley D, Yoder MC. Identification of a novel hierarchy of endothelial progenitor cells using human peripheral and umbilical cord blood. Blood. 2004;104:2752–60.
13. Boyer-Di Ponio J, El-Ayoubi F, Glacial FG, Ganeshamoorthy K, Driancourt C, Godet M, Perrière N, Guillevic O, Couraud PO, Uzan G. Instruction of circulating endothelial progenitors *in vitro* towards specialized blood-brain barrier and arterial phenotypes. PLoS One. 2014;9:e84179.
14. Gehling UM, Ergun S, Schumacher U, Wagener C, Pantel K, Otte M, Schuch G, Schafhausen P, Mende T, Kilic N, Kluge K, Schäfer B, Hossfeld DK, Fiedler W. *In vitro* differentiation of endothelial cells from AC133-positive progenitor cells. Blood. 2000;95:3106–12.
15. Peichev M, Naiyer AJ, Pereira D, Zhu Z, Lane WJ, Williams M, Oz MC, Hicklin DJ, Witte L, Moore MA, Rafii S. Expression of VEGFR-2 and AC133 by circulating human CD34(+) cells identifies a population of functional endothelial precursors. Blood. 2000;95:952–8.
16. Case J, Mead LE, Bessler WK, Prater D, White HA, Saadatzadeh MR, Bhavsar JR, Yoder MC, Haneline LS, Ingram DA. Human CD34+AC133+VEGFR-2+ cells are not endothelial progenitor cells but distinct, primitive hematopoietic progenitors. Exp Hematol. 2007;35:1109–18.
17. Timmermans F, Van Hauwermeiren F, De Smedt M, Raedt R, Plasschaert F, De Buyzere ML, Gillebert TC, Plum J, Vandekerckhove B. Endothelial outgrowth cells are not derived from CD133+ cells or CD45+ hematopoietic precursors. Arterioscler Thromb Vasc Biol. 2007;27:1572–9.
18. Yamanaka S, Takahashi K. Induction of pluripotent stem cells from mouse fibroblast cultures. Tanpakushitsu Kakusan Koso. 2006;51:2346–51.
19. Zangrossi S, Marabese M, Broggini M, Giordano R, D'Erasmo M, Montelatici E, Intini D, Neri A, Pesce M, Rebulla P, Lazzari L. Oct-4 expression in adult human differentiated cells challenges its role as a pure stem cell marker. Stem Cells. 2007;25:1675–80.
20. Gandhi NS, Mancera RL. The structure of glycosaminoglycans and their interactions with proteins. Chem Biol Drug Des. 2008;72:455–82.
21. Gallagher JT. Multiprotein signalling complexes: regional assembly on heparan sulphate. Biochem Soc Trans. 2006;34:38–441.
22. Papy-Garcia D, Barbier-Chassefière V, Rouet V, Kerros M-E, Klochendler Cb, Tournaire M-C, Barritault D, Caruelle J-P, Petit E. Nondegradative sulfation of polysaccharides. Synthesis and structure characterization of biologically active heparan sulfate mimetics. Macromolecules. 2005;38:4647–54.
23. Meddahi A, Bree F, Papy-Garcia D, Gautron J, Barritault D, Caruelle J-P. Pharmacological studies of RGTA 11, a heparan sulfate mimetic polymer, efficient on muscle regeneration. J Biomed Mater Res. 2002;62:525–31.
24. Rouet V, Hamma-Kourbali Y, Petit E, Panagopoulou P, Katsoris P, Barritault D, Caruelle J-P, Courty J. A synthetic glycosaminoglycan mimetic binds vascular endothelial growth factor and modulates angiogenesis. J Biol Chem. 2005;280:32792–800.
25. Albanese P, Caruelle D, Frescaline G, Delbé J, Petit-Cocault L, Huet E, Charnaux N, Uzan G, Papy-Garcia D, Courty J. Glycosaminoglycan mimetics-induced mobilization of hematopoietic progenitors and stem cells into mouse peripheral blood: structure/function insights. Exp Hematol. 2009;37:1072–83.
26. Frescaline G, Bouderlique T, Mansoor L, Carpentier G, Baroukh B, Sineriz F, Trouillas M, Saffar J-L, Courty J, Lataillade J-J, Papy-Garcia D, Albanese P. Glycosaminoglycan mimetic associated to human mesenchymal stem cell-based scaffolds inhibit ectopic bone formation, but induce angiogenesis in vivo. Tissue Eng. 2013;19:1641–53.
27. Yin AH, Miraglia S, Zanjani ED, Almeida-Porada G, Ogawa M, Leary AG, Olweus J, Kearney J, Buck DW. AC133, a novel marker for human hematopoietic stem and progenitor cells. Blood. 1997;90:5002–12.
28. Civin CI, Strauss LC, Brovall C, Fackler MJ, Schwartz JF, Shaper JH. Antigenic analysis of hematopoiesis. III. A hematopoietic progenitor cell surface antigen defined by a monoclonal antibody raised against KG-1a cells. J Immunol. 1984;133:157–65.
29. Fina L, Molgaard HV, Robertson D, Bradley NJ, Monaghan P, Delia D, Sutherland DR, Baker MA, Greaves MF. Expression of the CD34 gene in vascular endothelial cells. Blood. 1990;75:2417–26.
30. Hara T, Nakano Y, Tanaka M, Sekiguchi T, Minehata K, Copeland NG, Jenkins NA, Okabe M, Kogo H, Mukouyama Y, Miyajima A. Identification of podocalyxin-like protein 1 as a novel cell surface marker for hemangioblasts in the murine aorta-gonad-mesonephros region. Immunity. 1999;11:567–78.

31. Bayat H, Fathi F, Peyrovi H, Mowla SJ. Evaluating the expression of self-renewal genes in human endothelial progenitor cells. Cell J. 2013;14:298–305.
32. Kershaw DB, Beck SG, Wharram BL, Wiggins JE, Goyal M, Thomas PE, Wiggins RC. Molecular cloning and characterization of human podocalyxin-like protein. Orthologous relationship to rabbit PCLP1 and rat podocalyxin. J Biol Chem. 1997;272:15708–14.
33. Sassetti C, Tangemann K, Singer MS, Kershaw DB, Rosen SD. Identification of podocalyxin-like protein as a high endothelial venule ligand for L-selectin: parallels to CD34. J Exp Med. 1998;187:1965–75.
34. Pesce M, Scholer HR. Oct-4: control of totipotency and germline determination. Mol Reprod Dev. 2000;55:452–7.
35. Levine AJ, Brivanlou AH. GDF3, a BMP inhibitor, regulates cell fate in stem cells and early embryos. Development. 2006;133:209–16.
36. Chen C, Ware SM, Sato A, Houston-Hawkins DE, Habas R, Matzuk MM, Shen MM, Brown CW. The Vg1-related protein Gdf3 acts in a nodal signaling pathway in the pre-gastrulation mouse embryo. Development. 2006;133:319–29.
37. Moretti A, Caron L, Nakano A, Lam JT, Bernshausen A, Chen Y, Qyang Y, Bu L, Sasaki M, Martin-Puig S, Sun Y, Evans SM, Laugwitz KL, Chien KR. Multipotent embryonic isl1+ progenitor cells lead to cardiac, smooth muscle, and endothelial cell diversification. Cell. 2006;127:1151–65.
38. Thomson JA, Itskovitz-Eldor J, Shapiro SS, Waknitz MA, Swiergiel JJ, Marshall VS, Jones JM. Embryonic stem cell lines derived from human blastocysts. Science. 1998;282:1145–7.
39. Okano M, Xie S, Li E. Cloning and characterization of a family of novel mammalian DNA (cytosine-5) methyltransferases. Nat Genet. 1998;19:219–20.
40. Janic B, Guo AM, Iskander AS, Varma NR, Scicli AG, Arbab AS. Human cord blood-derived AC133 + progenitor cells preserve endothelial progenitor characteristics after long term in vitro expansion. PLoS One. 2010;5:e9173.
41. Kalka C, Masuda H, Takahashi T, Kalka-Moll WM, Silver M, Kearney M, Li T, Isner JM, Asahara T. Transplantation of ex vivo expanded endothelial progenitor cells for therapeutic neovascularization. Proc Natl Acad Sci U S A. 2000;97:3422–7.
42. Schlechta B, Wiedemann D, Kittinger C, Jandrositz A, Bonaros NE, Huber JC, Preisegger K-H, Kocher A. Ex-vivo expanded umbilical cord blood stem cells retain capacity for myocardial regeneration. Circ J. 2010;74:188–94.
43. Vanneaux V, El-Ayoubi F, Delmau C, Driancourt C, Lecourt S, Grelier A, Cras A, Cuccuini W, Soulier J, Lataillade J-J, Lebousse-Kerdiles M-C, Oury JF, Sibony O, Marolleau J-P, Benbunan M, Uzan G, Larghero J. In vitro and in vivo analysis of endothelial progenitor cells from cryopreserved umbilical cord blood: are we ready for clinical application? Cell Transplant. 2010;19:1143–55.
44. Kraushaar DC, Dalton S, Wang L. Heparan sulfate: a key regulator of embryonic stem cell fate. Biol Chem. 2013;394:741–51.
45. Ashikari-Hada S, Habuchi H, Kariya Y, Kimata K. Heparin regulates vascular endothelial growth factor165-dependent mitogenic activity, tube formation, and its receptor phosphorylation of human endothelial cells. Comparison of the effects of heparin and modified heparins. J Biol Chem. 2005;280:31508–15.
46. Gitay-Goren H, Soker S, Vlodavsky I, Neufeld G. The binding of vascular endothelial growth factor to its receptors is dependent on cell surface-associated heparin-like molecules. J Biol Chem. 1992;267:6093–8.
47. Zhao W, McCallum SA, Xiao Z, Zhang F, Linhardt RJ. Binding affinities of vascular endothelial growth factor (VEGF) for heparin-derived oligosaccharides. Biosci Rep. 2012;2:71–81.
48. Schenauer MR, Yu Y, Sweeney MD, Leary JA. CCR2 chemokines bind selectively to acetylated heparan sulfate octasaccharides. J Biol Chem. 2007;282:25182–8.
49. Hristov M, Zernecke A, Bidzhekov K, Liehn EA, Shagdarsuren E, Ludwig A, Weber C. Importance of CXC chemokine receptor 2 in the homing of human peripheral blood endothelial progenitor cells to sites of arterial injury. Circ Res. 2007;100:590–7.
50. Yamauchi H, Desgranges P, Lecerf L, Papy-Garcia D, Tournaire M-C, Mocza M, Loisance D, Barritault D. New agents for the treatment of infarcted myocardium. FASEB J. 2000;14:2133–4.

Proangiogenic Features of Perinatal Tissue-Derived Stem Cells in Cardiovascular Disease Therapy

11

Hongyan Tao and Zongjin Li

Abstract

Because of the terminal differentiation of cardiomyocytes in adult heart, damaged cardiomyocytes cannot be self-replicated and differentiated into new myocardium. Cardiac regeneration using stem cells represents as a promising therapeutic strategy for cardiovascular disease. However, available stem cells such as embryonic stem cells (ESCs) have the risk of teratoma formation and ethical problems. Thus, perinatal stem cells (PSCs) obtained from surplus pregnancy test samples taken for prenatal diagnostic procedures became an alternative. PSCs are broadly multipotent with strong angiogenic effect, which have emerged as promising therapeutic tools in regenerative medicine. In this chapter, we will highlight some of the basic problems in the therapy of cardiovascular disease by focusing on the proangiogenic features of stem cells from perinatal tissue and introducing the application of PSCs in cardiac regeneration.

H. Tao
Nankai University School of Medicine, Tianjin, China

Z. Li (✉)
School of Medicine and Key Laboratory of Bioactive Materials, Ministry of Education, College of Life Sciences, Nankai University, Tianjin, China
e-mail: zongjinli@nankai.edu.cn

11.1 Introduction

With cardiovascular disease (CAD) being the most common cause of global mortality, finding viable treatment methods to reduce the risk of deaths are aggressively addressed. Although coronary artery bypass grafting (CABG) and intracoronary percutaneous coronary intervention (PCI) technology are making great progress in disease progression controlling, problems such as potential infection and thrombosis have persisted. Therefore, ongoing research on regenerative medicine is in processing trying to find a safe and feasible alternative treatment. A novel therapeutic strategy in CAD therapy is therapeutic angiogenesis, which started from pre-existing vasculature, providing extra blood with new formation vessels for the injured region caused by endothelial dysfunction, myocardial ischemia, and myocardial infarction [1]. Currently, kinds of stem cells have been used in therapeutic angiogenesis [2–4] in combination with biomaterials and growth factors thereby facilitating vessel regeneration [5, 6].

Recently, evidence has reinforced the idea that stem cells from perinatal tissue provide advantages over other stem cell sources [7]. Given the unique cell biological properties and broad multipotent plasticity, perinatal stem cells (PSCs) are a novel stem cell compared with both embryonic stem cells (ESCs) and adult stem cells (ASCs). Studies proved that PSCs are capable of generating and replacing mature cells within their own

Z. C. Han et al. (eds.), *Perinatal Stem Cells*, https://doi.org/10.1007/978-981-13-2703-2_11

specific tissues, which are more flexible than other sources. Besides, PSCs exhibited high levels of telomerase and proliferation activity, allowing for greater regenerative potential before reaching senescence, but are not tumorigenic like ESCs. These stem cells can be isolated from extraembryonic tissues including amniotic fluid, Wharton's jelly, amniotic membrane, placenta, and umbilical cord blood [8] (Fig. 11.1) which are normally discarded after birth. Thus, these stem cells isolated from the abandoned fetus tissue are effectively avoid involved in ethical problems. These entire attractive characteristics make PSCs become promising and noncontroversial source of stem cells for extensive use in cardiovascular diseases therapy.

In this chapter, we will concentrate on four main stem/progenitor cells of PSCs which include umbilical cord blood-derived endothelial progenitor cells (EPCs), amniotic fluid-derived stem cells (AFSCs), amniotic epithelial cells (AECs), and perinatal mesenchymal stem cells (PMSCs). We are going to review the proangiogenic features of PSCs as a method for cardiovascular diseases therapy, addressing mechanisms of action and future clinical application.

11.2 Umbilical Cord Blood-Derived Endothelial Progenitor Cells

Endothelial progenitor cells (EPCs) derived from umbilical cord blood (UCB) have the potential to incorporate into the injury site while proliferating and differentiating into endothelial cells, thereby improving local blood perfusion and tissue regeneration. This improvement is achieved

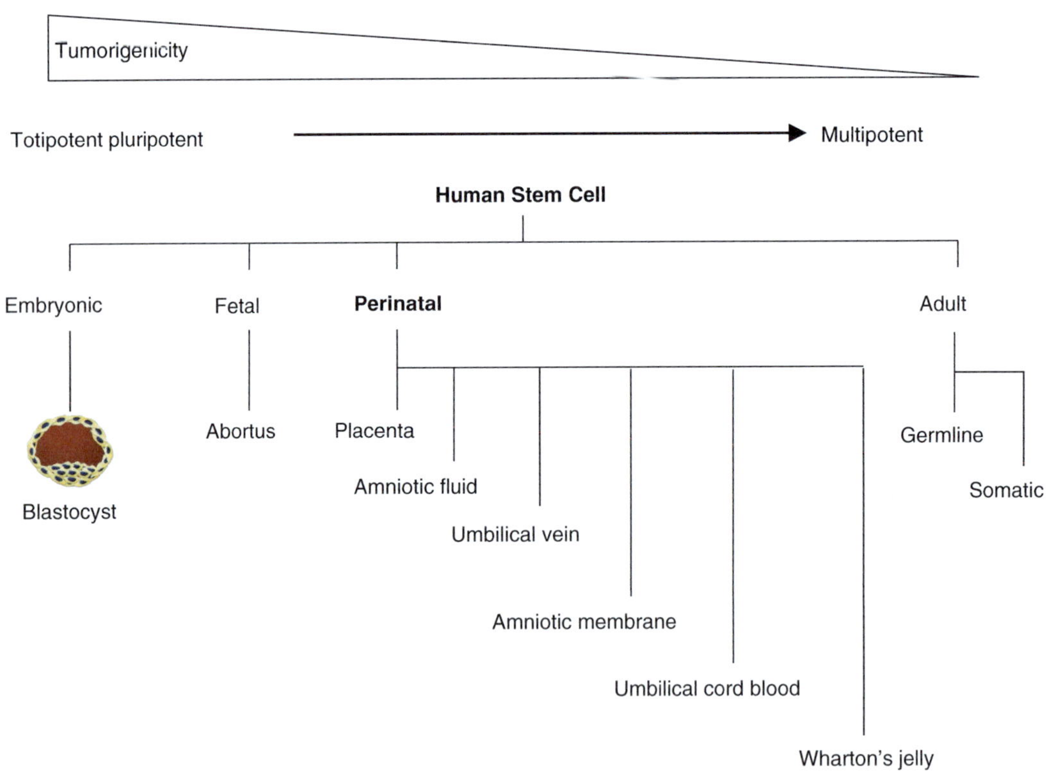

Fig. 11.1 Diagram of human stem cell classification. In mammals, there are two broad types of stem cells: embryonic stem cells (ESCs), which are isolated from the blastocysts, and adult stem cells (ASCs), which are found in various tissues. Compared with both ESCs and ASCs, perinatal stem cells (PSCs) represent a new paradigm isolated from perinatal tissue like umbilical cord blood, Wharton's jelly, amniotic membrane, placenta, and amniotic fluid

by modulating vascular regenerative processes such as angiogenesis and vasculogenesis [9, 10] and by replacing damaged tissue. The proangiogenic potential of EPCs isolated from UCB has garnered plenty of attention for therapeutic purposes. Compared with adult EPCs, EPCs derived from UCB have higher proliferative capacity, rapid self-renewal, and low apoptosis, which are able to enhance the angiogenesis ability after transplantation [11, 12]. Researchers described several antiapoptotic and proangiopoietic factors from UCB CD133^{+} cell-derived microvesicles, including kit ligand, insulin growth factor-1 (IGF-1), vascular endothelial growth factor (VEGF), basic fibroblast growth factor (bFGF), and interleukin (IL)-8 [13]. Therefore, we force on the isolation, proliferation, migration and survival capacities, and proangiogenic cytokine release of umbilical cord blood-derived EPCs.

Multiple cell surface antigens are used to identify human EPCs, including stem cell markers such as CD34, CD133 (AC133), c-Kit, and monocyte markers like endothelial growth factor receptor-2 (VEGFR-2, also known as platelet endothelial cell adhesion molecule or termed KDR in the human and flk-1 in the mouse), von Willebrand's factor (vWF), CD105 (endoglin), and VE-cadherin [14]. Other cell markers include CD146 (also known as the melanoma cell adhesion molecule) and CXCR4 (the receptor for stromal cell-derived factor), while the EPC-specific molecular profile has not been defined. Currently, it is generally agreed that human EPC are positive for CD34, CD133, and VEGFR-2 and negative for CD45, CD73, and CD90 [15]. As stem cells differentiate into mature postmitotic cells, CD133 represented the clonogenic, and proliferative potential of human PSCs is rapidly downregulated. Besides, CD31^{+} EPCs isolated from UCB were suggested contributing to the repair of ischemic vascular injuries introduced by femoral artery ligated [16]. Result showed that the foreign material-free EPCs injection significantly improved the limb blood flow compared with control group. Immunohistochemistry analysis also showed that the EPCs injection has increased the number of CD31^{+} capillaries. Altogether, the results indicate that with regard to therapeutic potential for ischemic vascular diseases, hUCB EPCs may offer an additional option as a cell therapeutic tool for ischemic diseases treatment [17–20].

However, the process of EPCs homing and migration to a target tissue still need to be explored. Sun et al. have used the EPCs isolated from UCB with fucosylation to research the homing of EPCs in the process of neovascularization [21]. They found few hUCB-derived PSCs expressing sLex antigen and capable of binding to E- and P-selectins, which appears necessary for selectin-dependent adhesion and transmigration in peripheral tissues. And fucose is a key element in sLex oligosaccharides, which is mediated by FucT activities. The expressed FucT VI in human UCB-derived EPCs lead in markedly increased sLex expression and E- and P-selectin binding in FucT VI-EPCs, indicated by markedly increased homing in the injured tissues and improved blood flow. EPC-mediated paracrine signaling has also been shown to be crucial in proangiogenesis by activating adjacent endothelial cells with proangiogenic growth factors [22, 23]. Thus, some researches try to improve the secretion of cytokines in the culture medium in order to find a significant increase in the expression of proangiogenic growth factors. Yin et al. have proved that the release of VEGF and SDF-1 was significantly increased in UCB-EPCs compared to MNCs and HUVECs [24]. Scanning electron microscopy analysis revealed that rough regenerated endothelium derived from EPCs has formation in the injury area. Moreover, this research also demonstrated more homing EPCs, and the transplantation of UCB-EPCs reduced neointima formation [24], and this result is also proved by other research [13].

In addition to soluble growth factors and chemokines, UCB-EPCs also secrete microvesicles (MVs) [25], which are small, spherical membrane fragments shed from the cell surface or secreted from the endosomal compartment and play a vital role in cell-to-cell communication. Unfortunately, a massive loss of transplanted cells has been reported, which often leads to the failure of the therapy. Environment and specific culture conditions may influence the survival rate, endothelial differentiation, as well as therapeutic benefit [26]. Numerous authors have proposed using biodegradable scaffold materials combined with UCB-derived cells to solve this problem. The application of three-dimensional (3D) microvascular system in angiogenesis therapy can be used as a positive control to validate novel biomaterial strategies that direct microvascular formation of EPCs. In particular, the model can help stem cell biologists integrate with other cell sources by screening candidate cells for their ability to support angiogenesis of EPCs. This strategy can also help promote EPC differentiation into tissue-specific endothelium. For example, ERICA et al., using polyethylene glycol (PEG) hydrogel systems containing covalently linked cell-binding agents and MMP-degradable peptides, can provide binders and degradation cues to support 3D angiogenesis [27]. hCB-EPC and vascular smooth muscle cell co-culture system can produce stable in vitro microvessels, showing the characteristics of physiological microvascular. This microvascular combined with multiple angiogenic cytokines and microvesicles could contribute to the improvement of ventricular function. In summary, these findings suggest that cell therapy with endothelial progenitor cells can serve as a new treatment for patients with endothelial dysfunction [28].

Currently, the main sources of EPCs include bone marrow, peripheral blood, and umbilical cord blood. Compared with umbilical cord blood, EPCs brought by marrow extraction when performed on patients with ischemic diseases are high risk of infection and formation of thrombus while the mobilization of peripheral blood stem cells. By comparison, the isolation process of UCB-EPCs is no risk to either the mother or the baby in the meantime with higher proliferation capacity. However, the number of cells is dependent on the volume of collected umbilical cord blood, which need further cryopreservation for future utilization. This limitation would affect the clinical application of umbilical cord blood-derived EPCs. In conclusion, the idea of using stem/progenitor cells, which aims to rebirth blood vessels, further reestablishes the structure and function of injured tissue are clearly promising.

11.3 Amniotic Fluid-Derived Stem Cells

Novel sources of stem/progenitor cells, the amniotic fluid of fetal tissues, provide a new tool for cell-based therapeutic applications of regenerative medicine. This tissue contains different cell types [29] and has rich source of stem/progenitor cells that maintain highly multipotent differentiation potential. Researchers have isolated and characterized highly multipotent cells with expression of pluripotency markers such as *Oct-3/Oct-4* from the amniotic fluid, which is called amniotic fluid-derived stem cells (AFSCs) [30]. The AFSCs could potentially provide an autologous cell source for clinical application due to their similar capacity for multilineage differentiation potential and self-renewal ability as compared with embryo stem cells (ESCs). Besides, the collection of amniotic fluid (AF) and AFSCs is less invasive than other sources of stem cells. These features represent AFSCs are in an intermediate stage between adult stem cells and ESCs [31]. The highly pluripotent and anti-inflammatory properties of these cells indicate the potential clinical application in the treatment of a range of diseases. Therefore, AFSCs have become promising tools for cardiovascular diseases therapy.

hAFSCs can be easily obtained using transabdominal approach between 14 and 21 weeks of gestation (Fig. 11.2). Second-trimester amniotic fluid contains hAFSCs with c-kit-positive (CD117) [31]. To date, there are numerous markers including CD133, CD105 (SH2), CD73 (SH3/4), CD90, and several adhesion molecules, for example, CD44 (HCAM-1 antigen) and CD29 [8], which can be used to identify AFSCs. Besides, the forming of embryoid bodies suggest that AFSCs may express embryonal markers and exhibit several features of ESCs [32]. However, unlike the ESCs, when implanted into immune-deficient mice, undifferentiated AFSCs expand extensively without feeders and are not tumorigenic, a basic safety features of a potential cell therapy. Current therapies in cardiovascular diseases mainly involve percutaneous revascularization and surgical or pharmaceutical approaches.

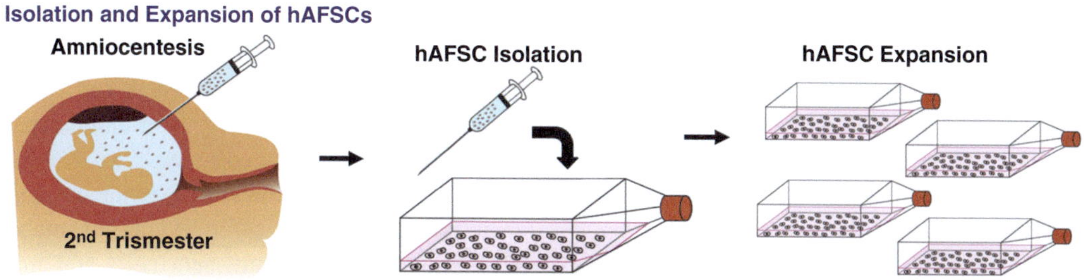

Fig. 11.2 Isolation and expansion of human amniotic fluid stem cells (hAFSCs). hAFSCs can be isolated from the second-trimester amniotic fluid less invasive than other sources of stem cells. The isolated hAFSCs were cultured in a modified minimum essential medium (a-MEM) supplemented with 20% fetal bovine serum and 4 ng/mL basic fibroblast growth factors and incubated at 37 C with 5% humidified CO_2. Reproduced with permission from [43]

Unfortunately, cardiovascular diseases are progressive, and the ischemia provokes further damages of cardiomyocytes that negatively affect the course of the disease [33]. Therefore, cell therapy using AFSCs is considered to be a promising strategy in tissue repair through angiogenesis or by modulation of the inflammatory process via paracrine signal.

VEGF plays a vital role in angiogenesis process, and hypoxic could enhance the secretion of VEGF compared with normoxia. To comfier the conclusion, researchers culture the AFSCs in intermittent hypoxia or continuous hypoxia. The results show that continuous hypoxia and intermittent hypoxia could enhance VEGF secretion. This result was further corroborated by the twofold increase in VEGF in intermittent hypoxia groups. Besides, AFSCs culture in continuous hypoxia was found to enhance endothelial gene expression and VEGF protein secretion [34]. Just as we know, hypoxic plays a central role in ischemia damage, and the secretion of VEGF by AFSCs is capable of stimulating angiogenesis. AFSCs have previously proved they can secrete several angiogenic factors like vWF MCP-1, Flk-1, and CD146, together with the absence of IL-1α [35], which may present alternative cell source for endothelial-like cells that could be used in clinical applications. In another research, it is considered that microRNAs (miRNAs) have an important role in transcriptional regulation of angiogenesis-related genes. Specifically, they chose the expression of miR210, 126, and132, which improve angiogenesis by targeting negative regulators in angiogenic pathways. Endothelial cell-specific miR126 has a pivotal role in angiogenesis following ischemia and was detected in AFSCs before endothelial induction and maintained during differentiation. In addition, the expression of miR210, which is a key regulator of angiogenesis, rapidly increased after induction [36].

According to above research conclusion, we know that the regeneration of endothelial cells has a vital role in tissue repair and regeneration, and therefore it could be used in regenerative medicine. However, the regeneration of abundant functional endothelium from stem cells is a slow process because endothelial cells are difficult to culture in many current strategies which have limited their proliferative potential and display vascular instability. Therefore, mid-gestation AFSCs can be a promising tool for vascular development and regeneration. In a study, mid-gestation amniotic fluid-derived cells (ACs) including AFSCs were utilized to transform into a stable and expandable population of vascular ECs (rAC-VECs). ETS (E26 transformation-specific or E-twenty-six [37]) family is one of the largest families of transcription factors [38] and important regulators of angiogenesis. The constitutive expression of ETS transcription factors FLI-1 and ERG1, meanwhile, transient expression of the ETS transcription factor ETV2, concomitant with TGF-β inhibition could reprogram AFSCs into functional rAC-VECs. Results show that these rAC-VECs maintain the typical morphology of mature ECs, as well as the expression of ECs surface markers, like CD31, VEGFR2, and VE-cadherin over numerous passages in vitro. What's more, rAC-VECs are able to form functional vessels after transplanted in vivo. Therefore, AFSCs can serve as an alternative to endothelial cells to reveal the molecular determinants of vascular development and heterogeneity and may represent the ideal ECs for treating tissue damage [39].

A study has used AFSCs transfected by a GFP-labeled plasmid system of an adenovirus vector carrying glial cell line-derived neurotrophic factor (GDNF) gene (Ad-GDNF-GFP) to confirm the differentiation of AFSCs. GDNF is a tissue morphogen to regulate stem cell proliferation and differentiation. Transfected AFSCs were cultured in endothelial growth medium-2 with VEGF. One week later, AFSCs were positive for markers of endothelial cells including vWF and CD31, and more importantly, the recombinant GDNF group was significantly higher than control groups. The result demonstrated that AFSCs are able to differentiate into vascular endothelial-like cells, and the differentiation can be promoted by recombination with GDNF [40]. In other study, AFSCs were co-cultured with human umbilical vein endothelial cell (HUVEC). The results of this study demonstrate that the AFSC-HUVEC co-culture was capable of prevascularising collagen chondroitin sulfate (CCS) scaffolds, and AFSCs are capable of behaving as pericytes while interacting with HUVECS to form microcapillary-like structures. Likewise, cultured in endothelial cell media supplemented with VEGF, AFSCs display an endothelial gene expression and adopted functional endothelial cell characteristics indicative of early endothelial differentiation [34]. Overall, these results demonstrate that AFSCs can function in the role of vascular endothelial-like cells in order to stabilize microcapillary-structure formation.

Tissue engineering is exploring the potential of developing tissue and organ substitutes, which has achieved tremendous progress in disease therapy [41]. However, there is a challenge in the design of tissue engineering scaffolds for their clinical applications, which is the lack of a perfusing blood supply to nourish the resident cells after implantation. This limits long-term engraftment of the transplanted cells, along with the ability of scaffolds, cell survival, and subsequent construct function [42]. As a unique stem cell source, AFSCs are considered as a useful tool in regeneration medicine. Given their angiogenic potential, AFSCs offer several advantages over embryonic and adult stem cells and may play a vital role in support of the recruitment of endothelial cells in vivo. Studies have shown that ectopic implants obtained with AFSCs and biomaterials have a strong angiogenic response in mouse models with collateral vessel infiltration grafts and significant blood perfusion [35]. Besides, in an immune-suppressed rat model with an ischemic injury in border area, following xenogeneic transplantation of AFSCs resulted in heart function improvement and engraftment efficiency. What's more, the transplantation leads to an increase in neovascularization and in situ proliferation with expression of cardiovascular markers [43]. With so much successful research in AFSCs-related tissue engineering, we believed these cells will be an important therapeutic tool in clinical. However, some questions still remain for the transplanted cells could not stay at the injured area for enough time. A method of ensuring a well preserved engraved AFSC is achieved using spherical porous beads of [poly (D, L-lactic acid-co-glycolic acid)] as a platform for in vivo cell transplantation. This biomaterial was used to design a micro-scaffold to form a composite fibrotic 3D structure while retaining endocrine-secreted extracellular matrix components. Fibrous-sized beads were injected intramyocardially into a rat model with myocardial infarction, and biomaterials in combination with AFSCs showed implantation at the injection site with cardiac function improvement. Histological analysis supports the delivered AFSCs and had differentiation into vascular and cardiomyocyte lineages with vWF, homeobox transcription factor Nkx2.5, sarcomeric actin, and cardiac troponin I expression [44].

AFSCs have shown promise as a cell therapy for cardiac disease and express the specific markers of cardiomyocyte-like cells including c-kit, myosin heavy chain, and cardiac troponin

I [45]. In a study, AFSCs was cultured with cardiac cells to induce its differentiation within the direction of cardiac fibroblast, smooth muscle cells, and endothelial cells through paracrine factors and mechanotransduction separately from the progenitor populations. The results showed that GATA4 and cTnT expression in the experimental groups were upregulated compared to the controls. Furthermore, compared to the controls and other co-culture conditions, the contact co-culture of these cells resulted in significantly higher expression of cTnT mRNA. This suggests AFSCs population exhibiting cardiac differentiation ability and cell-cell communication may have improved the expression of cardiac genes. Thus, the functional benefits of cell transplantation include reducing the progression of damaged tissue, increasing local wall motion, and improving global function. The ability of AFSCs to act directly on the myocardium is important for the long-term prognosis of patients with myocardial infarction [46].

In conclusion, AFSCs are a highly potent source of anti-inflammatory cells with little ethical and legal implications for their collection and use. Similarly, endothelial cells derived from hESCs and iPSCs proliferate poorly and drift into nonvascular lineages [47]. The shortcomings of existing strategies to generate mature and stable endothelial cells may be due to insufficient knowledge of the transcription factors and microenvironmental cues that establish long-lasting tissue-specific cells [48]. According to literature, since AFSCs have the ability to differentiate into different cell types and lack ethical concerns, they have so far been considered as promising tools for clinical treatment. Based on the data presented in this article, it can also be assumed that AFSCs will be one of the major therapy tools of the recent human cardiovascular disease model.

11.4 Human Amniotic Epithelial Cells

Amniotic membrane is the innermost layer of fetal membranes, located in the placenta of the baby surface, and composed of cubic or cylindrical epithelial cells. Stem/progenitor cells in amniotic membrane are mainly amniotic mesenchymal stem cells and amniotic epithelial stem cells. Both these two cell types are promising for cell-based therapy and tissue engineering. Recently, the isolation of amniotic epithelial cells (hAECs) from intraoperative or preoperative amniotic membranes has attracted the attention of many researchers and clinicians and has become a potential source of regenerative medicine. The reason for this interest is the evidence that these cells are pluripotent, immunomodulatory, and anti-inflammatory. hAECs were successfully isolated, showing fibroblast-like morphology and pebble-shaped epithelial morphology. Under certain conditions, hAECs are able to differentiate cell types from all three germ layers [49]. Therefore, the potential clinical applications of hAECs might include many different diseases therapy including stroke, multiple sclerosis, acute liver failure, diabetes, and chronic and acute lung diseases. For the low immunogenicity, immune regulation, and proangiogenesis feature, hAECs are considered to be promising cells for ischemia-related diseases therapy. We will introduce the proangiogenesis characteristics of hAECs and their application in clinical therapy.

The biomarker analysis showed that hAECs were positive for CD9, CD24, CD73, CD90, CD105, ATP-binding cassette transporter G2 (ABCG2), E-cadherin, c-met (also called tyrosine-protein kinase Met or hepatocyte growth factor receptor), stage-specific embryonic antigen 3 (SSEA-3), and SSEA-4 and are regularly reported to be negative for SSEA-1 and some markers used to identify MSCs, like CD44, CD90, CD133, HLA-DR and CD105, weakly expressed or do not express CD117 (c-kit), and C-C chemokine receptor type 4 (CCR4) [50]. HAECs do not express or express very low amounts of HLA-A, B, C, but with the passage of cells, their expression has increased. It has been reported that hAECs express keratan sulfate-associated antigens, including tumor rejection antigens (TRA) 1-60 and TRA1-81 [51]. HAECs also express some moleculars involved in the maintenance of pluripotency such as octamer-binding transcription factor (Oct)-4 and homeobox protein

Nanog. In addition, the immune-privileged and nontumorigenic features make them particularly attractive as a tool for cell-based therapy. These features have led researchers to investigate the potential use of hAECs in the treatment of various diseases in preclinical animal studies.

RT-PCR shows that telomerase is not expressed in naive hAECs, which means these cells are non-tumorigenic upon transplantation [52]. This property makes hAECs a safe and useful cell type for clinical cell-based therapy. Therefore, several studies have indicated that they are a promising substitute for embryonic stem cells as diseases therapy. Currently, AECs are considered to have potential application for the treatment of cardiovascular disease because of their proangiogenic feature. For safety and quality control considerations, the study on cell isolation protocols using animal product-free reagents is in progress to develop methods which allow researchers to isolate, cryopreserve, and culture hAECs [51]. The utilization of these cells could avoid delivering of potentially harmful animal pathogens to humans with this method while maintaining their regenerative potential, viabilities, and suitable cell yields. For researchers now moving from preclinical animal studies to clinical treatment, these approaches will significantly speed regulatory approvals, reduce risk, and improve the quality of their therapeutic cell population [53].

Although some study has gained the positive result of hAECs on its proangiogenic feature, the effect of hAECs on angiogenesis-related therapy remains controversial. For example, it is unclear whether the presence of hAEC donor inflammation and/or gestational age has an effect on the angiogenic effect. In a study, researchers try to examine the differences of hAECs between preterm and term on angiogenesis, and result shows that both preterm and term hAECs expressed VEGFA, platelet-derived growth factor subunit B (PDGFB), angiopoietin 1(ANGPT1), and forkhead box C1 (FOXC1). Stimulation with TNF-α and IFN-γ and all these angiogenesis-related gene expressions significantly increased. Besides, the formation of HUVEC tubules on Matrigel was significantly longer with the stimulation of conditioned media from term hAECs. In the presence of tumor necrosis factor alpha (TNF-α) and interferon gamma (IFN-γ), co-culture with term hAECs reduced gene transcription of Tie-2 (an endothelium-specific receptor tyrosine kinase) and FOXC1 in HUVEC, while co-culture with preterm hAECs increased gene transcription of PDGFRα and PDGFRβ and reduced gene transcription of FOXC1 in HUVEC. In summary, this results implied the presence of inflammation, and underlying pathology is able to influence the proangiogenesis ability of hAECs [54]. HAECs secreted higher levels of proangiogenic cytokines, including monocyte chemoattractant protein (MCP)-1, interleukin (IL)-6, and epidermal growth factor (EGF) in vitro. To figure out whether these proangiogenic cytokines can be expressed in vivo by transplanted hAECs, some researchers ligated the coronary artery of rats to cause MI and injected hAECs into the infarcted area. They found that the infarct area treated with hAECs had higher expression of MCP-1, EGF, and IL-6 compared to the tissues of saline-treated rats. In this study, hAECs are proved secreted proangiogenic cytokines after transplant into a rat model of myocardial infarction (MI). Besides, the improvement of cardiac function may confirm that the paracrine effect by hAECs could stimulate proangiogenesis as well as regenerate myocardial tissue.

HAECs, as a source of cells for cell therapy and tissue regeneration, have various advantages and are known to differentiate into endothelial cells. However, the ability of hAECs to directly differentiate into cardiomyocytes in vivo and their cell engraftment effect on myocardial infarction is still controversial. In a study, umbilical cord blood mesenchymal stem cells (UCB-MSCs) and adipose tissue-derived MSCs (AT-MSCs) were transplanted to differentiate into cardiomyocytes in vivo to reduce infarct size and improve cardiac function (Fig. 11.3). UCB-MSCs, AT-MSCs, hAECs, and saline were injected into areas of myocardial infarction in nude rats. The results showed that 3% of the hAECs expressed myosin heavy

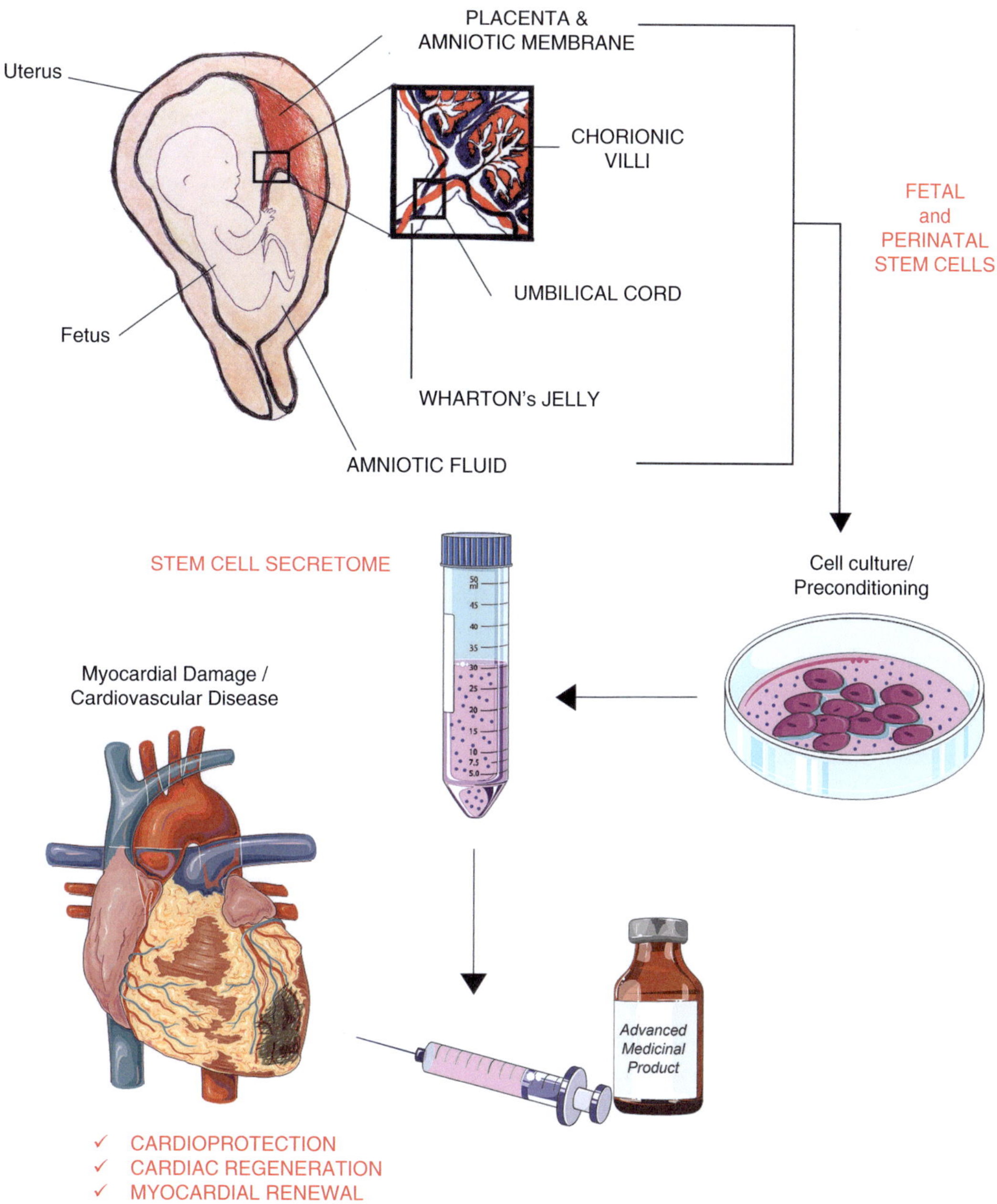

Fig. 11.3 Perinatal stem cells in cardiac regeneration. Perinatal stem cells (PSCs) isolated from extraembryonic tissues share features with both adult and embryonic stem cells and show high self-renewal. While fetal and perinatal stem cells demonstrated to improve cardiac function when injected in the injured heart. The paracrine potentials of perinatal stem cell secretome are able to provide cardiac repair and resurge the dormant mechanisms of cardiac regeneration for future therapy. Reproduced with permission from [110]

chain and hAECs-implanted groups showed a lower infarct area, higher ejection fraction, and more abundant myocardial gene and protein expression after 4 weeks culture, which indicated that hAECs could differentiate into cardiomyocyte-like cells. The ability of hAECs in cardiac function improvement as well as infarct size decrease was comparable to those of UCB-MSCs and AT-MSCs. These results support the need for further studies of hAECs as a cell source for myocardial regeneration due to their plentiful availability, low immunity, and lack of ethical issues [55]. Furthermore, co-culture has no impact on the phenotype of AT-MSCs and UCB-MSCs but increased the proliferation activity of hAECs and decreased the proliferation activity of AT-MSCs [56].

Based on the aforesaid study, hAECs may have the ability to improve the ischemic tissue regeneration. In response to incubation with transforming growth factor-β1 (TGF-β1), hAECs acquired fibroblastoid shape, associated with downregulation of E-cadherin, upregulation of N-cadherin, Akt phosphorylation, and intracellular periostin translocation. Epithelial-to-mesenchymal transition (EMT) in AECs showed significantly enhanced mobilization and secretion of gelatinase activity compared to untreated AECs. The surface presentation of CD105 and CD73 is diminished, and RNA microarray analysis reflects the loss of epithelial features and transcriptional profiling. Unimmunized AECs and EMT-AECs were then injected intramyocardially into fully immunocompetent mice after permanent left anterior descending (LAD) ligation, and cardiac function was followed by MRI and echocardiography. The infarcted heart treated with EMT-AEC showed better overall systolic function and improved longitudinal strain rate in the region of interest, while the vessel density increased EMT-AEC treatment of the heart. They conclude that EMT enhances the cardioprotective effects of hAECs [57].

Tissue regeneration is a highly orchestrated physiological process consisting in a complex interaction of cellular and biochemical events. HAECs have been shown to be an attractive resource for tissue regeneration because they are primitive stem cells. It has found that hAECs accelerate the migration of keratinocytes and induce a remarkable increase in the activity of phospho-ERK, phospho-JNK, and phospho-AKT, the blockade of which their specific inhibitors significantly inhibits migration induced by hAESC-conditioned medium (CM). Furthermore, the co-culture of keratinocytes with hAESCs upregulates the expression levels of cell proliferation proteins Cyclin D1, Cyclin D3, and Mdm2. Taken together, these results reveal that hAESCs promote wound healing by facilitating the migration and proliferation of keratinocytes via ERK, JNK, and AKT signaling pathways and might be a potential therapy in tissue regeneration [58]. The other study was also determining the impact of transplantation of hAECs on the pathologic changes of wound tissues in a rat ulcer model. Ninety-six SD rats were randomly divided into the model group, hAEC transplantation group, traditional treatment group, and the control group. HE staining proved that the healing time of hAEC transplantation group was significantly shorter than the other three groups. Moreover, pathological observations revealed that hAEC transplantation group exhibited significant proliferation of fibrous tissues and vessels in the dermal layer, and the appearance time and degree of skin appendages were significantly greater than that observed in the other groups. Pathological observations showed that hAEC transplantation could significantly speed up the healing of ulcer [59].

11.5 Human Mesenchymal Stem Cells

Mesenchymal stem cells (MSCs) have been used for therapeutic angiogenesis and tissue regenerative for decades. Under numerous researches in animal models and clinical trials, MSCs have been utilized as a potential therapeutic agent for several diseases. According to an international agreement, human MSCs are defined by three minimal criteria: (1) adherence to plastic ability; (2) positive expression of specific surface antigens (CD29, CD44, CD73, CD90, CD105) and

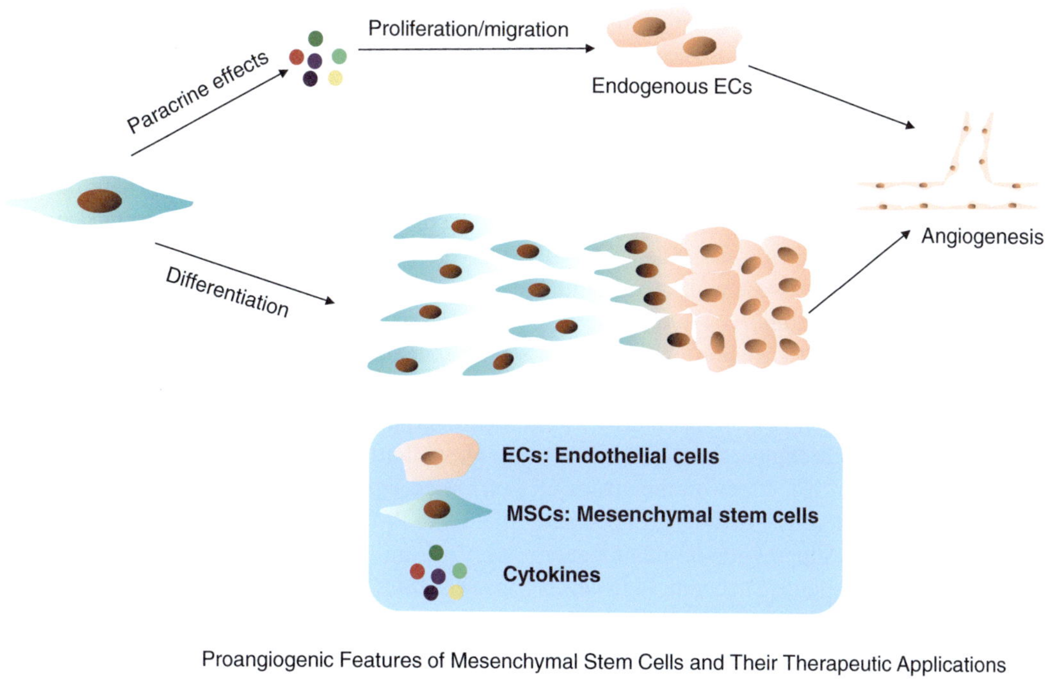

Fig. 11.4 Angiogenic potency of MSCs. MSCs could promote angiogenesis either by paracrine effects or by transdifferentiation. The secretion of cytokines could enhance the proliferation and migration of endogenous endothelial cells. Moreover, MSCs may differentiate into lineage with functional and phenotypic features of ECs and participate in angiogenesis directly

negative expression of hematopoietic/endothelial markers CD45, CD34, CD14, CD19, and human leukocyte antigen (HLA)-DR; and (3) multilineage differentiation potential in vitro [60]. The main source of MSCs in clinical investigation is adult bone marrow. Bone marrow mesenchymal stem cells (BMSCs) have been proven and significantly promote improvement of patients with critical ischemia in clinical application for therapeutic angiogenesis [61]. Recently, due to their accessibility, painless procedures to donors, and lower risk of viral contamination, perinatal stem cells are considered to be valuable alternatives to BM as a source of MSCs.

There are several researches that verified the proangiogenic feature of MSCs derived from perinatal tissue. Studies by Du et al. have suggested that MSCs isolated from placental chorionic villi are preferred in therapeutic angiogenesis for their eNOS producing and acetylated-low density lipoprotein (Ac-LDL) uptaking ability during endothelial differentiation [62]. Their studies showed that compared with AT-MSCs and UCB-MSCs, the conditioned medium of placental chorionic villi (PMSCs) showed significant proangiogenic bioactivities on endothelial cells (Fig. 11.4). These cells could directly form tubular structures in vitro with several angiogenic genes that were upregulated compared with AT-MSCs and UCB-MSCs. Enzyme-linked immunosorbent assay showed that PMSCs secreted much more PGE2 and HGF than another group. Furthermore, MSCs isolated from placental chorionic villi have the weakest expression of major histocompatibility complex class I (MHC I) genes and major histocompatibility complex class II (MHC II) genes, while they do not express immune-related genes [63], which means that fetal tissue such as amniotic, cord blood, umbilical cord Wharton's jelly (WT), and placental chorionic villi-derived MSCs might be preferred in clinical application [30]. This part explores the potential clinical use and benefit of MSCs and analogous regenerative medicine therapies sourced from the amniotic fluid, UCB, umbilical cord Wharton's Jelly (WJ), and placental membrane.

PMSCs isolated from the fetal tissue possess widespread clinical applications—from cell-based regenerative medicine to tissue engineering. PMSCs possess many advantages over stem cells, and these cells have been shown to preferentially accumulate at sites of tissue damage and/or inflammation caused by ischemia. The feature of MSCs homing to the injury area may provide an avenue for clinical translation of MSCs as a cellular tool for ischemia diseases therapeutics in the future [64]. After homing to the injured sites, here are some possible explanations and evidence correspond to the MSCs effect on the stages of tissue healing, which can be broadly grouped into three categories: (1) suppression of inflammation and immunomodulation; (2) secrete high levels of tissue repair factors such as PDGF TGF-β, EGF (epidermal growth factor), bFGF, and KGF (keratinocyte growth factor) on cell survival, proliferation, migration, and augmentation of angiogenesis; and (3) generation and remodeling of new tissue including dermal fibroblasts, keratinocytes, and epithelial cells [65].

PMSCs are pluripotent progenitors that can differentiate into multiple types of cells including cell lineages with phenotypic and functional features of endothelial cells (ECs) [66]. Genetic modification of PMSCs to mimic the function of ECs has been reported. Further in vivo research also confirmed that engrafted PMSCs were positive for mature endothelial marker vWF, suggesting their differentiation into ECs [67]. Meanwhile, endothelial dysfunction precedes, and initial denudations are major contributing factors to the pathogenesis and clinical course of CADs by promoting the mechanisms of vasoconstriction, neointima formation, thrombosis, and atherosclerosis[28, 68]. These results proved that MSCs might be a viable source for the regeneration of ECs in therapeutic angiogenesis application. Recently, study shows that the diminished production or availability of nitric oxide (NO) with an imbalance of endothelium-derived relaxing and contracting factors might cause the decrease of ECs [69]. NO is normally released by ECs to offer feasible effects of angiogenic on the native coronary arterial, as well as prevent thrombogenesis and improve self-patency [70]. It has been revealed that NO signaling is a necessary regulator of proangiogenic in PMSCs, and this finding offers exciting insights for PMSC-based clinical therapy. Therefore, the production and regulation of NO are the key factors of the therapy in endothelial dysfunction. Previous studies showed that MSCs derived from human Wharton's jelly were negative for typical endothelial-related markers like pericyte markers (like NG2- and platelet-derived growth factor receptor β), VE-cadherin, vWF, and eNOS [71], while the expression of these markers was increased significantly after endothelial differentiation of PMSCs [72], which has been proved in MSCs from other sources as well. On the other hand, MSCs could be utilized as a useful tool in CADs therapy. In a study, MSCs with transfection of eNOS gene could produce significant amounts of NO, which indicate the possibility of stem cell production of an engineered vascular [73]. This hybrid vascular prosthesis is expected to provide a therapeutic advantage by extended production of NO from the inner surfaces. By considering these phenomenon, what we are interested in is how the NO signaling impacts the behavior of PMSCs during the process of differentiating them into ECs [74]. MSC-conditioned media have been shown to contain soluble factors that exert profound biological effects. These studies indicate that MSCs may exert their function in a paracrine fashion.

In addition to the endothelial differentiation potential of PMSCs, studies in recent years are mainly focused on the role of PMSC paracrine in diseases therapy. Numerous numbers of studies have revealed that the mechanism underlying the benefit achieved by transplanting PMSCs appears to relate to their paracrine effect, numbers of angiogenic enzymes and cytokines secreted by PMSCs, including VEGFR-1 (FLT-1) and VEGFR-2 (KDR), along with the expression of TGF-β, hepatocyte growth factor (HGF), and matrix metalloproteinases (MMPs) [75, 76]. Besides, angiogenic factors such as EGF, VEGF, IL-8, IGF-1, and angiopoietin have also been demonstrated a direct link with PMSCs [77].

Therefore, based on their proangiogenic properties, PMSCs are attractive in various clinical trials. Du et al., for example, reported that MSCs derived from human placenta chorionic villi are able to contribute to proangiogenic including promote vessel maturation and stabilization via paracrine effect [78]. Vascular cell adhesion molecule 1 (VCAM-1), also known as cluster of differentiation 106 (CD106) or vascular cell adhesion protein 1, is extensively expressed on endothelial cells and some stromal cells. Soluble VCAM-1 (sVCAM-1) has shown evidence of mediating angiogenesis in rat cornea [79], and the sVCAM-1/α4 integrin pathway plays a central role in inflammatory stimuli-induced angiogenesis. VCAM-1$^+$-MSC subpopulation displayed a unique angiogenic property and exerted enhanced therapeutic efficacy on regeneration after ischemia in comparison with the VCAM-1$^-$-MSC subpopulation. Angiogenic genes were highly expressed in VCAM-1$^+$CV-MSCs. The gene profile result indicated that the CXC chemokine family, MMPs, several growth factors (VEGFA, HGF, bFGF, TGFβ1, and TGFβ3), hypoxia-induced factor-1α (HIF-1α), and angiopoietin-like protein 2 (ANGPTL2) were highly expressed than VCAM-1$^-$ -MSCs. Meanwhile, the expressions of lymphangiogenesis-related VEGF-C and intercellular cell adhesion molecule-1 (ICAM-1) were lower in VCAM-1$^+$-MSCs. Results showed that HGF, angiogenin (ANG), VEGFA, TGFβ, and bFGF expressed in VCAM-1$^+$-MSCs were upregulated compared with VCAM-1$^-$-MSCs. Besides, VCAM-1$^+$CV-MSC conditioned mediums are able to effectively promote endothelial cell proliferation and migration.

Moreover, as other fate tissue, human amnion or amniotic membrane has emerged as an alternative source of MSCs, which are easily to obtain and can be expanded in vitro for several passages. hAMSCs possess a lot of properties including immune-privileged, antimicrobial, and antitumorigenic, which are considered to be suitable for cell transplantation, regenerative medicine, and diseases therapy. As a promising clinically tool, hAMSCs were widely studied due to these properties and to their pluripotent capacity. Besides, numerous growth factors has been determined that hAMSCs are capable of secreting, which can support angiogenesis, tissue remodeling, and decrease inflammation [80, 81]. This expression of cytokines related to different cell types including VEGF, PDGF, and TGF-β [82, 83]. VEGF is a major factor in the process of endothelial differentiation of MSCs [84–86]. A study implied that VEGF induces human MSCs differentiation into ECs by Rho/ROCK signaling pathway, which could suppress the upregulation of tube formation, migration, and proliferation of ECs [87, 88]. Moreover, PDGF could induce endothelial differentiation of MSCs [89], whereas TGF-β signals are SMC specification [90]. It has revealed that TGF-β could increase the expression of smooth muscle β-actin while decrease the expression of gelsolin according to proteomic profiling analysis [91], which will lead to more focus and in-depth research on the influences of TGF-β on endothelial differentiation of MSCs.

Considering endothelial differentiation ability and paracrine effect of MSCs, optimized biomaterials, gene transfection, and culture conditions can be used to enhance therapeutic effect of MSCs. Tissue-engineered vascular grafts (TEVGs) are commonly used in the clinic for bypass surgery and dialysis [92, 93] but have high failure rates because of neointima formation and thrombosis. Common models of living vascular grafts have been developed to some different types like matrix alone, cells alone, or the hybrid vascular grafts combined with cells, matrix, and soluble factors [94]. In clinical application, TEVGs are supposed to have mechanically durable structure and anti-thrombogenic ability. Most approaches of TEVGs fabrication rely on the forms of scaffold to provide mechanical integrity upon implantation to the arterial circulation. In vivo transplantation of acellular conduits or MSC-seeded as interpositional grafts suggested that MSCs could block the inflammatory responses induced by nanofibrous scaffolds [95]. The combined application of polymer grafts and MSCs for developing vascular conduits confers TEVGs with mechanical strength, good tissue compatibility, and ability of remodeling. Thus, seeding with genetically modified MSCs vascular grafts improved long-term small caliber graft patency [96]. The biosynthetic grafts may provide a novel therapeutic modality for patients with cardiovascular disease.

The angiogenic cytokines are crucial factors of angiogenesis, which participate in cell proliferation, migration, differentiation, and communication. However, plenty of studies show that the communication between MSCs and other cells is not only through cytokines but also exerts their function via exosomes. Exosomes are small particles dynamically composed by lipids, proteins, and a large amount of mRNA as well as microRNA that can convey biological materials to surrounding cells [97–101], and these exosomes can be functional for recipient cells. Results suggest that exosomes may serve as an essential mediator of angiogenesis by transferring genetic materials and angiogenic molecules. Exosomes with miRNAs can be secreted by MSCs in culture [102], and the altered expression of miRNAs has been correlated with CVDs [103, 104]. Recently, the discovery of miRNA/vesicle-mediated communication between endothelial cells and cardiovascular cells was confirmed [105, 106]. Moreover, miRNAs have been detected in the supernatant of human MSCs that may be associated with exosomes [107]. These exosomes were injected into a rat model of ischemia/reperfusion and revealed cardioprotective effect [108]. The result indicated that exosomes secreted by MSCs might possess therapeutic potential. Although significant progresses have been made in CVDs therapy, further investigations are needed on the exploration of MSCs-derived exosomes. Future clinical use of exosomes in diagnosis, monitoring disease progress, and treatment and monitoring treatment efficacy is promising.

11.6 Conclusion and Future Perspectives

In the past, the standards of care are limited to extenuating loss and preventing further damage, rather than facilitating improvement and regeneration of heart tissue. Currently, cell-based therapies using autologous stem/progenitor cells from fetal tissue have been evaluated in preclinical models, and several clinical trials have demonstrated that these cells can be used safely in humans [109]. Many kind of stem/progenitor cells were utilizing include human embryonic stem cells (hESCs), induced pluripotent stem cells (IPSCs), mesenchymal stem cells (MSCs), myocyte progenitor cells, and myocardium stem cells in cell therapy, however, the cultivation of stable cardiomyocytes and ECs to clinically relevant scales have not been achieved.

Given the lack of successful treatments available for CADs patients, there is an urgent need to develop therapies to treat the endothelial dysfunction and myocardial ischemia that are involved in CADs. As outlined in this review, the involvement of the stem/progenitor cell therapy with perinatal stem cells is promising and has thus far been the primary tool for current clinical treatments. However, as we have highlighted, the ethical difficulties and technical hurdles of ESCs have limited their application despite their unlimited self-renewal and have the potential to give rise to all tissue types. Meanwhile, perinatal stem cells offer the ability to repair the injury tissue caused by ischemia in humans, which might represent a valuable alternative.

Although many clinical trials have demonstrated the undisputed safety and viability of stem cell therapy, there are still many important issues, including the best cell type, delivery route, and time point of administration. Therefore, additional multicenter studies are under way to try to address these issues and reveal the true benefits of clinical outcomes. These cells have demonstrated the ability to differentiate into functional organ-specific cell types and are implanted in damaged tissues to restore disease or post-injury function. It is noteworthy that increasing the number of introduced cells remaining in the damaged myocardium is a key challenge in developing more effective treatments. In addition, modulating this response with cellular therapy may have a significant impact on healing and tissue regeneration where inflammation plays an important role in disease and injury. These cells have important immunomodulatory activity that appears to be maintained in differentiated populations as well as more promising for cell therapy applications in cardiovascular disease.

References

1. Du W, Tao H, Zhao S, He ZX, Li Z. Translational applications of molecular imaging in cardiovascular disease and stem cell therapy. Biochimie. 2015;116:43–51. https://doi.org/10.1016/j.biochi.2015.06.021.
2. Nair V, Madan H, Sofat S, Ganguli P, Jacob MJ, Datta R, et al. Efficacy of stem cell in improvement of left ventricular function in acute myocardial infarction—MI3 Trial. Indian J Med Res. 2015;142(2):165–74. https://doi.org/10.4103/0971-5916.164245.
3. Roura S, Pujal JM, Bayes-Genis A. Umbilical cord blood for cardiovascular cell therapy: from promise to fact. Ann N Y Acad Sci. 2012;1254:66–70. https://doi.org/10.1111/j.1749-6632.2012.06515.x.
4. Roura S, Soler-Botija C, Bago JR, Llucia-Valldeperas A, Fernandez MA, Galvez-Monton C, et al. Postinfarction functional recovery driven by a three-dimensional engineered fibrin patch composed of human umbilical cord blood-derived mesenchymal stem cells. Stem Cells Transl Med. 2015;4(8):956–66. https://doi.org/10.5966/sctm.2014-0259.
5. Galvez-Monton C, Prat-Vidal C, Roura S, Soler-Botija C, Bayes-Genis A. Update: innovation in cardiology (IV). Cardiac tissue engineering and the bioartificial heart. Rev Esp Cardiol (Engl Ed). 2013;66(5):391–9. https://doi.org/10.1016/j.rec.2012.11.012.
6. Wang L, Wang Y, Li Z. Nanoparticle-based tumor theranostics with molecular imaging. Curr Pharm Biotechnol. 2013;14(7):683–92.
7. Si JW, Wang XD, Shen SG. Perinatal stem cells: a promising cell resource for tissue engineering of craniofacial bone. World J Stem Cells. 2015;7(1):149–59. https://doi.org/10.4252/wjsc.v7.i1.149.
8. Bajek A, Olkowska J, Gurtowska N, Kloskowski T, Walentowicz-Sadlecka M, Sadlecki P, et al. Human amniotic-fluid-derived stem cells: a unique source for regenerative medicine. Expert Opin Biol Ther. 2014;14(6):831–9. https://doi.org/10.1517/14712598.2014.898749.
9. Asahara T, Murohara T, Sullivan A, Silver M, van der Zee R, Li T, et al. Isolation of putative progenitor endothelial cells for angiogenesis. Science. 1997;275(5302):964–7.
10. Guillevic O, Ferratge S, Pascaud J, Driancourt C, Boyer-Di-Ponio J, Uzan G. A novel molecular and functional stemness signature assessing human cord blood-derived endothelial progenitor cell immaturity. PLoS One. 2016;11(4):e0152993. https://doi.org/10.1371/journal.pone.0152993.
11. Kim H, Huang L, Critser PJ, Yang Z, Chan RJ, Wang L, et al. Notch ligand delta-like 1 promotes in vivo vasculogenesis in human cord blood-derived endothelial colony forming cells. Cytotherapy. 2015;17(5):579–92. https://doi.org/10.1016/j.jcyt.2014.12.003.
12. Samuel R, Duda DG, Fukumura D, Jain RK. Vascular diseases await translation of blood vessels engineered from stem cells. Sci Transl Med. 2015;7(309):309rv6. https://doi.org/10.1126/scitranslmed.aaa1805.
13. Ratajczak J, Kucia M, Mierzejewska K, Marlicz W, Pietrzkowski Z, Wojakowski W, et al. Paracrine proangiopoietic effects of human umbilical cord blood-derived purified CD133+ cells—implications for stem cell therapies in regenerative medicine. Stem Cells Dev. 2013;22(3):422–30. https://doi.org/10.1089/scd.2012.0268.
14. Wils J, Favre J, Bellien J. Modulating putative endothelial progenitor cells for the treatment of endothelial dysfunction and cardiovascular complications in diabetes. Pharmacol Ther. 2017;170:98–115. https://doi.org/10.1016/j.pharmthera.2016.10.014.
15. Sandler VM, Lis R, Liu Y, Kedem A, James D, Elemento O, et al. Reprogramming human endothelial cells to haematopoietic cells requires vascular induction. Nature. 2014;511(7509):312–8. https://doi.org/10.1038/nature13547.
16. Yoon JW, Jang IH, Heo SC, Kwon YW, Choi EJ, Bae K-H, et al. Isolation of foreign material-free endothelial progenitor cells using CD31 aptamer and therapeutic application for ischemic injury. PLoS One. 2015;10(7):e0131785. https://doi.org/10.1371/journal.pone.0131785.
17. Fadini GP, Losordo D, Dimmeler S. Critical reevaluation of endothelial progenitor cell phenotypes for therapeutic and diagnostic use. Circ Res. 2012;110(4):624–37. https://doi.org/10.1161/circresaha.111.243386.
18. Iskander A, Knight RA, Zhang ZG, Ewing JR, Shankar A, Varma NR, et al. Intravenous administration of human umbilical cord blood-derived AC133+ endothelial progenitor cells in rat stroke model reduces infarct volume: magnetic resonance imaging and histological findings. Stem Cells Transl Med. 2013;2(9):703–14. https://doi.org/10.5966/sctm.2013-0066.
19. Raval Z, Losordo DW. Cell therapy of peripheral arterial disease: from experimental findings to clinical trials. Circ Res. 2013;112(9):1288–302. https://doi.org/10.1161/circresaha.113.300565.
20. Segers VFM, Lee RT. Stem-cell therapy for cardiac disease. Nature. 2008;451(7181):937–42. https://doi.org/10.1038/nature06800.
21. Sun S, Liu Z, Zhou H, Li G, Liu M, Yao J, et al. The role of fucosylation in the promotion of endothelial progenitor cells in neovascularization and bone repair. Biomaterials. 2014;35(12):3777–85. https://doi.org/10.1016/j.biomaterials.2014.01.025.
22. Rehman J, Li J, Orschell CM, March KL. Peripheral blood "endothelial progenitor cells" are derived from monocyte/macrophages and secrete angiogenic growth factors. Circulation. 2003;107(8):1164–9.
23. Urbich C, Aicher A, Heeschen C, Dernbach E, Hofmann WK, Zeiher AM, et al. Soluble factors released by endothelial progenitor cells promote migration of endothelial cells and cardiac resident progenitor cells. J Mol Cell Cardiol. 2005;39(5):733–42. https://doi.org/10.1016/j.yjmcc.2005.07.003.
24. Yin Y, Liu H, Wang F, Li L, Deng M, Huang L, et al. Transplantation of cryopreserved human umbili-

cal cord blood-derived endothelial progenitor cells induces recovery of carotid artery injury in nude rats. Stem Cell Res Ther. 2015;6:37. https://doi.org/10.1186/s13287-015-0022-4.
25. Li X, Chen C, Wei L, Li Q, Niu X, Xu Y, et al. Exosomes derived from endothelial progenitor cells attenuate vascular repair and accelerate reendothelialization by enhancing endothelial function. Cytotherapy. 2016;18(2):253–62. https://doi.org/10.1016/j.jcyt.2015.11.009.
26. Santini MP, Forte E, Harvey RP, Kovacic JC. Developmental origin and lineage plasticity of endogenous cardiac stem cells. Development. 2016;143(8):1242–58. https://doi.org/10.1242/dev.111591.
27. Peters EB, Christoforou N, Leong KW, Truskey GA, West JL. Poly(ethylene glycol) hydrogel scaffolds containing cell-adhesive and protease-sensitive peptides support microvessel formation by endothelial progenitor cells. Cell Mol Bioeng. 2016;9(1):38–54. https://doi.org/10.1007/s12195-015-0423-6.
28. Pias-Peleteiro J, Perez-Mato M, Lopez-Arias E, Rodriguez-Yanez M, Blanco M, Campos F, et al. Increased endothelial progenitor cell levels are associated with good outcome in intracerebral hemorrhage. Sci Rep. 2016;6:28724. https://doi.org/10.1038/srep28724.
29. Di Trapani M, Bassi G, Fontana E, Giacomello L, Pozzobon M, Guillot PV, et al. Immune regulatory properties of CD117(pos) amniotic fluid stem cells vary according to gestational age. Stem Cells Dev. 2015;24(1):132–43. https://doi.org/10.1089/scd.2014.0234.
30. Zani A, Cananzi M, Fascetti-Leon F, Lauriti G, Smith VV, Bollini S, et al. Amniotic fluid stem cells improve survival and enhance repair of damaged intestine in necrotising enterocolitis via a COX-2 dependent mechanism. Gut. 2014;63(2):300–9. https://doi.org/10.1136/gutjnl-2012-303735.
31. De Coppi P, Bartsch G Jr, Siddiqui MM, Xu T, Santos CC, Perin L, et al. Isolation of amniotic stem cell lines with potential for therapy. Nat Biotechnol. 2007;25(1):100–6. https://doi.org/10.1038/nbt1274.
32. Petsche Connell J, Camci-Unal G, Khademhosseini A, Jacot JG. Amniotic fluid-derived stem cells for cardiovascular tissue engineering applications. Tissue Eng Part B Rev. 2013;19(4):368–79. https://doi.org/10.1089/ten.TEB.2012.0561.
33. Behfar A, Crespo-Diaz R, Terzic A, Gersh BJ. Cell therapy for cardiac repair—lessons from clinical trials. Nat Rev Cardiol. 2014;11(4):232–46. https://doi.org/10.1038/nrcardio.2014.9.
34. Lloyd-Griffith C, Duffy GP, O'Brien FJ. Investigating the effect of hypoxic culture on the endothelial differentiation of human amniotic fluid-derived stem cells. J Anat. 2015;227(6):767–80. https://doi.org/10.1111/joa.12283.
35. Mirabella T, Cilli M, Carlone S, Cancedda R, Gentili C. Amniotic liquid derived stem cells as reservoir of secreted angiogenic factors capable of stimulating neo-arteriogenesis in an ischemic model. Biomaterials. 2011;32(15):3689–99. https://doi.org/10.1016/j.biomaterials.2011.01.071.
36. Schiavo AA, Franzin C, Albiero M, Piccoli M, Spiro G, Bertin E, et al. Endothelial properties of third-trimester amniotic fluid stem cells cultured in hypoxia. Stem Cell Res Ther. 2015;6:209. https://doi.org/10.1186/s13287-015-0204-0.
37. Leprince D, Gegonne A, Coll J, de Taisne C, Schneeberger A, Lagrou C, et al. A putative second cell-derived oncogene of the avian leukaemia retrovirus E26. Nature. 1983;306(5941):395–7.
38. Nunn MF, Seeburg PH, Moscovici C, Duesberg PH. Tripartite structure of the avian erythroblastosis virus E26 transforming gene. Nature. 1983;306(5941):391–5.
39. Ginsberg M, Schachterle W, Shido K, Rafii S. Direct conversion of human amniotic cells into endothelial cells without transitioning through a pluripotent state. Nat Protoc. 2015;10(12):1975–85. https://doi.org/10.1038/nprot.2015.126.
40. Zhang R, Lu Y, Li J, Wang J, Liu C, Gao F, et al. Glial cell line-derived neurotrophic factor induced the differentiation of amniotic fluid-derived stem cells into vascular endothelial-like cells in vitro. J Mol Histol. 2016;47(1):9–19. https://doi.org/10.1007/s10735-015-9649-9.
41. Khademhosseini A, Langer R. A decade of progress in tissue engineering. Nat Protoc. 2016;11(10):1775–81. https://doi.org/10.1038/nprot.2016.123.
42. Harrison RH, St-Pierre JP, Stevens MM. Tissue engineering and regenerative medicine: a year in review. Tissue Eng Part B Rev. 2014;20(1):1–16. https://doi.org/10.1089/ten.TEB.2013.0668.
43. Yeh YC, Lee WY, Yu CL, Hwang SM, Chung MF, Hsu LW, et al. Cardiac repair with injectable cell sheet fragments of human amniotic fluid stem cells in an immune-suppressed rat model. Biomaterials. 2010;31(25):6444–53. https://doi.org/10.1016/j.biomaterials.2010.04.069.
44. Huang CC, Wei HJ, Yeh YC, Wang JJ, Lin WW, Lee TY, et al. Injectable PLGA porous beads cellularized by hAFSCs for cellular cardiomyoplasty. Biomaterials. 2012;33(16):4069–77. https://doi.org/10.1016/j.biomaterials.2012.02.024.
45. Guan X, Delo DM, Atala A, Soker S. In vitro cardiomyogenic potential of human amniotic fluid stem cells. J Tissue Eng Regen Med. 2011;5(3):220–8. https://doi.org/10.1002/term.308.
46. Peng SY, Chou CJ, Cheng PJ, Tseng TY, Cheng WT, Shaw SW, et al. Intramuscular transplantation of pig amniotic fluid-derived progenitor cells has therapeutic potential in a mouse model of myocardial infarction. Cell Transplant. 2015;24(6):1003–12. https://doi.org/10.3727/096368914x680109.
47. James D, Nam HS, Seandel M, Nolan D, Janovitz T, Tomishima M, et al. Expansion and maintenance of human embryonic stem cell-derived endothelial cells by TGFbeta inhibition is Id1 dependent.

Nat Biotechnol. 2010;28(2):161–6. https://doi.org/10.1038/nbt.1605.

48. Ginsberg M, James D, Ding BS, Nolan D, Geng F, Butler JM, et al. Efficient direct reprogramming of mature amniotic cells into endothelial cells by ETS factors and TGFbeta suppression. Cell. 2012;151(3):559–75. https://doi.org/10.1016/j.cell.2012.09.032.
49. Strom SC, Gramignoli R. Human amnion epithelial cells expressing HLA-G as novel cell-based treatment for liver disease. Hum Immunol. 2016;77(9):734–9. https://doi.org/10.1016/j.humimm.2016.07.002.
50. Silini AR, Cargnoni A, Magatti M, Pianta S, Parolini O. The long path of human placenta, and its derivatives, in regenerative medicine. Front Bioeng Biotechnol. 2015;3:162. https://doi.org/10.3389/fbioe.2015.00162.
51. Gramignoli R, Srinivasan RC, Kannisto K, Strom SC. Isolation of human amnion epithelial cells according to current good manufacturing procedures. Curr Protoc Stem Cell Biol. 2016;37:1e.10.1–1e.10.13. https://doi.org/10.1002/cpsc.2.
52. Miki T, Strom SC. Amnion-derived pluripotent/multipotent stem cells. Stem Cell Rev. 2006;2(2):133–42. https://doi.org/10.1007/s12015-006-0020-0.
53. Murphy SV, Kidyoor A, Reid T, Atala A, Wallace EM, Lim R. Isolation, cryopreservation and culture of human amnion epithelial cells for clinical applications. J Vis Exp. 2014;(94). https://doi.org/10.3791/52085.
54. Zhu D, Muljadi R, Chan ST, Vosdoganes P, Lo C, Mockler JC, et al. Evaluating the impact of human amnion epithelial cells on angiogenesis. Stem Cells Int. 2016;2016:4565612. https://doi.org/10.1155/2016/4565612.
55. Fang CH, Jin J, Joe JH, Song YS, So BI, Lim SM, et al. In vivo differentiation of human amniotic epithelial cells into cardiomyocyte-like cells and cell transplantation effect on myocardial infarction in rats: comparison with cord blood and adipose tissue-derived mesenchymal stem cells. Cell Transplant. 2012;21(8):1687–96. https://doi.org/10.3727/096368912x653039.
56. Koo H, Choi M, Kim E, Hahn SK, Weissleder R, Yun SH. Bioorthogonal click chemistry-based synthetic cell glue. Small. 2015;11(48):6458–66. https://doi.org/10.1002/smll.201502972.
57. Roy R, Kukucka M, Messroghli D, Kunkel D, Brodarac A, Klose K, et al. Epithelial-to-mesenchymal transition enhances the cardioprotective capacity of human amniotic epithelial cells. Cell Transplant. 2015;24(6):985–1002. https://doi.org/10.3727/096368913x675151.
58. Zhao B, Liu JQ, Zheng Z, Zhang J, Wang SY, Han SC, et al. Human amniotic epithelial stem cells promote wound healing by facilitating migration and proliferation of keratinocytes via ERK, JNK and AKT signaling pathways. Cell Tissue Res. 2016;365(1):85–99. https://doi.org/10.1007/s00441-016-2366-1.
59. Zheng X, Jiang Z, Zhou A, Yu L, Quan M, Cheng H. Pathologic changes of wound tissue in rats with stage III pressure ulcers treated by transplantation of human amniotic epithelial cells. Int J Clin Exp Pathol. 2015;8(10):12284–91.
60. Huber-Lang M, Wiegner R, Lampl L, Brenner RE. Mesenchymal stem cells after polytrauma: actor and target. Stem Cells Int. 2016;2016:6289825. https://doi.org/10.1155/2016/6289825.
61. Huang PP, Yang XF, Li SZ, Wen JC, Zhang Y, Han ZC. Randomised comparison of G-CSF-mobilized peripheral blood mononuclear cells versus bone marrow-mononuclear cells for the treatment of patients with lower limb arteriosclerosis obliterans. Thromb Haemost. 2007;98(6):1335–42.
62. Du WJ, Chi Y, Yang ZX, Li ZJ, Cui JJ, Song BQ, et al. Heterogeneity of proangiogenic features in mesenchymal stem cells derived from bone marrow, adipose tissue, umbilical cord, and placenta. Stem Cell Res Ther. 2016;7(1):163. https://doi.org/10.1186/s13287-016-0418-9.
63. Tao H, Han Z, Han ZC, Li Z. Proangiogenic features of mesenchymal stem cells and their therapeutic applications. Stem Cells Int. 2016;2016:1314709. https://doi.org/10.1155/2016/1314709.
64. Li F, Huang Q, Chen J, Peng Y, Roop DR, Bedford JS, et al. Apoptotic cells activate the "phoenix rising" pathway to promote wound healing and tissue regeneration. Sci Signal. 2010;3(110):ra13. https://doi.org/10.1126/scisignal.2000634.
65. Krawiec JT, Weinbaum JS, St Croix CM, Phillippi JA, Watkins SC, Rubin JP, et al. A cautionary tale for autologous vascular tissue engineering: impact of human demographics on the ability of adipose-derived mesenchymal stem cells to recruit and differentiate into smooth muscle cells. Tissue Eng Part A. 2015;21(3–4):426–37. https://doi.org/10.1089/ten.TEA.2014.0208.
66. Oswald J, Boxberger S, Jørgensen B, Feldmann S, Ehninger G, Bornhäuser M, et al. Mesenchymal stem cells can be differentiated into endothelial cells in vitro. Stem Cells. 2004;22(3):377–84. https://doi.org/10.1634/stemcells.22-3-377.
67. Thakker R, Yang P. Mesenchymal stem cell therapy for cardiac repair. Curr Treat Options Cardiovasc Med. 2014;16(7):323. https://doi.org/10.1007/s11936-014-0323-4.
68. Verma S, Anderson TJ. Fundamentals of endothelial function for the clinical cardiologist. Circulation. 2002;105(5):546–9.
69. Rammos C, Zeus T, Balzer J, Kubatz L, Hendgen-Cotta UB, Veulemans V, et al. Percutaneous mitral valve repair in mitral regurgitation reduces cell-free hemoglobin and improves endothelial function. PLoS One. 2016;11(3):e0151203. https://doi.org/10.1371/journal.pone.0151203.
70. Kanki-Horimoto S, Horimoto H, Mieno S, Kishida K, Watanabe F, Furuya E, et al. Implantation of mesenchymal stem cells overexpressing endothelial nitric oxide synthase improves right ventricular

impairments caused by pulmonary hypertension. Circulation. 2006;114(1 suppl):I-181–5. https://doi.org/10.1161/circulationaha.105.001487.
71. Ball SG, Shuttleworth CA, Kielty CM. Vascular endothelial growth factor can signal through platelet-derived growth factor receptors. J Cell Biol. 2007;177(3):489–500.
72. Aguilera V, Briceno L, Contreras H, Lamperti L, Sepulveda E, Diaz-Perez F, et al. Endothelium trans differentiated from Wharton's jelly mesenchymal cells promote tissue regeneration: potential role of soluble pro-angiogenic factors. PLoS One. 2014;9(11):e111025. https://doi.org/10.1371/journal.pone.0111025.
73. McIlhenny S, Zhang P, Tulenko T, Comeau J, Fernandez S, Policha A, et al. eNOS transfection of adipose-derived stem cells yields bioactive nitric oxide production and improved results in vascular tissue engineering. J Tissue Eng Regen Med. 2015;9(11):1277–85. https://doi.org/10.1002/term.1645.
74. Gomes SA, Rangel EB, Premer C, Dulce RA, Cao Y, Florea V, et al. S-nitrosoglutathione reductase (GSNOR) enhances vasculogenesis by mesenchymal stem cells. Proc Natl Acad Sci U S A. 2013;110(8):2834–9. https://doi.org/10.1073/pnas.1220185110.
75. Lee MY, Huang JP, Chen YY, Aplin JD, Wu YH, Chen CY, et al. Angiogenesis in differentiated placental multipotent mesenchymal stromal cells is dependent on integrin alpha5beta1. PLoS One. 2009;4(10):e6913. https://doi.org/10.1371/journal.pone.0006913.
76. Nazarov I, Lee JW, Soupene E, Etemad S, Knapik D, Green W, et al. Multipotent stromal stem cells from human placenta demonstrate high therapeutic potential. Stem Cells Transl Med. 2012;1(5):359–72. https://doi.org/10.5966/sctm.2011-0021.
77. Kim SW, Zhang HZ, Guo L, Kim JM, Kim MH. Amniotic mesenchymal stem cells enhance wound healing in diabetic NOD/SCID mice through high angiogenic and engraftment capabilities. PLoS One. 2012;7(7):e41105. https://doi.org/10.1371/journal.pone.0041105.
78. Du W, Li X, Chi Y, Ma F, Li Z, Yang S, et al. VCAM-1+ placenta chorionic villi-derived mesenchymal stem cells display potent pro-angiogenic activity. Stem Cell Res Ther. 2016;7:49. https://doi.org/10.1186/s13287-016-0297-0.
79. Koch AE, Halloran MM, Haskell CJ, Shah MR, Polverini PJ. Angiogenesis mediated by soluble forms of E-selectin and vascular cell adhesion molecule-1. Nature. 1995;376(6540):517–9. https://doi.org/10.1038/376517a0.
80. Alcaraz A, Mrowiec A, Insausti CL, Bernabe-Garcia A, Garcia-Vizcaino EM, Lopez-Martinez MC, et al. Amniotic membrane modifies the genetic program induced by TGFss, stimulating keratinocyte proliferation and migration in chronic wounds. PLoS One. 2015;10(8):e0135324. https://doi.org/10.1371/journal.pone.0135324.
81. Koob TJ, Lim JJ, Massee M, Zabek N, Rennert R, Gurtner G, et al. Angiogenic properties of dehydrated human amnion/chorion allografts: therapeutic potential for soft tissue repair and regeneration. Vasc Cell. 2014;6:10. https://doi.org/10.1186/2045-824x-6-10.
82. Gholizadeh-Ghaleh Aziz S, Pashaei-Asl F, Fardyazar Z, Pashaiasl M. Isolation, characterization, cryopreservation of human amniotic stem cells and differentiation to osteogenic and adipogenic cells. PLoS One. 2016;11(7):e0158281. https://doi.org/10.1371/journal.pone.0158281.
83. Tamagawa T, Oi S, Ishiwata I, Ishikawa H, Nakamura Y. Differentiation of mesenchymal cells derived from human amniotic membranes into hepatocyte-like cells in vitro. Hum Cell. 2007;20(3):77–84. https://doi.org/10.1111/j.1749-0774.2007.00032.x.
84. Xu Y, Liu L, Zhang L, Fu S, Hu Y, Wang Y, et al. Efficient commitment to functional CD34+ progenitor cells from human bone marrow mesenchymal stem-cell-derived induced pluripotent stem cells. PLoS One. 2012;7(4):e34321. https://doi.org/10.1371/journal.pone.0034321.
85. Chen MY, Lie PC, Li ZL, Wei X. Endothelial differentiation of Wharton's jelly-derived mesenchymal stem cells in comparison with bone marrow-derived mesenchymal stem cells. Exp Hematol. 2009;37(5):629–40. https://doi.org/10.1016/j.exphem.2009.02.003.
86. Wang M, Su Y, Sun H, Wang T, Yan G, Ran X, et al. Induced endothelial differentiation of cells from a murine embryonic mesenchymal cell line C3H/10T1/2 by angiogenic factors in vitro. Differentiation. 2010;79(1):21–30. https://doi.org/10.1016/j.diff.2009.08.002.
87. He QW, Xia YP, Chen SC, Wang Y, Huang M, Huang Y, et al. Astrocyte-derived sonic hedgehog contributes to angiogenesis in brain microvascular endothelial cells via RhoA/ROCK pathway after oxygen-glucose deprivation. Mol Neurobiol. 2013;47(3):976–87. https://doi.org/10.1007/s12035-013-8396-8.
88. Wang N, Zhang R, Wang SJ, Zhang CL, Mao LB, Zhuang CY, et al. Vascular endothelial growth factor stimulates endothelial differentiation from mesenchymal stem cells via Rho/myocardin-related transcription factor—a signaling pathway. Int J Biochem Cell Biol. 2013;45(7):1447–56. https://doi.org/10.1016/j.biocel.2013.04.021.
89. Lin H, Shabbir A, Molnar M, Yang J, Marion S, Canty JM Jr, et al. Adenoviral expression of vascular endothelial growth factor splice variants differentially regulate bone marrow-derived mesenchymal stem cells. J Cell Physiol. 2008;216(2):458–68. https://doi.org/10.1002/jcp.21414.
90. Wang D, Park JS, Chu JS, Krakowski A, Luo K, Chen DJ, et al. Proteomic profiling of bone marrow mesenchymal stem cells upon transforming growth factor beta1 stimulation. J Biol Chem.

2004;279(42):43725–34. https://doi.org/10.1074/jbc.M407368200.
91. Abnaof K, Mallela N, Walenda G, Meurer SK, Sere K, Lin Q, et al. TGF-beta stimulation in human and murine cells reveals commonly affected biological processes and pathways at transcription level. BMC Syst Biol. 2014;8:55. https://doi.org/10.1186/1752-0509-8-55.
92. Wang Z, Cui Y, Wang J, Yang X, Wu Y, Wang K, et al. The effect of thick fibers and large pores of electrospun poly(ε-caprolactone) vascular grafts on macrophage polarization and arterial regeneration. Biomaterials. 2014;35(22):5700–10. https://doi.org/10.1016/j.biomaterials.2014.03.078.
93. Wang K, Chen X, Pan Y, Cui Y, Zhou X, Kong D, et al. Enhanced vascularization in hybrid PCL/gelatin fibrous scaffolds with sustained release of VEGF. Biomed Res Int. 2015;2015:865076. https://doi.org/10.1155/2015/865076.
94. L'Heureux N, Paquet S, Labbe R, Germain L, Auger FA. A completely biological tissue-engineered human blood vessel. FASEB J. 1998;12(1):47–56.. http://www.ncbi.nlm.nih.gov/pubmed/9438410.
95. Hashi CK, Zhu Y, Yang GY, Young WL, Hsiao BS, Wang K, et al. Antithrombogenic property of bone marrow mesenchymal stem cells in nanofibrous vascular grafts. Proc Natl Acad Sci U S A. 2007;104(29):11915–20. https://doi.org/10.1073/pnas.0704581104.
96. Zhang ZY, Teoh SH, Hui JH, Fisk NM, Choolani M, Chan JK. The potential of human fetal mesenchymal stem cells for off-the-shelf bone tissue engineering application. Biomaterials. 2012;33(9):2656–72. https://doi.org/10.1016/j.biomaterials.2011.12.025.
97. Emanueli C, Shearn AI, Angelini GD, Sahoo S. Exosomes and exosomal miRNAs in cardiovascular protection and repair. Vasc Pharmacol. 2015;71:24–30. https://doi.org/10.1016/j.vph.2015.02.008.
98. Kim DK, Nishida H, An SY, Shetty AK, Bartosh TJ, Prockop DJ. Chromatographically isolated CD63+CD81+ extracellular vesicles from mesenchymal stromal cells rescue cognitive impairments after TBI. Proc Natl Acad Sci U S A. 2016;113(1):170–5. https://doi.org/10.1073/pnas.1522297113.
99. Phinney DG, Di Giuseppe M, Njah J, Sala E, Shiva S, St Croix CM, et al. Mesenchymal stem cells use extracellular vesicles to outsource mitophagy and shuttle microRNAs. Nat Commun. 2015;6:8472. https://doi.org/10.1038/ncomms9472.
100. Valadi H, Ekstrom K, Bossios A, Sjostrand M, Lee JJ, Lotvall JO. Exosome-mediated transfer of mRNAs and microRNAs is a novel mechanism of genetic exchange between cells. Nat Cell Biol. 2007;9(6):654–9. https://doi.org/10.1038/ncb1596.
101. Yeh YY, Ozer HG, Lehman AM, Maddocks K, Yu L, Johnson AJ, et al. Characterization of CLL exosomes reveals a distinct microRNA signature and enhanced secretion by activation of BCR signaling. Blood. 2015;125(21):3297–305. https://doi.org/10.1182/blood-2014-12-618470.
102. Waldenstrom A, Ronquist G. Role of exosomes in myocardial remodeling. Circ Res. 2014;114(2):315–24. https://doi.org/10.1161/CIRCRESAHA.114.300584.
103. Sahoo S, Losordo DW. Exosomes and cardiac repair after myocardial infarction. Circ Res. 2014;114(2):333–44. https://doi.org/10.1161/CIRCRESAHA.114.300639.
104. Zhang H, Xiang M, Meng D, Sun N, Chen S. Inhibition of myocardial ischemia/reperfusion injury by exosomes secreted from mesenchymal stem cells. Stem Cells Int. 2016;2016:4328362. https://doi.org/10.1155/2016/4328362.
105. Halkein J, Tabruyn SP, Ricke-Hoch M, Haghikia A, Nguyen NQ, Scherr M, et al. MicroRNA-146a is a therapeutic target and biomarker for peripartum cardiomyopathy. J Clin Invest. 2013;123(5):2143–54. https://doi.org/10.1172/jci64365.
106. Hergenreider E, Heydt S, Treguer K, Boettger T, Horrevoets AJ, Zeiher AM, et al. Atheroprotective communication between endothelial cells and smooth muscle cells through miRNAs. Nat Cell Biol. 2012;14(3):249–56. https://doi.org/10.1038/ncb2441.
107. Chen TS, Lai RC, Lee MM, Choo AB, Lee CN, Lim SK. Mesenchymal stem cell secretes microparticles enriched in pre-microRNAs. Nucleic Acids Res. 2010;38(1):215–24. https://doi.org/10.1093/nar/gkp857.
108. Lai RC, Arslan F, Lee MM, Sze NS, Choo A, Chen TS, et al. Exosome secreted by MSC reduces myocardial ischemia/reperfusion injury. Stem Cell Res. 2010;4(3):214–22. https://doi.org/10.1016/j.scr.2009.12.003.
109. Kim J, Shapiro L, Flynn A. The clinical application of mesenchymal stem cells and cardiac stem cells as a therapy for cardiovascular disease. Pharmacol Ther. 2015;151:8–15. https://doi.org/10.1016/j.pharmthera.2015.02.003.
110. Balbi C, Bollini S. Fetal and perinatal stem cells in cardiac regeneration: moving forward to the paracrine era. Placenta. 2017;59:96–106. https://doi.org/10.1016/j.placenta.2017.04.008.

12 Perinatal Stem Cells in Kidney Regeneration: Current Knowledge and Perspectives

Guowei Feng, Xin Yao, and Zongjin Li

Abstract

Perinatal stem cells (PSCs) are gaining increasing attention with regard to their therapeutic potential in kidney disease. Many proof-of-principle experiments have demonstrated that PSCs could improve renal recovery in the setting of acute and chronic insults. Deciphering the mechanisms involved in PSC-based therapy, mounting evidence favors the paracrine actions (e.g., immune-modulatory properties) rather than the multi-differentiating capacity of PSCs during renal restoration. In particular, PSCs have been shown to promote kidney regeneration via releasing soluble factors and extracellular vesicles into damaged renal cells. In this chapter, we summarize current understanding of the characteristics and advantages of PSCs in regenerative medicine, discuss the underlying mechanisms of their therapeutic actions, and examine the strategies for increasing protective effects of PSCs. Finally, we focus on the recent progress in the use of PSCs to alleviate kidney injury and provide suggestions for accelerating the implementation of cellular therapy to prevent and treat kidney diseases.

G. Feng · X. Yao
Key Laboratory of Cancer Prevention and Therapy, Department of Genitourinary Oncology, Tianjin Medical University Cancer Institute and Hospital, National Clinical Research Center for Cancer, Tianjin's Clinical Research Center for Cancer, Tianjin, China

Z. Li (✉)
School of Medicine and Key Laboratory of Bioactive Materials, Ministry of Education, College of Life Sciences, Nankai University, Tianjin, China
e-mail: zongjinli@nankai.edu.cn

12.1 Introduction

Acute kidney injury (AKI) is a global health problem with rapidly increasing incidence due to the growing rates of cardiovascular disease, hypertension, diabetes, and the aged population [1]. When severe and persistent, AKI is associated with high morbidity/mortality and poor quality of life. Among AKI patients who survive, long-term outcomes include the development of chronic kidney disease (CKD) and end-stage renal disease (ESRD) [2]. In this incurable condition, dialysis and kidney transplantation remain the most effective treatments; however, these approaches are limited by significant complications, shortage of donor organs, immunological rejection, and high medical costs [3]. Decades of research have identified numerous chemical and biological agents (e.g., antioxidants, hormones, cytokines, growth factors, and molecular-targeted agents) with renoprotective effects in experimental models (reviewed in [4]). Most have yet to be translated into clinical use. This is due, in great part, to the fact that pharmacological therapies target only a limited number of the pathogenetic

Z. C. Han et al. (eds.), *Perinatal Stem Cells*, https://doi.org/10.1007/978-981-13-2703-2_12

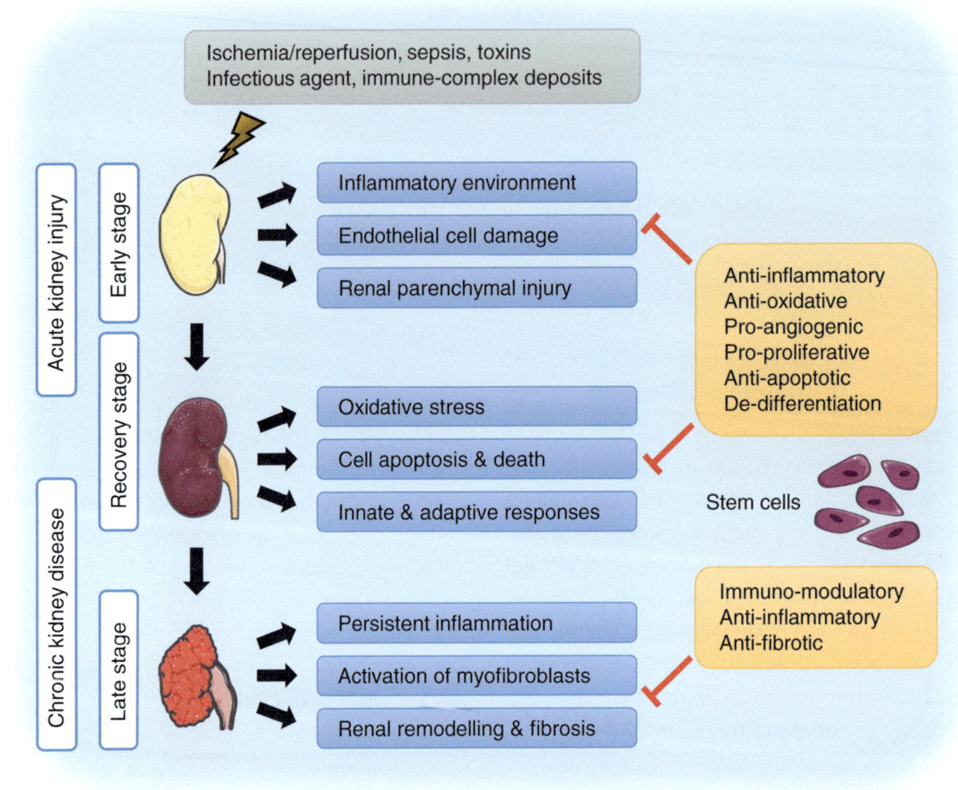

Fig. 12.1 Stem cell-based therapy targets all major components of the pathophysiology of kidney disease. Various types of insults (e.g., ischemia, endotoxin) instigate primary tissue injury that results in renal inflammation and vascular and tubular damage at early stage. The renal damage subsequently stimulates the production of pro-inflammatory cytokines and chemokines from renal cells, thereby inducing oxidative stress (the release of ROS, RNS). Sustained oxidative stress, in turn, exacerbates renal inflammation and cause substantial tissue damage. At recovery stage, regulative T cells are recruited to the site of injury, and innate or adaptive immune responses are triggered. Severe damage together with continually expressed pathogenic factors leads to progressive injury and persistent inflammation, during which stage resident or recruited macrophages release profibrotic mediators (i.e., TGF-β1, Galectin-3) in response to the inflammatory microenvironment. The secreted factors drive myofibroblast activation and accumulation, thereby promoting renal fibrosis. The pathophysiological process involved at the early and the late stages could be blocked by stem cell actions via diverse mechanisms

mechanisms. Consequently, the rising disease burden, complex disease pathogenesis, and treatment dilemma serve as substantial motivation for pursuit of novel therapeutic strategies to prevent or mitigate renal disorders.

Stem cell-based therapy has emerged as an attractive treatment alternative for kidney diseases. On one hand, stem cells possess a broad enough differentiation potential to produce the entire spectrum of mature progeny [5]. On the other hand, they can target the underlying disease process and mobilize endogenous regenerative capacity (Fig. 12.1). Although various stem cells have shown promise in renal repair, the optimal cell population remains disputed. Embryonic stem cells are recognized as a likely candidate because of their ability to differentiate into a wide range of cell types [6]. However, the ethical, technical, and safety concerns greatly limit their application. Induced pluripotent stem cells have similar pluripotency as embryonic stem cells, but their use is constrained by the risk of tumorigenesis and immunological rejection [7]. Multiple studies have therefore examined the therapeutic application of mesenchymal stem cells (MSCs) isolated from the bone marrow, adipose, and other adult tissues/organs, such as the kidney [8], whereas adult MSCs have unsatisfying availability, and their number declines with age [9]. Besides, these cells have relatively low expansive potential and restricted differentiation capability. Hence, recently considerable attention has been focused on stem cells that have intermediate properties between embryonic and adult progenitors. MSCs isolated from birth-associated tissue (i.e., fetal, cord blood, and extraembryonic tissues) are named perinatal stem cells (PSCs). PSCs are a potential therapeutic candidate for multiple diseases with advantages over both embryonic and adult stem cells [10]. First, the retrieval of PSCs is ethically acceptable and does not involve the destruction of living human embryos. Second, they do not form tumor when injected in vivo despite their multi-differentiation potential. Of note, these cells have unique advantages in the treatment of kidney injury, because they are more plastic and stable than their adult counterparts [11].

The use of PSCs as a cure for kidney diseases began in the last decade with pioneering preclinical studies describing the ability of PSCs to regenerate several renal compartments [12]. Afterwards, accumulating evidence favors the concept that PSCs protect the kidney via paracrine and endocrine activity rather than direct incorporation into renal components [13]. Indeed, extracellular vesicles (EVs) secreted by PSCs have emerged as new mediators of cell-to-cell communication that may induce target cell reprogramming and tissue regeneration via active transfer of proteins and functional mRNAs and miRNAs. Thus, in this chapter, we will first discuss the biological characteristics (including phenotype and secretome) of PSCs that are related to their therapeutic actions. Moreover, we will summarize recent insight into how PSCs confer renoprotective effects and the modulation of cell behavior to enhance therapeutic efficacy. Finally, we will describe the preclinical data on the therapeutic applications of PSCs in a wide spectrum of conditions as varied as AKI, CKD, and immune-related renal diseases and dissect the mechanisms, benefits, and limits involved.

12.2 Biological Characteristics of PSCs

12.2.1 Phenotypic Profile and Pluripotency

Various stem cell types have been identified from fetal, cord blood, and extraembryonic tissues and are collectively named PSCs. Among them, stem cells isolated from amniotic fluid (AF-MSCs), umbilical cord (UC-MSCs), umbilical cord blood (UCB-MSCs), Wharton's jelly (WJ-MSCs), and placenta (including chorionic villi, fetal membrane) are most commonly used for regenerative medicine. It should be noted that birth-associated tissue is not one single source of MSCs but rather a comprising name for several subpopulations [14]. Most of them can be further categorized into even more subpopulations. AF-MSCs are a very promising cell population with a high differentiation potential. They can differentiate into cells from all three germ layers, such as cardiomyocyte, osteoblast, chondrocyte, and insulin-producing cells [15–17]. The plasticity of AF-MSCs was corroborated by the fact that these cells were fully reprogrammed to pluripotency without ectopic factors [18]. They have higher proliferation rates than adult MSCs and express mesenchymal (CD29, CD44, CD73, CD90, and CD105) and embryonic (Oct-4, Nanog, SSEA-1/4) markers. Despite their high pluripotency, AF-MSCs display a normal karyotype in vitro and do not induce teratoma formation in vivo, which represents an advantage for their use in translational medicine [19]. UC-MSCs possess potent self-renewal and multi-differentiation properties [20]. They have greater proliferative capacity and decreased immunological reactivity as compared with bone marrow-derived MSCs. Apart from this, UC-MSCs support the expansion of hematopoietic stem cells and serve as an accessory cell population for cell engraftment. This effect could be related to their secretion of hematopoietic growth factors, including granulocyte-colony stimulating factor (G-CSF), interleukin (IL)-6, and IL-11 [21]. Like UC-MSCs, UCB-MSCs outweigh adult MSCs in terms of immunogenicity, ease of harvesting, risk of viral contamination, and potential to form tumors [22]. Wharton's jelly is the connective tissue within the umbilical cord; accordingly, WJ-MSCs share similar characteristics with UC-MSCs in terms of phenotypic properties, differentiation capacity, as well as immune modulation [23]. WJ-MSCs were also found to support hematopoiesis when used as a feeder layer for hematopoietic stem cell ex vivo expansion. Moreover, WJ-MSCs have been successfully induced to differentiate into mature cell types such as neurons and osteocyte-like cells [24, 25]. Placenta-derived MSCs share with adult MSCs many biological properties except that they expand quicker and are both less immunogenic and more immunosuppressive than their adult counterpart [26].

12.2.2 Secretome in Angiogenesis and Immunomodulation

Generally, PSCs secrete a variety of bioactive compounds to exert their therapeutic effects. According to previous reports, the secretion pattern of PSCs differs significantly from that of adult MSCs. For instance, the production of pro-angiogenic chemokines including C-X-C motif ligand (CXCL)1, CXCL5, and CXCL8 and growth factors including basic fibroblast growth factor (bFGF), platelet-derived growth factor (PDGF)-AA, and hepatocyte growth factor (HGF) by PSCs are increased compared with MSCs derived from the bone marrow and adipose tissue [27, 28]. In contrast, the level of vascular endothelial growth factor (VEGF)-A which is highly expressed by bone marrow-derived MSCs is almost undetectable in UC-MSCs, suggesting that they promoted angiogenesis by a VEGF-A-independent pathway [29]. This can be explained by the presence of soluble VEGF receptors that neutralize exogenous VEGF. In parallel, placenta-derived MSCs secreted higher levels of HGF, prostaglandin E2 (PGE2), and showed stronger pro-angiogenic bioactivities on endothelial cells in vitro compared with bone marrow-derived MSCs [30]. The immunosuppressive activity of PSCs has been regarded as an important property by which these cells can evade the host immune response which is associated with rejection of the transplant. Evidence is accumulating to indicate that the immunomodulatory action of PSCs can be attributed to their paracrine mechanism. Indoleamine 2, 3-dioxygenase (IDO) is an interferon (IFN)-γ-inducible enzyme that catalyzes the conversion of tryptophan, an essential molecule in the activation of T cells. Both AF-MSCs and UC-MSCs have been shown to secrete IDO to induce T cell apoptosis [31, 32]. Similarly, WJ-MSCs were found to have higher expression of immunosuppressive factors, for instance, IDO1 and human leukocyte antigen (HLA)-G, but lower expression of vascular cell adhesion molecule-1 as compared with adipose-derived MSCs [33]. In addition, when co-cultured with peripheral blood mononuclear cells, WJ-MSCs showed an immunosuppressive function by inhibiting the proliferative response of Th1 and Th17 cells but augmenting Th2 and regulatory T (Treg) cells. This finding is in agreement with an earlier report showing that WJ-MSCs possessed the strongest inhibitory effect on T cell proliferation when comparing to counterparts derived from adipose, the bone marrow, and placenta [34].

12.3 How Do PSCs Contribute to Kidney Regeneration?

The knowledge regarding how stem cells stimulate kidney regeneration is mounting at a fast pace. Despite the pioneering work by different groups demonstrating a critical role of mobilized stem cells of extrarenal origin (e.g., bone marrow) in renal repair [35, 36], subsequent studies challenged this notion [37, 38] and focused on the use of exogenous stem cells, such as MSCs (including PSCs) [39]. MSCs possess robust multi-lineage differentiation capacity; however, the realization that the replacement of damaged renal cells by MSCs was a very rare and late phenomenon that could not account for the effects of the treatment led to the understanding that paracrine and endocrine actions exerted by MSCs were principal mechanism responsible for the observed therapeutic effects [40].

12.3.1 Renal Commitment

Upon ischemic or toxic insults, the kidney regenerates itself via re-proliferation of uninjured tubules or stimulation of resident progenitors. However, this process is usually ineffective because the kidney is a terminally differentiated organ with relatively low proliferative potential [41]. In addition, the kidney is comprised of more than 20 cell types, and its repair is associated with regeneration of functional tissue unit rather than individual cells. Accordingly, the renal architectural recovery which is critical to overall restoration requires exogenous pluripotent stem cells. Several studies have shown that AF-MSCs were capable of replenishing renal components by direct differentiation. To recapitulate the specific niche of developing renal compartment, Perin et al. microinjected C-kit$^+$ human AF-MSCs into murine embryonic kidneys and tracked the differentiation of AF-MSCs within developing fetal kidneys [12]. After 10 days, the injected cells were found to express early renal markers such as zonula occludens-1, glial cell-derived neurotrophic factor (GDNF), and claudin. Moreover, AF-MSCs integrated into renal vesicle and C- and S-shaped bodies, suggesting that they contributed to the development of primordial nephrogenic structures. This finding was attested by another group showing AF-MSCs differentiated toward mesenchymal and epithelial lineages finally expressing markers (CD2-associated protein and NPHS2) for podocytes [42]. Further, they cultivated the labeled AF-MSCs with embryonic kidneys cells and demonstrated the induced expression of specific renal marker (paired box gene 2, calbindin, E-cadherin, laminin, and Wilms' tumor 1) in AF-MSCs. In addition, they revealed that the mammalian target of rapamycin pathway played a critical role in mediating the contribution of AF-MSCs to renal development [43]. In parallel, Perin's group has identified metanephric mesenchyma-like cells that express CD24 and OB-cadherin from amniotic fluid [44]. This cell population shares characteristics of specific kidney progenitor cells, in particular nephron precursors. Recently, Xinaris et al. mixed AF-MSCs with mouse embryonic kidney cells to establish three-dimensional chimeric organoids which form vascularized glomeruli and tubular structures [45]. After transplantation into mature rat kidneys, AF-MSCs within the organoids generated functional podocyte-like cells expressing several podocyte markers, including podocin, nestin, and α-actinin-4, and organized foot processes with intact interposed slit diaphragms. Of interest, the propensity of AF-MSCs to form podocytes was markedly enhanced by GDNF modification of stem cells. Collectively, these findings imply that progenitor cells or organoids committed to renal lineage may represent an invaluable source of cells for regenerative medicine and drug discovery purposes (Fig. 12.2).

12.3.2 Paracrine Actions

Despite early evidence supporting integration of MSCs into renal components, subsequent studies have questioned the ability of MSCs to directly transdifferentiate into resident cells and the significance of renal replenishment by MSCs [46, 47]. Transplanted MSCs were only transiently present in injured kidney, and they predominantly localized in the peritubular area. Besides, the transdifferentiation rate is too low to explain their therapeutic effects. In this context, a growing body of evidence indicates that MSCs (including PSCs) may confer beneficial effects via the release of a broad repertoire of renotrophic and bioactive proteins including growth factors, cytokines, and chemokines [11]. These bioactive factors act to promote cell proliferation, dedifferentiation, and angiogenesis and attenuate renal apoptosis, inflammation, and fibrosis, thereby facilitating kidney regeneration. Much progress has been made in defining the factors

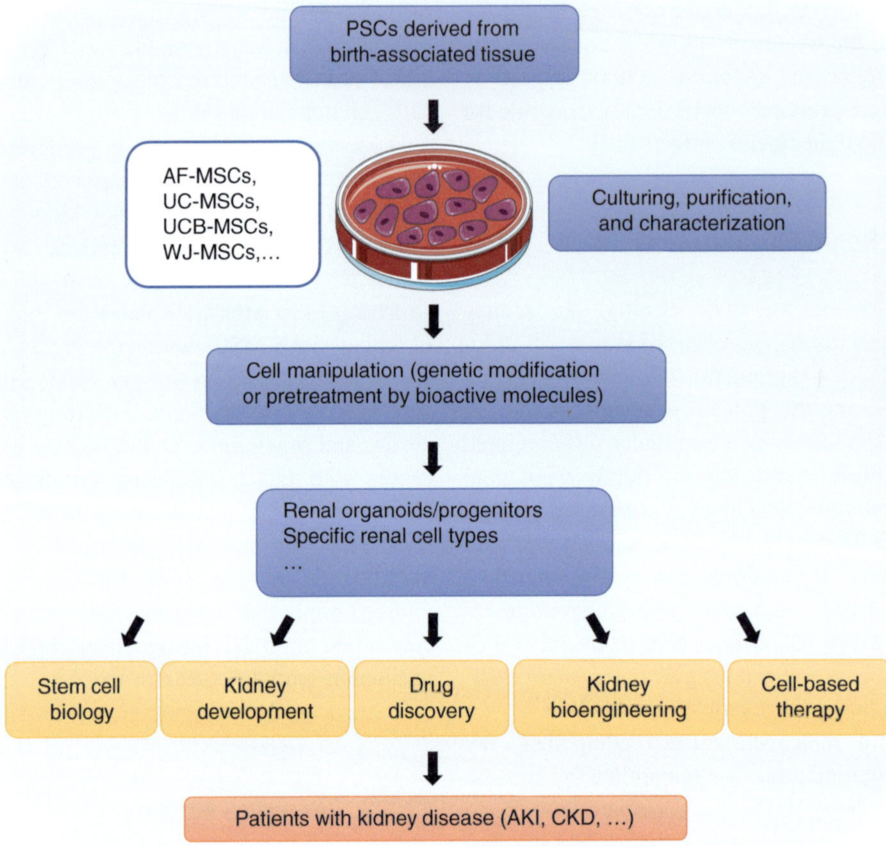

Fig. 12.2 Therapeutic application of renal components generated from PSCs in kidney disease. PSCs can be isolated from a broad range of birth-associated tissues. After in vitro expansion and characterization, these cells can be manipulated via genetic modification or pretreatment by bioactive molecules to allow directed differentiation into renal progenitors, organoids, and specific renal cell types. The generation of renal components can be used to understand stem cell biology and kidney development, to discover drugs for specific disease conditions, and to employ bioengineering and cellular therapy for kidney diseases

produced by MSCs that might be critical in driving the repair response. It has been proposed that UCB-MSCs mitigated ischemic AKI by humoral effects and secretion of VEGF, which was reinforced by the fact that VEGF inhibitor abolished the renoprotective effects of UCB-MSCs [48]. Recently, Du et al. showed that systemic delivery of WJ-MSCs ameliorated renal dysfunction in a rat model of unilateral I/R injury, which was associated with increased cell proliferation and phosphor-Akt expression and decreased cell apoptosis and fibrosis [49]. WJ-MSCs were shown to produce HGF in vitro and upregulate renal expression of HGF, IL-10, and heme oxygenase-1. The same group further demonstrated that WJ-MSCs prevented epithelial-mesenchymal transition (EMT) and inhibited renal fibrosis [50]. In search of the underlying mechanism, the researchers found that WJ-MSCs induced in vivo both native and foreign HGF syntheses in damaged renal tubules. Moreover, conditioned medium derived from WJ-MSCs stimulated in vitro HGF upregulation of both rat and human origin in rat tubular epithelial cells subjected to a hypoxic insult. These results indicate that human HGF gene transcript, released by WJ-MSCs, is delivered to tubular cells and then translated into protein [51]. Another attractive finding is that the secretion profile of PSCs appears to be dependent on external stimulation; in other words, the paracrine action may be regulated by the cross talk between PSCs and local microenvironment. For instance, treatment with hypoxia/reoxygenation in vitro markedly increased HGF gene expression and protein release in human UC-MSCs [52]. Likewise, co-culture with peroxide hydrogen-injured renal tubular cells significantly promoted the expression of HGF and insulin-like growth factor-1 (IGF-1) in UC-MSCs, indicating that the interaction between UC-MSCs and resident renal cells could be responsible for the pro-survival and pro-angiogenic effects.

12.3.3 Extracellular Vesicles (EVs)

Finding few MSCs engrafting into the injured tissue in the face of robust functional recovery has raised the interest to unveil additional mechanisms that act in concert with the release of soluble factors to explain the therapeutic effects of MSCs [53]. Evidence is mounting to indicate that MSCs produce complex structures called EVs, responsible for the cell-to-cell communication that allows the transfer of membrane receptors, functional proteins, organelles, or genetic material [54]. EVs are comprised of exosomes and microvesicles (MVs). Exosomes are membrane fragments of 30–90 nm diameter originated from the endosomal membrane compartment after fusion of secretory granules with the plasma membrane. MVs are relatively large vesicles (100 nm to 1 μm diameter) released from the surface membrane of activated cells. What is not clear is the mechanism of action through which PSCs-derived EVs confer protective effects. A very recent study showed that renal tubular cells lacking IL-10 expression and exposed to UCB-MSC-derived EVs acquired the IL-10 mRNA, which was efficiently translated into the corresponding protein, leading to improved cell viability [55]. This implies that horizontal mRNA shuttling via EVs is a critical mechanism involved in in vitro restoring ability of UCB-MSCs. Of interest, the content and functional property of EVs is influenced by the environment in which they are cultured. For instance, MVs derived from IFN-γ-stimulated UC-MSCs contained tetraspanins and more complete proteasome complex accompanied with MHC-I, suggesting that the inflammatory conditioning regulates protein packaging into EVs and thereby modulating EVs function [56]. Confirming this, the protective effect of UC-MSCs on renal function was abrogated by IFN-γ stimulation.

Exciting recent discoveries have implicated PSC-derived EVs in the amelioration of renal disorders (Table 12.1). Oxidative stress is regarded as a prime therapeutic target due to its pivotal role in AKI progression by activation of inflammatory response. In rodent models of AKI, the anti-oxidative and anti-inflammatory effects of EVs have been documented. For instance, in a rat model of cisplatin-induced AKI, renal subcapsular injection of human UC-MSC-derived exosomes promoted kidney repair through alleviating oxidative stress as evidenced by increased glutathione (GSH) and decreased malondialdehyde (MDA) levels [57]. Furthermore, in vitro studies demonstrated an inhibitory effect of exosomes on the activation of p38 mitogen-activated protein kinase pathway induced by cisplatin, together with a stimulatory effect of exosomes on extracellular signal-regulated kinase pathway. Similar results have been observed in human WJ-MSC-derived EVs where systemically delivered EVs exhibited anti-oxidative activities via reducing the expression of NADPH oxidase 2 and reactive oxygen species [58], suppressing renal and in vitro expression of CX3CL1 [59] and enhancing nuclear factor E2-related factor 2 pathway [60]. Recently, VEGF has been found

Table 12.1 Studies focusing on the protective effects of PSC-derived EVs on kidney diseases

Ref.	Animal, model	EV type, source	No. of EVs, delivery, dose	Functional and structural recovery	Therapeutic effects and mechanism
2013 Stem Cell Res Ther	Rat, cisplatin (6 mg/kg wt)	Exosomes, human UC-MSCs	200 μg, renal capsular, 1	Improved renal function (BUN, SCr), decreased renal tubular necrosis, and protein cast formation	Suppressed oxidative stress (reduced renal expression of 8-OHdG and MDA, increased GSH) and tubular cell apoptosis, increased cell proliferation
2015 PLoS One	Rat, I/R (left, 60 min)	MVs, human UC-MSCs	30 μg, tail vein, 1	Improved renal function (BUN, SCr), attenuated collagen deposition, myofibroblast accumulation	HGF mRNA delivery by exosome to tubular cells led to increased cell dedifferentiation, proliferation, and renal expression of HGF but reduced tubular cell apoptosis
2016 Am J Transl Res	Rat, I/R (left, 45 min)	EVs, human UC-MSCs	100 μg, tail vein, 1	Improved renal function (BUN, SCr), alleviated tubular necrosis, dilatation, and cast formation	VEGF transferred to tubular cells by EVs resulted in decreased tubular cell apoptosis and renal fibrosis but increased cell proliferation and capillary density
2016 Biomater Res	Patient, CKD (stage III and IV)	EVs, human UCB-MSCs	100 μg/kg/dose, median cubital vein, first dose; renal artery, second dose	Improved renal function (BUN, SCr, estimated glomerular filtration rate, urine albumin-to-creatinine ratio)	Immuno-regulative effects of exosome (increased TGF-β1 and IL-10, decreased TNF-α), enhanced tubular epithelial cell proliferation and differentiation
2017 Stem Cell Res Ther	Rat, cisplatin (5 mg/kg wt)	Exosomes, human UC-MSCs	200 μg, intra-capsular, 1	Improved renal function (BUN, SCr), reduced number of renal tubules with edema and structural damage	Activated autophagy led to inhibited mitochondrial apoptosis, reduced secretion of inflammatory cytokines, enhanced tubular cell proliferation, and decreased cell death
2014 PLoS One	Rat, I/R (left, 60 min)	MVs, human WJ-MSCs	100 μg, tail vein, 1	Improved renal function (BUN, SCr), decreased number of necrotic proximal tubules and protein casts	Alleviated oxidative stress (reduced levels of MDA and ROS, expression of 8-OHdG and NOX2 in kidney), renal fibrosis (myofibroblast accumulation) and cell apoptosis, increased tubular cell proliferation
2014 Stem Cell Res Ther	Rat, I/R (left, 60 min)	MVs, human WJ-MSCs	100 μg, tail vein, 1	Improved renal function (BUN, SCr), alleviated tubular necrosis and cast formation	Reduced expression of CX3CL1 in the kidney and human umbilical vein endothelial cells; decreased renal cell death, fibrosis (myofibroblast accumulation), and macrophage recruitment; enhanced tubular cell proliferation

Table 12.1 (continued)

Ref.	Animal, model	EV type, source	No. of EVs, delivery, dose	Functional and structural recovery	Therapeutic effects and mechanism
2016 Kidney Blood Press Res	Rat, I/R (left, 45 min)	EVs, human WJ-MSCs	100 μg, tail vein, 1	Improved renal function (BUN, SCr), attenuated tubular detachment, foamy degeneration and necrosis	Weakened oxidative stress (reduced levels of MDA and 8-OHdG, increased expression of SOD, Nrf2, and HO-1 in renal tissue), activated Nrf2/ARE pathway, decreased tubular cell apoptosis
2016 Stem Cells Int	Rat, I/R (left, 45 min, plus right nephrectomy)	EVs, human WJ-MSCs	100 μg, tail vein, 1	Improved renal function (BUN, SCr), alleviated mitochondrial fragmentation	Delivered and restored expression of miR-30 resulted in inhibited mitochondrial fission and apoptosis, reduced renal tubular cell apoptosis
2017 J Cell Biochem	Rat, renal transplantation	EVs, human WJ-MSCs	100 μg, tail vein, 1	Improved renal function (BUN, SCr) and animal survival, reduced collagen deposition and myofibroblast accumulation	Abrogated fibrosis (reduced expression of α-SMA and TGF-β1, upregulated expression of HGF), increased tubular cell proliferation, decreased cell apoptosis and macrophage infiltration

PSCs perinatal stem cells; *EVs* extracellular vesicles; *MVs* microvesicles; *I/R* ischemia and reperfusion; *CKD* chronic kidney disease; *MSCs* mesenchymal stem cells; *UC* umbilical cord; *UCB* umbilical cord blood; *WJ* Wharton jelly; *BUN* blood urea nitrogen; *SCr* serum creatinine; *8-OHdG* 8-hydroxy-2′ -deoxyguanosine; *MDA* malondialdehyde; *GSH* glutathione; *HGF* hepatocyte growth factor; *VEGF* vascular endothelial growth factor; *TGF-β1* transforming growth factor-β1; *IL-10* interleukin-10; *TNF-α* tumor necrosis factor-α; *ROS* reactive oxygen species; *NOX2* NADPH oxidase 2; *CX3CL1* C-X3-C motif ligand 1; *Nrf2* nuclear factor E2-related factor 2; *HO-1* heme oxygenase-1; *ARE* antioxidant response element; *α-SMA* alpha-smooth muscle actin

to be transferred by UC-MSC-derived EVs to tubular cells to induce capillary growth and protect against renal ischemia [61]. Notably, these reparative effects of EVs can be attributed to RNA delivery, as inactivation of RNA by RNAse abrogated their therapeutics. Besides, it has been proposed that administration of PSC-derived EVs in AKI model led to functional restoration mediated by induction of HGF synthesis [62], inhibition of mitochondrial fission [63], and activation of autophagy [64]. An additional important renoprotective property of EVs is to decrease fibrosis. For instance, a very recent study revealed the anti-fibrotic effect of WJ-MSC-derived MVs in a rat model of syngeneic kidney transplantation where intravenous injection of MVs mitigated inflammatory cell infiltration and reduced alpha-smooth muscle actin (α-SMA) expression [65]. The therapeutic potential of human UCB-MSC-derived EVs in CKD patients has been addressed in a clinical study [66]. EVs treated patients demonstrated a significant improvement in renal function compared with patients receiving placebo. Moreover, treatment with EVs increased plasma levels of IL-10 and transforming growth factor (TGF)-β1 while decreased tumor necrosis factor (TNF)-α level, indicating their anti-inflammatory role in counteracting CKD. To sum up, these studies disclose that the diverse beneficial functions of PSCs should be ascribed to the transfer of EVs containing immune modulators and trophic growth factors to injured renal tissue (Fig. 12.3). Insights from these studies might provide a fundamental understanding of the therapeutic actions of EVs and avoid the complex production and regulatory processes involved in developing PSC therapies.

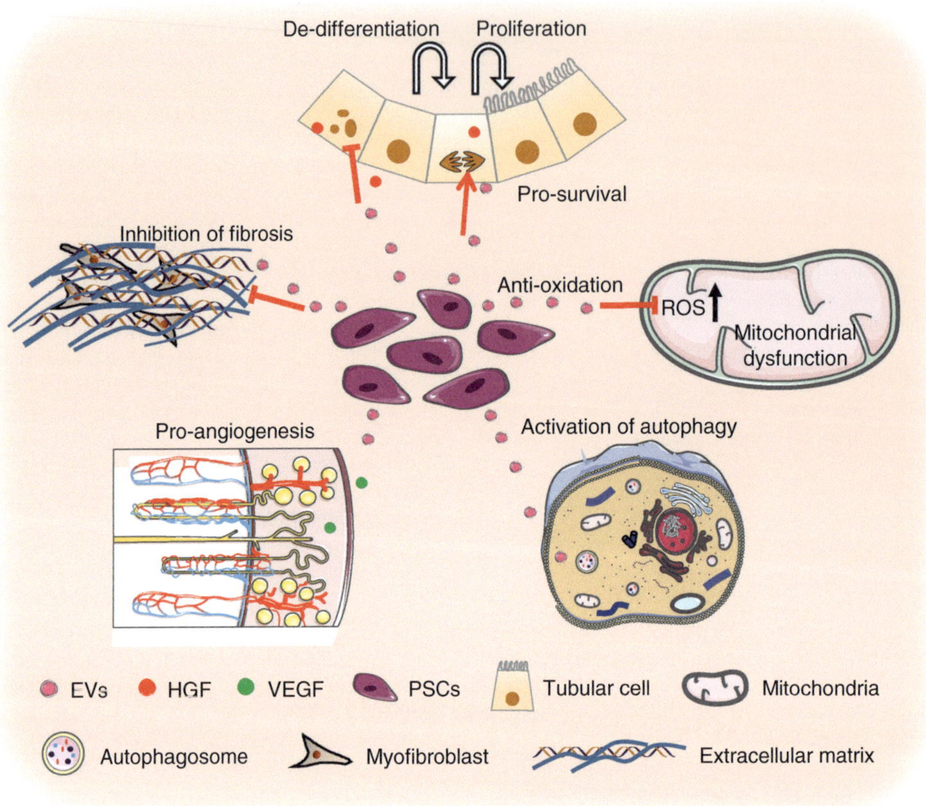

Fig. 12.3 PSC-derived EVs promote kidney regeneration via multiple mechanisms. The therapeutic effects of PSCs are attributed to EV-induced paracrine/endocrine mechanism based upon pro-proliferative, immunomodulatory, pro-angiogenic, pro-autophagic, and anti-fibrotic activities. PSC-derived EVs contain HGF mRNA that promotes endogenous reparative regenerative by inducing dedifferentiation, proliferation, and redifferentiation of surviving tubular cells. Besides, EVs exert anti-oxidative and anti-inflammatory effects via reducing renal expression of CXCL3 and activating Nrf2/ARE pathway. Moreover, miR-30 delivered by EVs is responsible for the inhibition of mitochondrial fission. Similarly, activated autophagy by EVs acts to alleviate mitochondrial apoptosis. In addition, EVs stimulate local angiogenesis by transferring VEGF. What's more, EVs could also upregulate renal expression of HGF, thereby contributing to the reduction of kidney fibrosis

12.4 Strategies to Enhance Cellular Therapy

The discovery of stem cells has fueled intense research interest in the field of regenerative medicine. Although numerous studies substantiate the therapeutic benefits of stem cells in renal diseases, non-specific engraftment and poor survival significantly undermine the functional benefits of transplanted cells [67]. Therefore, various strategies are being developed to increase PSC homing capacity, to promote cell engraftment and retention, and to enhance their regenerative potential by cell pretreatment/preconditioning with growth factors, bioactive compounds, or genetic manipulation (Fig. 12.4).

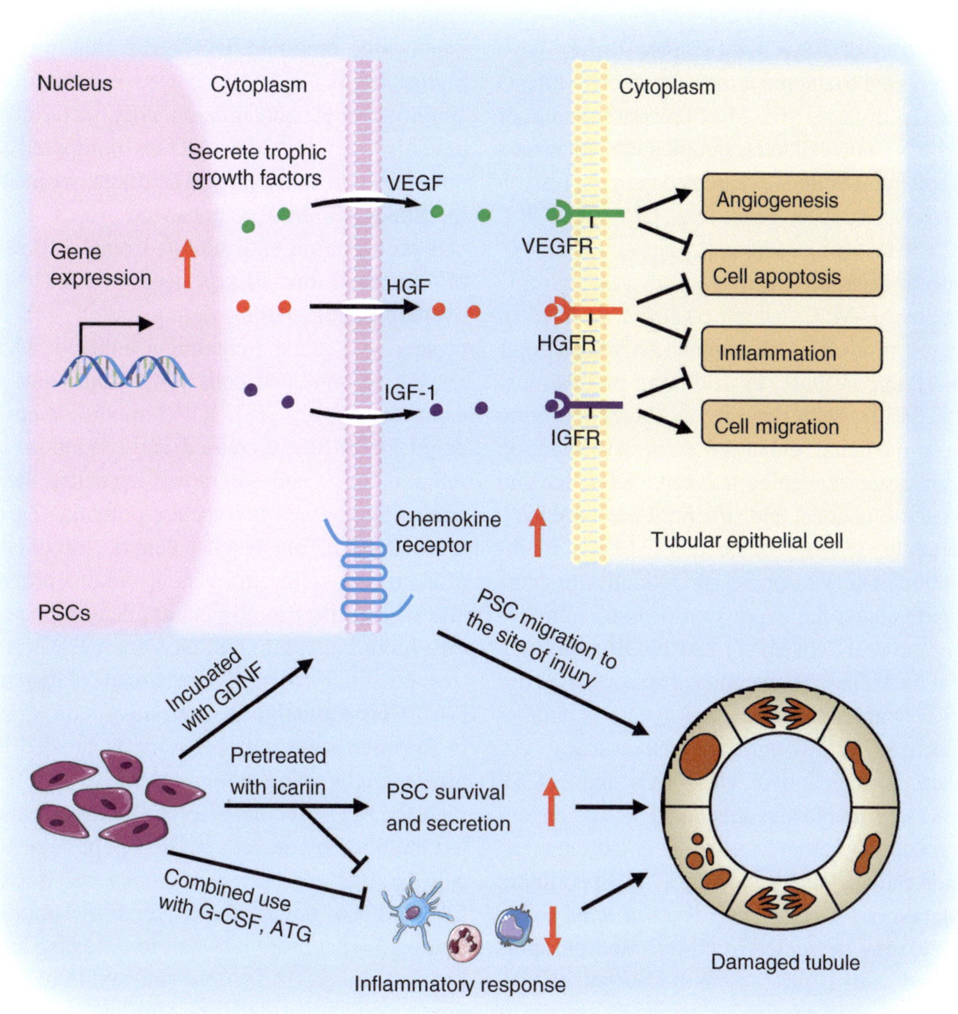

Fig. 12.4 Strategies to enhance therapeutic effects of PSCs in kidney disease. The therapeutic efficacy of PSCs can be potentiated in various ways to generate renoprotective capabilities. Genetically modified PSCs secrete therapeutic proteins that function directly on tubular epithelial cells. For instance, VEGF, HGF, and IGF-1 produced by PSCs can bind to their corresponding receptors to exert diverse protective (i.e., anti-apoptotic, anti-inflammatory, mitogenic, pro-angiogenic) effects. The migratory property of PSCs can be enhanced using cell pretreatment approach. For instance, incubation with GDNF leads to upregulated expression of chemokine receptors, including CD44, CXCR4, and CX3CR1, thereby increasing stem cell motility. In addition, stimulation with icariin could improve PSC survival and promote their secretion of pro-regenerative factors (e.g., BMP-7 and bFGF). PSC-based therapeutics can also be strengthened by the combined use of pharmacological agents, for instance, G-CSF and anti-thymocyte globulin (ATG). The observed benefits can be ascribed to attenuated inflammation and oxidative stress

12.4.1 Cell Preconditioning

Pretreatment of PSCs with trophic factors have been reported to augment their therapeutic effects in kidney diseases. Ex vivo preconditioning of AF-MSCs with GDNF significantly increased stem cell motility and upregulated their expression of chemotaxis-related genes (e.g., CD44, CXCR4, and CX3CR1) [68]. Moreover, exposure to GDNF enhanced migratory activity and vasculotropic effects of AF-MSCs which could be ascribed to increased production of stromal-derived factor-1 (SDF-1) and VEGF. In addition, pretreatment of UC-MSCs with icariin, a kidney-tonifying Chinese medicine, enhanced stem cell retention in renal tissue of adenine-induced CKD mice and facilitated functional and structural recovery [69]. Importantly, icariin-pretreated UC-MSCs inhibited inflammatory response and oxidative damage and upregulated the expression of bone morphogenetic protein 7 (BMP-7) and bFGF which are critical for kidney regeneration. The combined use of PSCs with certain bioactive agents is another approach to improving stem cell efficacy. For instance, co-injection of UC-MSCs and G-CSF reduced serum urea and attenuated tubular degeneration and dilation in a rat model of carbon tetrachloride-induced nephrotoxicity [70]. In addition, the glutathione S-transferase level in renal tissue was markedly decreased in co-injection group as compared with control or monotherapy groups. Similarly, in a recent study using murine model of cisplatin-induced AKI, antithymocyte globulin (a kind of immunosuppressive drug) was employed to provide UC-MSCs a compromised immune system as evidenced by diminished inflammation and oxidative stress in local environment [71]. As a result, combined use of antithymocyte globulin and UC-MSCs induced a synergistic effect on renal recovery and animal survival.

12.4.2 Genetic Modification

Evidence favoring the use of genetic manipulation in cell-based therapy for renal disorders has accumulated in recent years. AF-MSCs overexpressing VEGF exhibited potent pro-regenerative actions by mitogenic, pro-angiogenic, and anti-inflammatory mechanisms [72]. Furthermore, genetically modified UC-MSCs that produced higher levels of HGF showed enhanced anti-apoptotic and anti-inflammatory activities in animals with ischemic AKI as compared with untransfected cells [73]. Treatment with HGF-modified UC-MSCs enhanced renal tubular cell proliferation and inhibited renal expression of caspase-3 and IL-1β. Recently, UC-MSCs overexpressing IGF-1 demonstrated enhanced protection against gentamicin-induced AKI in respect to untreated cells [74]. Consistent with a previous study [75], IGF-1 modification rendered UC-MSCs increased anti-oxidation, anti-inflammation, and migratory capacity, thereby strengthening their therapeutic potential. Despite these encouraging results, genetic modification of stem cells with viral vector remains problematic with respect to clinical application, because virus could integrate into the genome which carries potential risks of insertional mutagenesis [76]. Consequently, an alternative safe strategy to improve stem cell behavior is needed. Our recent study using adipose-derived MSCs has reported the beneficial effects of chitosan hydrogel modified by bioactive IGF-1 peptide on stem cell survival and therapeutics in renal ischemia [77]. MSCs encapsulated in IGF-1-modified hydrogel were more resistant to oxidative stress-induced apoptosis in vitro and exhibited more potent immunomodulatory effects in vivo. This implicates that engineered microenvironment that mimics native stem cell niche may provide an exciting and promising approach to improving cellular therapy for renal disease.

12.5 Therapeutic Use of PSCs in Kidney Diseases

Much of the recent encouragement and excitement surrounding PSCs is connected with the hope that they can serve as a cell-based therapeutic approach for a wide array of kidney diseases. PSCs hold great potential in the treatment of AKI and CKD by virtue of their complex paracrine and endocrine activities including anti-apoptotic, anti-inflammatory, and pro-survival effects, together with the stimulation of angiogenesis and vasculogenesis. In addition, the immunomodulatory effects enable PSCs to play a role in immune-mediated renal diseases, e.g., kidney transplantation, lupus nephritis, glomerulosclerosis, etc.

12.5.1 Acute Kidney Injury (AKI)

AKI is a common and serious medical condition associated with significant morbidity and mortality in critically ill patients. Aside from ischemia, toxicity, shock, and sepsis, the use of invasive procedures and increasing age of the patient population also account for the ever-increasing incidence of AKI. Despite the advances in elucidating the pathogenesis of AKI, effective treatments that improve survival are still lacking. In this regard, identification and implementation of novel therapeutic options are urgently needed. PSCs have emerged as a promising candidate for cellular therapeutics for AKI. The overall rationale for using PSCs is that single-agent therapies usually affect only a limited number of various pathways that characterize renal pathogenesis [78]. PSC-based therapy is a new treatment modality that targets the complex pathophysiology of AKI. Thus, the potential of PSCs to repair acutely damaged kidney has been the subject of intense investigations (Table 12.2).

UC-MSCs are the most widely studied cell population with respect to PSC-based therapy for AKI. The mechanisms responsible for the renoprotective capacity of UC-MSCs have been extensively investigated. Previous in vivo work revealed the capability of UC-MSCs to promote tubular cell proliferation and to reduce cell apoptosis by downregulating renal expression of caspase-3 [79] and upregulating renal expression of Bcl-2 [80]. Fang et al. reported that systemic infusion of human UC-MSCs protected renal function and structure in nonobese diabetic-severe combined immune deficiency (NOD-SCID) mice with AKI induced by folic acid [81]. UC-MSCs increased the expression of p-Akt/IGF-1R in damaged kidney via secretion of IGF-1, thereby inducing proliferation of tubular epithelial cells. In addition, preclinical studies have highlighted the potential of UC-MSCs to promote kidney regeneration by conferring anti-inflammatory and immunomodulatory effects. Cao and colleagues observed a significant reduction in renal gene expression of pro-inflammatory cytokines such as IL-1β, IL-6, and TNF-α in I/R-injured rats receiving human UC-MSCs [79]. This result was in consonance with a recent study showing that intravenously administered fetal membrane-derived MSCs engrafted into ischemic kidney, increased serum level of IL-10, and decreased renal expression of monocyte chemoattractant protein-1 (MCP-1) and IL-6, therefore protecting animals from function impairment and tubular injury [82]. Besides, UC-MSCs were also shown to decrease intercellular cell adhesion molecule-1 expression and reduce neutrophil infiltration in I/R-injured kidney [83]. What's more, two recent reports unveil the importance of UC-MSCs in improving oxidative status in rats suffering renal ischemia [84, 85]. Macrophages serve as key mediators in renal inflammation in AKI [86]. They exhibit distinct phenotypic and functional characteristics in response to various stimuli in the local microenvironment. Li and co-workers injected human UC-MSCs by tail vein into I/R-injured mice and observed reduced macrophage infiltration and an increased proportion of anti-inflammatory M2 macrophage recruited [87]. Additionally, macrophages when co-cultured with UC-MSCs exhibited a high level of CD206, a marker of alternatively activated M2 macrophages, indicating that UC-MSCs can switch macrophages to an anti-inflammatory phenotype. Similar to UC-MSCs, UCB-MSCs have also been documented to exert immunoregulatory effects in murine model of AKI. In NOD-SCID mice with cisplatin-induced AKI, intravenous delivery of human UCB-MSCs alleviated acute tubular necrosis by inducing the pro-survival factor Akt in tubular cells, resulting in ameliorated deficit in function and reduced recipient mortality [88]. Moreover, UCB-MSCs acted by attenuating oxidative damage and inhibiting expression of IL-1β and TNF-α in renal tubules, suggesting the anti-inflammatory action of UCB-MSCs treatment.

Table 12.2 Studies focusing on the protective effects of PSCs on AKI

Ref.	Animal, model	Cell, source	No. of cells, delivery, dose	Functional and structural recovery	Therapeutic effects and mechanism
2010 Am J Pathol	SCID mouse, glycerol (8 mL/kg wt)	AF-MSCs, human	3.5×10^5, intravenous, 1	Improved renal function (BUN, SCr), reduced tubular necrosis, vacuolization, cast formation, and brush border loss	Increased proliferation and reduced apoptosis of tubular cells, engraftment and accumulation of AF-MSCs within peritubular capillary and interstitium, secretion of leukemia inhibitory factor
2010 PloS One	Nu/nu mouse, glycerol (10 mL/kg wt)	AF-MSCs, human	1×10^6, intrarenal, 1	Improved renal function (BUN, SCr), decreased number of damaged tubules and casts	Engraftment and integration of AF-MSCs into injured kidney, stimulated tubular cell proliferation and reduced cell apoptosis, immune-modulatory actions
2016 Int J Stem Cells	Rat, cisplatin (5 mg/kg wt)	AF-MSCs, human	5×10^6, tail vein, 1	Improved renal function (BUN, SCr), reduced tubular necrosis, vacuolization, shedding, and inflammatory cell infiltration	Antioxidant activity (decreased renal expression of MDA, increased GSH and SOD), attenuated interstitial fibrosis
2016 Stem Cell Res Ther	Rat, cisplatin (5 mg/kg wt)	AF-MSCs, human	5×10^6, tail vein, 1	Improved renal function (BUN, SCr, creatinine clearance); weakened renal tubular necrosis, dilation, and epithelial shedding; inhibited inflammation	Antioxidant activity (decreased renal expression of MDA, increased GSH and SOD), reduced interstitial fibrosis, and increased renal cell proliferation
2010 Biotech	Rat, I/R (bilateral, 60 min)	UC-MSCs, human	1×10^6, carotid artery, 1	Improved renal function (BUN, SCr), reduced tubular necrosis, dilation, inflammatory cell infiltration, and brush border disruption	Increased renal cell proliferation, inhibited renal expression of caspase-3 and inflammatory genes (i.e., IL-1β, IL-6, TNF-α)
2014 Cell Transplant	Rat, I/R (left, 60 min, plus right NPX)	Fetal membrane-derived MSCs, mismatched ACI rat	5×10^5, intravenous, 1	Improved renal function (BUN, SCr), attenuated tubular degeneration, dilatation, swelling, and luminal congestion with loss of the brush border	Reduced tubular cell apoptosis; inhibited T cell and macrophage infiltration; decreased expression of MCP-1, IL-6, α-SMA, TGF-β, and collagen 1; engraftment of MSCs and their secretion of IL-10
2012 PLoS One	NOD-SCID mouse, folic acid (250 mg/kg wt)	UC-MSCs, human	1×10^6, external jugular vein, 1	Improved renal function (BUN, SCr), alleviated renal tubular cell damage	Promoted tubular cell proliferation, reduced cell apoptosis, IGF-1 secretion by UC-MSCs increased the expression of p-Akt/IGF-1R

Table 12.2 (continued)

Ref.	Animal, model	Cell, source	No. of cells, delivery, dose	Functional and structural recovery	Therapeutic effects and mechanism
2013 Int J Low Extrem Wounds	Rat, burn	UC-MSCs, rat	1×10^6, intravenous, 1	Improved renal function (BUN, SCr) and animal survival, reduced tubular cell necrosis, blebbing, and hyaline casts	Attenuated tubular cell apoptosis, increased renal expression of Bcl-2
2013 Stem Cell Res	Mouse, I/R (bilateral, 30 min)	UC-MSCs, human	2×10^6, tail vein, 1	Improved renal function (BUN, SCr), attenuated renal tubular necrosis, dilation, cast formation, and inflammatory cell infiltration	Promoted renal cell proliferation, reduced cell apoptosis and macrophage infiltration, enhanced M2 macrophage recruitment, and activated M2 phenotype
2014 Int Braz J Urol	Rat, I/R (left, 60 min, plus right NPX)	UC-MSCs, human	1×10^6, tail vein, 1	Weakened tubular cell necrosis, swelling, cast formation, and infiltration of inflammatory cells	Decreased ICAM-1 expression and infiltrating polymorphonuclear neutrophils in renal tissue, engraftment of UC-MSCs
2017 Stem Cell Res Ther	Rat, I/R (bilateral, 45 min)	UC-MSCs, human	1×10^6, intraperitoneal, 1	Improved renal function (BUN, SCr, creatinine clearance), attenuated renal damage, and macrophage infiltration	Inhibited premature senescence (increased Klotho) and oxidative state (increased SOD), decreased renal fibrosis and β-galactosidase expression
2017 Tissue Cell	Rat, I/R (left, 60 min)	UC-MSCs, human	1×10^6, tail vein, 1	Improved renal function (BUN, SCr, uric acids, urine albumin), alleviated tubular degeneration, interstitial edema, and blood vessel dilation	Improved oxidative status, (decreased the level of MDA, increased the levels of GSH, catalase, and GST in the liver)
2010 Stem Cells	NOD-SCID mouse, cisplatin (12.7 mg/kg wt)	UCB-MSCs, human	5×10^5, intravenous, 1	Improved renal function (BUN) and animal survival, reduced number of necrotic tubules and casts	Promoted tubular cell proliferation, alleviated cell apoptosis (induction of Akt), oxidative damage, and inflammation (inhibited expression of IL-1β and TNF-α in tubular cells)
2014 Am J Physiol Renal Physiol	Mouse, I/R (bilateral, 27 min)	UCB-MSCs, human	1×10^6, intraperitoneal, 1	Improved renal function (BUN, SCr), attenuated renal tubular necrosis	Suppressed IFN-γ expression and preserved VEGF expression, the immunoregulatory effects of UCB-SCs (reduced NK T cell but increased T cell infiltration)

(continued)

Table 12.2 (continued)

Ref.	Animal, model	Cell, source	No. of cells, delivery, dose	Functional and structural recovery	Therapeutic effects and mechanism
2017 Am J Physiol Renal Physiol	Mouse, cisplatin (22 mg/kg wt)	UCB-MSCs, human	1×10^6, intravenous or intraperitoneal, 1	Improved renal function (BUN, SCr), attenuated tubular dilatation, brush border loss, cast formation, and vacuolization	Decreased tubular cell apoptosis and renal expression of MCP-1, IL-6, and TNF-α, increased TReg cell infiltration and expression of IL-10 and VEGF in the kidney
2016 Stem Cells Transl Med	Rat, sepsis (cecal ligation and puncture)	WJ-MSCs, human	1×10^6, intraperitoneal, 1	Improved renal function (GFR, inulin clearance, fractional excretion of sodium, urea) and animal survival	Reduced macrophage infiltration, tubular cell apoptosis, and fractional interstitial area; decreased expression of NF-kB and IL-1α, IL-6, and IFN-γ; increased expression of eNOS, VEGF, Klotho

PSCs perinatal stem cells; *SCID* severe combined immune deficiency; *NOD* non-obese diabetic; *I/R* ischemia and reperfusion; *NPX* nephrectomy; *MSCs* mesenchymal stem cells; *AF* amniotic fluid; *UC* umbilical cord; *UCB* umbilical cord blood; *WJ* Wharton jelly; *BUN* blood urea nitrogen; i serum creatinine; *GFR* glomerular filtration rate; *MDA* malondialdehyde; *GSH* glutathione; *SOD* superoxide dismutase; *TNF-α* tumor necrosis factor-α; *MCP-1* monocyte chemoattractant protein-1; *α-SMA* alpha-smooth muscle actin; *TGF-β* transforming growth factor; *IGF-1* insulin-like growth factor-1; *Bcl-2* B-cell lymphoma-2; *ICAM-1* intercellular cell adhesion molecule-1; *GST* glutathione S-transferase; *IFN-γ* interferon-γ; *VEGF* vascular endothelial growth factor; *NK* natural killer; *TReg cells* regulatory T cells; *eNOS* endothelial nitric oxide synthase

In addition, intraperitoneal infusion of human UCB-MSCs protected against renal functional deterioration via reducing NK T cell infiltration and increasing Treg cell recruitment in either cisplatin or I/R model of AKI, respectively [48, 89].

The observation that AF-MSCs protect renal tubules from acute injury has been supported by data obtained in different models of AKI. In a glycerol-induced murine model of AKI, Perin et al. reported that intraparenchymally injected AF-MSCs migrated and integrated into the damaged kidney, promoted tubular cell proliferation, and expedited renal structural functional recovery [90]. In a rat model of cisplatin-induced nephrotoxicity, Al-Husseiny and colleagues confirmed the protective effects of AF-MSCs [91]. Delivery of AF-MSCs via the tail vein resulted in a significant improvement of renal function. Examining the expression of oxidation-related indices in the kidneys of AF-MSC-treated rats revealed an antioxidant activity of AF-MSCs as evidenced by a decrease in the level of MDA and an increase in the levels of GSH and superoxide dismutase (SOD). Moreover, recent findings from the same group showed that human AF-MSCs exhibited greater architectural improvement and pro-proliferative activity in cisplatin-induced AKI compared with MSCs derived from a rat bone marrow and human adipose [92]. Of interest, it has been reported that intraperitoneal injection of WJ-MSCs resulted in improved renal function and prolonged animal survival in a rat model of sepsis-induced AKI [93]. The therapeutic effects were associated with reduced macrophage infiltration and decreased expression of a plethora of pro-inflammatory chemokines. These findings are concordant with the observations in studies using other types of PSCs as described before.

12.5.2 Chronic Kidney Disease (CKD)

For patient whose renal function is already impaired, AKI is often severe, and functional recovery may be incomplete or irreversible, eventually leading to CKD and ESRD. Besides, the aging population and the increasing prevalence of noncommunicable diseases such as diabetes and hypertension also lead to an increased prevalence of CKD [94]. The use of traditional treatment modalities including dialysis and organ transplantation has been limited by adverse events, small number of organ donors, and rejection across xenogenic barrier [95]. Thus, innovative interventions are required to delay the progression of CKD. In the view of pathophysiology, the kidney has an inherent capacity for self-regeneration in the presence of acute injury, but the cellular repair is ineffective in the face of chronic insult. In this context, administration of exogenous stem cells has been regarded as a promising approach to potentiate the regenerative ability of the kidney. Although PSCs have been shown to improve the outcome of experimental AKI, the use of PSCs to treat CKD remains a relatively new area of investigation. To date, only limited data are available with respect to the role of PSCs in models of CKD (Table 12.3).

Several groups have attempted to determine whether PSCs could improve disease status in CKD. In the CKD model of Alport syndrome, mice deficient for collagen alpha 5-chain type IV collagen (Col4a5$^{-/-}$) developed progressive glomerular sclerosis leading to renal failure. Single injection of AF-MSCs preserved glomerular basement membrane structure, attenuated renal interstitial fibrosis, and improved renal function in Col4a5$^{-/-}$ mice [96]. Of note, treatment with AF-MSCs resulted in downregulated expression of TNF-α, CCL2, and CXCL2 (M1 macrophage markers) and an elevated expression of IL1-R2 (M2 macrophage marker). Moreover, in vitro experiments confirmed the role of AF-MSCs in inducing an M1-to-M2 phenotype switch in macrophages, suggesting the strong immunoregulatory effects of PSCs. Intra-arterial infusion of human AF-MSCs into I/R rats decreased renal expression of iNOS and inhibited myofibroblast infiltration and renal fibrosis [97]. In an experimental model of CKD induced by unilateral ureteral obstruction (UUO), AF-MSCs alleviated tubular damage and interstitial fibrosis [98]. In the same model, human UC-MSCs injected via tail vein engrafted into renal interstitium and reduced inflammatory cell infiltration [99]. Interestingly, Liu and colleagues reported that infusion of conditioned medium derived from UC-MSCs into UUO rats attenuated fibrosis by inhibiting oxidative stress as evidenced by decreased expression of TGF-β1, α-SMA, and TNF-α and increased expression of E-cadherin in renal tissue [100]. Similar findings have been observed in cisplatin-induced CKD model, where UC-MSCs protected against tubular necrosis and functional deterioration through inhibiting collagen deposition, TGF-β1 expression, and renal EMT [101].

Diabetic nephropathy (or diabetic CKD), the leading cause of ESRD, is a complex clinical condition with a sequence of phases from increased GFR, microalbuminuria, proteinuria, and hypertension to reduced GFR with increasing proteinuria [102]. The main feature of diabetic CKD is glomerular and interstitial fibrosis characterized by excessive accumulation of ECM and the development of EMT, which have been recognized as therapeutic targets. The ability of PSCs to exert a pro-regenerative effect via local release of renotrophic and anti-inflammatory factors has also been documented in diabetic CKD. Park et al. examined the therapeutic use of human UCB-MSCs in preventing CKD in streptozotocin-induced diabetic rats [103]. They found that intravenously delivered UCB-MSCs engrafted into diabetic kidneys, reduced fibronectin expression, and prevented renal EMT. In addition, the same group further revealed that administration of UCB-MSCs resulted in ameliorated renal function and structure, together with inhibited renal expression of TGF-β1, collagen 1, and α-SMA, while increased the expression of BMP-7 [104]. Of note, in vitro studies demonstrated an inhibitory effect of UCB-MSC-derived conditioned medium on TGF-β1-induced ECM upregulation and EMT in NRK-52E cells, a renal epithelial cell line. These results suggest that PSCs could prevent diabetic CKD possibly through a paracrine action.

Table 12.3 Studies focusing on the protective effects of PSCs on CKD

Ref.	Animal, model	Cell, source	No. of cells, delivery, dose	Functional and structural recovery	Therapeutic effects and mechanism
2012 J Am Soc Nephrol	Col4a5−/− mouse, Alport syndrome	AF-MSCs, mouse	1×10^6, intraventricular, 1	Improved renal function (BUN, SCr, proteinuria) and animal survival, delayed interstitial fibrosis and glomerular sclerosis, attenuated macrophage recruitment	Activation of M2 macrophages (increased expression of IL1-R2), inhibition of fibrosis (decreased expression of TGF-β, α-SMA, and collagen 4 α1), preserved glomerular basement membrane structure and podocyte number
2013 PloS One	Nu/nu mouse, UUO (left)	AF-MSCs, human	3.5×10^5, tail vein, 1	Attenuated tubular atrophy, cast formation, brush border disruption, and interstitial fibrosis	Increased proliferation and reduced apoptosis of tubular cells; stimulated renal expression of VEGF, E-cadherin; decreased expression of MCP-1, HIF-1α, TGF-β1, and collagen 1
2015 PloS One	Rat, I/R (bilateral, 50 min)	AF-MSCs, human	1×10^6, intra-arterial, 1	Improved renal function (SCr, proteinuria), reduced tubular necrosis, hyaline cast formation, loss of brush border, and macrophage recruitment	Increased proliferation of tubular cells, decreased renal expression of iNOS, attenuated myofibroblast infiltration and interstitial fibrosis
2013 Exp Biol Med	Rat, cisplatin (6.0 mg/kg)	UC-MSCs, human	2×10^6, tail vein, 1	Improved renal function (BUN, SC), reduced tubular necrosis and inflammatory cell accumulation	Increased tubular cell proliferation; reduced cell apoptosis, Bax/Bcl-2 ratio, MDA level, and TGF-β1 expression; inhibited collagen deposition and EMT
2013 Ren Fail	Rat, UUO (left, 7 days)	UC-MSCs, human	5×10^6, tail vein, 1	Alleviated tubular cell atrophy, cystic expansion of distal tubules, and collecting ducts, reduced inflammatory cell infiltration	Attenuated cellular proliferation and fibrosis, engraftment of UC-MSCs to renal interstitium
2017 Nephrology	Rat, UUO (left)	UC-MSC-derived conditioned medium, human	500 μL from 5×10^6 cells, left renal artery, 1	Alleviated tubular dilation and atrophy, inflammatory cell infiltration, vascular collapse, extracellular matrix deposition, and interstitial fibrosis	Attenuated oxidative stress (increased GSH, reduced MDA and ROS levels), inhibited fibrosis (increased E-cadherin, decreased TGF-β1, α-SMA, TNF-α, and collagen 1), promoted tubular cell proliferation, reduced cell apoptosis

Table 12.3 (continued)

Ref.	Animal, model	Cell, source	No. of cells, delivery, dose	Functional and structural recovery	Therapeutic effects and mechanism
2012 Diabetes Res Clin Pract	Rat, diabetic renal injury (type 1 DM, 4 weeks)	UCB-MSCs, human	5×10^5, intravenous, 1	Improved renal function (proteinuria), decreased fractional mesangial area	Prevented renal EMT; inhibited expression of TGF-β1, HSP47, collagen 1, α-SMA, collagen accumulation; increased BMP-7 expression
2012 Transplant Proc	Rat, diabetic nephropathy (type 1 DM, 4 weeks)	UCB-MSCs, human	1×10^6, tail vein, 1	Improved renal function (SCr, proteinuria)	Reduced fibronectin and α-SMA expression, prevented EMT, and homing of UCB-MSCs to the injured kidney

PSCs perinatal stem cells; *UUO* unilateral ureteral obstruction; *I/R* ischemia and reperfusion; *DM* diabetes mellitus; *MSCs* mesenchymal stem cells; *AF* amniotic fluid; *UC* umbilical cord; *UCB* umbilical cord blood; *BUN* blood urea nitrogen; *SCr* serum creatinine; *TGF-β* transforming growth factor; *α-SMA* alpha-smooth muscle actin; *VEGF* vascular endothelial growth factor; *MCP-1* monocyte chemoattractant protein-1; *HIF-1* hypoxia-inducible factor-1; *iNOS* induced nitric oxide synthase; *Bax* bcl-2-associated X; *Bcl-2* B-cell lymphoma-2; *MDA* malondialdehyde; *EMT* epithelial-mesenchymal transitions; *GSH* glutathione; *ROS* reactive oxygen species; *TNF-α* tumor necrosis factor-α; *HSP 47* heat shock protein 47; *BMP-7* bone morphogenetic protein

12.5.3 Kidney Transplantation

Kidney transplantation is the treatment of choice for patients with ESRD. However, lifelong systemic immunosuppression required to prevent allograft rejection in transplant recipient can result in life-threatening complications [e.g., nephrotoxicity [105], infection [106], and tumorigenesis [107]]. In addition, non-specific immunosuppressive drugs are ineffective in prolonging long-term graft survival [108]. Accordingly, novel strategies to promote allograft tolerance and avoid the use of immunosuppressive drugs are urgently needed. MSCs are a promising therapeutic candidate in kidney transplantation, given their low immunogenicity and immune-modulatory properties [109]. The absence of MHC class II allows MSCs to evade allogeneic rejection. Multiple studies disclose that MSCs could regulate alloimmune response and improve kidney transplantation. These tolerance-inducing effects of MSCs could be ascribed to their capacity of downregulating the effector functions of immune cells and generating T cells and antigen-presenting cells [110]. In a porcine model of autologous renal transplantation, injection of autologous AF-MSCs via renal artery 6 days after transplantation ameliorated glomerular and tubular injury, hastened renal functional recovery, and prevented the development of interstitial fibrosis and tubular atrophy at 3 months [111]. This structural and functional improvement was associated with upregulated expression of VEGF-A and Ang-1 and decreased expression of α-SMA and TGF-β in the kidney. Notably, in an in vitro model simulating organ preservation conditions, AF-MSCs were found to confer protective effects on endothelial cells, as indicated by increased survival and tube-formation capacity of endothelial cells treated by conditioned medium of AF-MSCs than that of hypoxic endothelial cells. Moreover, AF-MSCs have also been shown to induce immune tolerance in a heterogenous kidney graft model. In Sprague-Dawley rat recipients of Wistar rat kidneys, posttransplant infusion of Wistar rat-derived AF-MSCs led to improved creatinine clearance rate, reduced inflammatory infiltrate, and attenuated pathological injury in the kidney transplant [112]. Tolerance induction was associated with a dramatic decrease in the serum cytokine (IL-2 and IFN-γ) as well as peripheral blood $CD4^+$ and $CD8^+$ lymphocytes. Of note, delivery of AF-MSCs significantly inhibited oxidative stress as evidenced by lower level of MDA in urine and renal tissue in AF-MSCs-treated rats than in untreated controls. These results implicate that AF-MSCs protect from graft dysfunction and modulate the host alloimmune response. Nonetheless, further studies are required to provide evidence for the safety and long-term efficacy of PSC therapy.

12.5.4 Lupus Nephritis (LN)

LN is one of the most common and severe manifestations of systemic lupus erythematosus and is associated with significant morbidity and mortality. This disease is characterized by autoantibody production, immune complex deposition, and complement activation. The use of conventional therapies based on nonselective immunosuppressive drugs is limited due to treatment-related toxicity and teratogenicity [113]. Thus, novel agents specifically interfere with underlying pathogenic mechanisms of LN are needed to minimize side effects and improve diseasc outcomes. The immunomodulatory potential of MSCs, along with their low immunogenicity, offers a promising treatment for refractory LN [114]. In the MRL/lpr mouse model of LN, transplantation of UC-MSCs induced therapeutic effect comparable to that of bone marrow MSCs [115]. This effect was associated with decreased expression of MCP-1 and HMGB-1 in the kidney. Moreover, UC-MSCs inhibited production of pro-inflammatory Th1 cytokines (i.e., IFN-γ, IL-2) but increased secretion of Th2 cytokines (i.e., IL-4, IL-10) [116]. Interestingly, these findings were supported by a recent work showing that UC-MSCs reduced renal injury by downregulating a broad range of pro-inflammatory cytokines (e.g., IFN-γ, IL-2, MCP-1, MIP-1α, M-CSF) in a tree shrew model of systemic lupus erythematosus [117]. Together, these preclinical studies indicate that PSCs inhibit renal inflammation via their immunoregulatory actions, thus supporting further exploration of the possibility to use PSCs from mismatched donors in LN treatment. The clinical translation of PSCs remains a work in progress, results of which are varied, with some showing benefit and others not. Very recently, infusion of UC-MSCs into patients with systemic lupus erythematosus led to alleviated renal inflammation and functional impairment [118]. This effect was owing to secretion of soluble HLA-G by UC-MSCs and subsequent upregulation of Treg cells. By contrast, a randomized double-blind, placebo-controlled trial assessing the therapeutic efficacy of human UC-MSCs in patients with LN reported that intravenously administrated UC-MSCs had no apparent additional effect over and above standard immunosuppression [119]. Thus, further studies will be required to address the safety, effectiveness, and efficacy of PSCs in LN treatment in clinical settings.

12.5.5 Focal Segmental Glomerulosclerosis (FSGS)

FSGS, the leading cause of CKD, is a primary glomerulopathy that is characterized by decreased podocyte number and glomerulosclerosis [120]. FSGS progresses from podocyte damage to loss of integrity of the glomerular filtration barrier, leading to renal dysfunction and proteinuria. Traditional pharmacological approaches (i.e., corticosteroids and calcineurin inhibitors) have failed to achieve a durable remission of proteinuria [121]. Thus, there is a pressing need to develop alternative interventions to treat this glomerular disease. PSCs are emerging as a promising candidate for the treatment of FSGS owing to their immune-modulatory and anti-inflammatory properties. It has been observed that human UC-MSCs attenuated the progression of adriamycin-induced FSGS by ameliorating renal fibrosis and modulating inflammatory response in rats [122]. Repeated infusions of 2×10^6 UC-MSCs via tail vein resulted in reduced collagen deposition; decreased serum level of IL-6, TNF-α, TGF-β, and connective tissue growth factor; and upregulated serum level of IL-10, as well as inhibited leukocyte infiltration in renal interstitium. This finding was in agreement with a previous study reporting that intravenous delivery of human UCB-MSCs significantly alleviated glomerular damage and accelerated functional recovery in rats with adriamycin-induced FSGS [123].

12.5.6 Other Kidney Diseases

Owing to their robust immunomodulatory effects, PSCs hold therapeutic promise in other renal disorders. IgA nephropathy is a common cause of CKD and is characterized by circulating IgA that forms immune complexes, which are deposited in the glomerular mesangium, leading to inflammation and renal dysfunction. Systemic infusion of UC-MSCs into mice with IgA nephropathy ameliorated proteinuria and hematuresis, alleviated fibrosis, and protected against the symptoms of nephropathy [124]. Anti-Thy1 glomerulonephritis is characterized by complement-mediated mesangial cell injury and endothelial cell loss. Allogenic fetal membrane-derived MSCs have been shown to protect against renal function impairment in experimental anti-Thy1 nephritis [125]. The protective effect was associated with reduced mesangial cell proliferation, glomerular macrophage infiltration, mesangial matrix accumulation, as well as glomerular expression of inflammatory or extracellular matrix-related genes including TNF-α, MCP-1, type I collagen, and TGF-β. Besides, PSCs hold therapeutic potential in inherited renal diseases. Intrauterine transplantation of MSCs derived from human fetal blood led to reduction of abnormal homotrimeric type I collagen deposition in the glomeruli and improved renal glomerulopathy in collagen type Iα2-deficient mice [126]. Similar results were observed in 129Sv-Col4a3^{tm1Dec}/J (−/−) mice model of Alport syndrome, where intraperitoneal injection of human fetal chorionic stem cells significantly delayed the progression of glomerulopathy by their anti-fibrotic and anti-inflammatory effects as well as replacement of the defective resident podocytes [127].

12.6 Future Challenges and Perspectives

Despite enormous progress made in PSC-based kidney repair, concerns remain regarding effectiveness and long-term safety based upon results from recent animal studies. Gheisari et al. examined the therapeutic effects of UC-MSCs and bone marrow-derived MSCs on cisplatin-induced AKI [128]. MSC-derived conditioned media were infused into animals via tail vein 24 h after cisplatin injection. Surprisingly, none of the conditioned medium from these two types of stem cells demonstrated preventive or reversal actions in terms of functional impairment and tubular injury. Moreover, even repetitive injection of conditioned medium was not effective for ameliorating renal damage. Considering that these negative findings were indeed on the grounds of functional and histological assessment, this does not rule out a paracrine mechanism for renotrophic activities of MSCs at the cellular and molecular levels. What's more, it is conceivable that experimental conditions (e.g., animal models, type, and dosing of cells) might explain the discrepancy between this study and other reports. In a different study using a mouse model of renal I/R injury, delivery of human UCB-derived CD133$^+$ cells at reperfusion exacerbated renal inflammation and apoptosis [129], whereas administration of a separate population of UCB-derived CD133$^-$ cells had no impact on kidney damage, suggesting a cell-specific effect. Hence, the untoward effect observed in this study highlights the need for caution in selecting appropriate cell subpopulations for the treatment of AKI. Studies performed to date have already led to the identification of a wide variety of renoprotective properties of PSCs; however, challenges remain in developing PSCs as a clinical therapy. Several important issues including the selection of the cell type, the number of cells needed, the route of cell delivery, and the timing of intervention are still unaddressed. In addition, a much better understanding of the mechanism of action of PSCs in renal homeostasis and rejuvenation is also critical.

12.7 Conclusion

PSCs exhibit great promise for cellular therapy without raising ethical and safety concerns associated with embryonic stem cells. Additionally, they have advantages of relatively easy isolation and culture conditions over adult MSCs. Given the robust evidence favoring a paracrine mechanism of treatment with PSCs, clearly more work is needed to identify the secreted soluble factors that mediate paracrine effects. What is intriguing, multiple proof-of-concept studies have shed light on the potential use of EVs to avoid direct injection of the parental cells and to circumvent safety concerns of aberrant stem cell differentiation, transformation, or antigenicity. Adverse effects of PSC-based therapy have been described in animal studies, and advances in our understanding of stem cell biology may contribute to elucidate repair mechanism and minimize unwanted outcomes. Despite the absence of data regarding feasibility and safety of PSCs in the clinical setting, this cell population holds enormous potential to treat renal disorders and to facilitate cell-based therapies that will eventually be used in regenerative medicine.

References

1. Vanmassenhove J, et al. Management of patients at risk of acute kidney injury. Lancet. 2017;389(10084):2139–51.
2. Zuk A, Bonventre JV. Acute kidney injury. Annu Rev Med. 2016;67:293–307.
3. Hickson LJ, Eirin A, Lerman LO. Challenges and opportunities for stem cell therapy in patients with chronic kidney disease. Kidney Int. 2016;89(4):767–78.
4. Yang Y, et al. Renoprotective approaches and strategies in acute kidney injury. Pharmacol Ther. 2016;163:58–73.
5. Little MH, Combes AN, Takasato M. Understanding kidney morphogenesis to guide renal tissue regeneration. Nat Rev Nephrol. 2016;12(10):624–35.
6. Takasato M, et al. Directing human embryonic stem cell differentiation towards a renal lineage generates a self-organizing kidney. Nat Cell Biol. 2014;16(1):118–26.
7. Neofytou E, et al. Hurdles to clinical translation of human induced pluripotent stem cells. J Clin Invest. 2015;125(7):2551–7.
8. Leuning DG, et al. A novel clinical grade isolation method for human kidney perivascular stromal cells. J Vis Exp. 2017;(126).
9. Kuang W, et al. Functional and molecular changes of MSCs in aging. Curr Stem Cell Res Ther. 2015;10(5):384–91.
10. La Rocca G, Anzalone R. Perinatal stem cells revisited: directions and indications at the crossroads between tissue regeneration and repair. Curr Stem Cell Res Ther. 2013;8(1):2–5.
11. Morigi M, De Coppi P. Cell therapy for kidney injury: different options and mechanisms--mesenchymal and amniotic fluid stem cells. Nephron Exp Nephrol. 2014;126(2):59.
12. Perin L, et al. Renal differentiation of amniotic fluid stem cells. Cell Prolif. 2007;40(6):936–48.
13. Hauser PV, et al. Stem cells derived from human amniotic fluid contribute to acute kidney injury recovery. Am J Pathol. 2010;177(4):2011–21.
14. Mattar P, Bieback K. Comparing the immunomodulatory properties of bone marrow, adipose tissue, and birth-associated tissue mesenchymal stromal cells. Front Immunol. 2015;6:560.
15. Zavatti M, et al. Estrogen receptor signaling in the ferutinin-induced osteoblastic differentiation of human amniotic fluid stem cells. Life Sci. 2016;164:15–22.
16. Preitschopf A, et al. Rapamycin-induced hypoxia inducible factor 2A is essential for chondrogenic differentiation of amniotic fluid stem cells. Stem Cells Transl Med. 2016;5(5):580–90.
17. Mu XP, et al. Enhanced differentiation of human amniotic fluid-derived stem cells into insulin-producing cells in vitro. J Diabetes Investig. 2017;8(1):34–43.
18. Moschidou D, et al. Valproic acid confers functional pluripotency to human amniotic fluid stem cells in a transgene-free approach. Mol Ther. 2012;20(10):1953–67.
19. Dziadosz M, Basch RS, Young BK. Human amniotic fluid: a source of stem cells for possible therapeutic use. Am J Obstet Gynecol. 2016;214(3):321–7.
20. Ding DC, et al. Human umbilical cord mesenchymal stem cells: a new era for stem cell therapy. Cell Transplant. 2015;24(3):339–47.
21. Friedman R, et al. Umbilical cord mesenchymal stem cells: adjuvants for human cell transplantation. Biol Blood Marrow Transplant. 2007;13(12):1477–86.
22. Olson AL, McNiece IK. Novel clinical uses for cord blood derived mesenchymal stromal cells. Cytotherapy. 2015;17(6):796–802.
23. Beeravolu N, et al. Isolation and comparative analysis of potential stem/progenitor cells from different regions of human umbilical cord. Stem Cell Res. 2016;16(3):696–711.
24. Frausin S, et al. Wharton's jelly derived mesenchymal stromal cells: biological properties, induction of neuronal phenotype and current applications in neurodegeneration research. Acta Histochem. 2015;117(4–5):329–38.
25. Mechiche Alami S, et al. Harnessing Wharton's jelly stem cell differentiation into bone-like nodule on calcium phosphate substrate without osteoinductive factors. Acta Biomater. 2017;49:575–89.
26. Murphy SV, Atala A. Amniotic fluid and placental membranes: unexpected sources of highly multipotent cells. Semin Reprod Med. 2013;31(1):62–8.
27. Amable PR, et al. Protein synthesis and secretion in human mesenchymal cells derived from bone marrow, adipose tissue and Wharton's jelly. Stem Cell Res Ther. 2014;5(2):53.

28. Wegmeyer H, et al. Mesenchymal stromal cell characteristics vary depending on their origin. Stem Cells Dev. 2013;22(19):2606–18.
29. Kuchroo P, et al. Paracrine factors secreted by umbilical cord-derived mesenchymal stem cells induce angiogenesis in vitro by a VEGF-independent pathway. Stem Cells Dev. 2015;24(4):437–50.
30. Du WJ, et al. Heterogeneity of proangiogenic features in mesenchymal stem cells derived from bone marrow, adipose tissue, umbilical cord, and placenta. Stem Cell Res Ther. 2016;7(1):163.
31. Romani R, et al. Stem cells from human amniotic fluid exert immunoregulatory function via secreted indoleamine 2,3-dioxygenase1. J Cell Mol Med. 2015;19(7):1593–605.
32. Li X, et al. Umbilical cord tissue-derived mesenchymal stem cells induce T lymphocyte apoptosis and cell cycle arrest by expression of indoleamine 2, 3-dioxygenase. Stem Cells Int. 2016;2016:7495135.
33. Wang Q, et al. Comparative analysis of human mesenchymal stem cells from fetal-bone marrow, adipose tissue, and Warton's jelly as sources of cell immunomodulatory therapy. Hum Vaccin Immunother. 2016;12(1):85–96.
34. Li X, et al. Comprehensive characterization of four different populations of human mesenchymal stem cells as regards their immune properties, proliferation and differentiation. Int J Mol Med. 2014;34(3):695–704.
35. Kale S, et al. Bone marrow stem cells contribute to repair of the ischemically injured renal tubule. J Clin Invest. 2003;112(1):42–9.
36. Lin F, et al. Hematopoietic stem cells contribute to the regeneration of renal tubules after renal ischemia-reperfusion injury in mice. J Am Soc Nephrol. 2003;14(5):1188–99.
37. Lin F, Moran A, Igarashi P. Intrarenal cells, not bone marrow-derived cells, are the major source for regeneration in postischemic kidney. J Clin Invest. 2005;115(7):1756–64.
38. Humphreys BD, et al. Intrinsic epithelial cells repair the kidney after injury. Cell Stem Cell. 2008;2(3):284–91.
39. Humphreys BD, Bonventre JV. Mesenchymal stem cells in acute kidney injury. Annu Rev Med. 2008;59:311–25.
40. Togel FE, Westenfelder C. Mesenchymal stem cells: a new therapeutic tool for AKI. Nat Rev Nephrol. 2010;6(3):179–83.
41. Borges FT, Schor N. Regenerative medicine in kidney disease: where we stand and where to go. Pediatr Nephrol. 2018;33(9):1457–65.
42. Siegel N, et al. Induction of mesenchymal/epithelial marker expression in human amniotic fluid stem cells. Reprod Biomed Online. 2009;19(6):838–46.
43. Siegel N, et al. Contribution of human amniotic fluid stem cells to renal tissue formation depends on mTOR. Hum Mol Genet. 2010;19(17):3320–31.
44. Da Sacco S, et al. Human amniotic fluid as a potential new source of organ specific precursor cells for future regenerative medicine applications. J Urol. 2010;183(3):1193–200.
45. Xinaris C, et al. Functional human podocytes generated in organoids from amniotic fluid stem cells. J Am Soc Nephrol. 2016;27(5):1400–11.
46. Duffield JS, Bonventre JV. Kidney tubular epithelium is restored without replacement with bone marrow-derived cells during repair after ischemic injury. Kidney Int. 2005;68(5):1956–61.
47. Duffield JS, et al. Restoration of tubular epithelial cells during repair of the postischemic kidney occurs independently of bone marrow-derived stem cells. J Clin Invest. 2005;115(7):1743–55.
48. Jang HR, et al. Effect of preemptive treatment with human umbilical cord blood-derived mesenchymal stem cells on the development of renal ischemia-reperfusion injury in mice. Am J Physiol Renal Physiol. 2014;307(10):F1149–61.
49. Du T, et al. The alleviation of acute and chronic kidney injury by human Wharton's jelly-derived mesenchymal stromal cells triggered by ischemia-reperfusion injury via an endocrine mechanism. Cytotherapy. 2012;14(10):1215–27.
50. Du T, et al. Human Wharton's jelly-derived mesenchymal stromal cells reduce renal fibrosis through induction of native and foreign hepatocyte growth factor synthesis in injured tubular epithelial cells. Stem Cell Res Ther. 2013;4(3):59.
51. Bruno S, Bussolati B. Therapeutic effects of mesenchymal stem cells on renal ischemia-reperfusion injury: a matter of genetic transfer? Stem Cell Res Ther. 2013;4(3):55.
52. Cao H, et al. Pretreatments with injured microenvironmental signals altered the characteristics of human umbilical cord mesenchymal stem cells. Biotechnol Lett. 2016;38(1):157–65.
53. Morigi M, Benigni A. Mesenchymal stem cells and kidney repair. Nephrol Dial Transplant. 2013;28(4):788–93.
54. Karpman D, Stahl AL, Arvidsson I. Extracellular vesicles in renal disease. Nat Rev Nephrol. 2017;13(9):545–62.
55. Ragni E, et al. Extracellular vesicle-shuttled mRNA in mesenchymal stem cell communication. Stem Cells. 2017;35(4):1093–105.
56. Kilpinen L, et al. Extracellular membrane vesicles from umbilical cord blood-derived MSC protect against ischemic acute kidney injury, a feature that is lost after inflammatory conditioning. J Extracell Vesicles. 2013;2.
57. Zhou Y, et al. Exosomes released by human umbilical cord mesenchymal stem cells protect against cisplatin-induced renal oxidative stress and apoptosis in vivo and in vitro. Stem Cell Res Ther. 2013;4(2):34.
58. Zhang G, et al. The anti-oxidative role of microvesicles derived from human Wharton-jelly mesenchymal stromal cells through NOX2/gp91(phox) suppression in alleviating renal ischemia-reperfusion injury in rats. PLoS One. 2014;9(3):e92129.

59. Zou X, et al. Microvesicles derived from human Wharton's jelly mesenchymal stromal cells ameliorate renal ischemia-reperfusion injury in rats by suppressing CX3CL1. Stem Cell Res Ther. 2014;5(2):40.
60. Zhang G, et al. Mesenchymal stromal cell-derived extracellular vesicles protect against acute kidney injury through anti-oxidation by enhancing Nrf2/ARE activation in rats. Kidney Blood Press Res. 2016;41(2):119–28.
61. Zou X, et al. Human mesenchymal stromal cell-derived extracellular vesicles alleviate renal ischemic reperfusion injury and enhance angiogenesis in rats. Am J Transl Res. 2016;8(10):4289–99.
62. Ju GQ, et al. Microvesicles derived from human umbilical cord mesenchymal stem cells facilitate tubular epithelial cell dedifferentiation and growth via hepatocyte growth factor induction. PLoS One. 2015;10(3):e0121534.
63. Gu D, et al. Mesenchymal stromal cells derived extracellular vesicles ameliorate acute renal ischemia reperfusion injury by inhibition of mitochondrial fission through miR-30. Stem Cells Int. 2016;2016:2093940.
64. Wang B, et al. Pre-incubation with hucMSC-exosomes prevents cisplatin-induced nephrotoxicity by activating autophagy. Stem Cell Res Ther. 2017;8(1):75.
65. Wu X, et al. Micro-vesicles derived from human Wharton's jelly mesenchymal stromal cells mitigate renal ischemia-reperfusion injury in rats after cardiac death renal transplantation. J Cell Biochem. 2018;119(2):1879–88.
66. Nassar W, et al. Umbilical cord mesenchymal stem cells derived extracellular vesicles can safely ameliorate the progression of chronic kidney diseases. Biomater Res. 2016;20:21.
67. Roushandeh AM, Bahadori M, Roudkenar MH. Mesenchymal stem cell-based therapy as a new horizon for kidney injuries. Arch Med Res. 2017;48(2):133–46.
68. Rota C, et al. Human amniotic fluid stem cell preconditioning improves their regenerative potential. Stem Cells Dev. 2012;21(11):1911–23.
69. Li W, et al. Icariin combined with human umbilical cord mesenchymal stem cells significantly improve the impaired kidney function in chronic renal failure. Mol Cell Biochem. 2017;428(1–2):203–12.
70. Koc Y, et al. Effects of human umbilical cord stem cells and granulocyte colony- stimulating factor (G-CSF) on carbon tetrachloride-induced nephrotoxicity. Nephrourol Mon. 2012;4(3):545–50.
71. Veceric-Haler Z, et al. Improved protective effect of umbilical cord stem cell transplantation on cisplatin-induced kidney injury in mice pretreated with antithymocyte globulin. Stem Cells Int. 2016;2016:3585362.
72. Mori da Cunha MG, et al. Vascular endothelial growth factor up-regulation in human amniotic fluid stem cell enhances nephroprotection after ischemia-reperfusion injury in the rat. Crit Care Med. 2017;45(1):e86–96.
73. Chen Y, et al. Hepatocyte growth factor modification promotes the amelioration effects of human umbilical cord mesenchymal stem cells on rat acute kidney injury. Stem Cells Dev. 2011;20(1):103–13.
74. Liu P, et al. Enhanced renoprotective effect of IGF-1 modified human umbilical cord-derived mesenchymal stem cells on gentamicin-induced acute kidney injury. Sci Rep. 2016;6:20287.
75. Imberti B, et al. Insulin-like growth factor-1 sustains stem cell mediated renal repair. J Am Soc Nephrol. 2007;18(11):2921–8.
76. Feng G, et al. Identification, characterization and biological significance of very small embryonic-like stem cells (VSELs) in regenerative medicine. Histol Histopathol. 2012;27(7):827–33.
77. Feng G, et al. IGF-1 C domain-modified hydrogel enhances cell therapy for AKI. J Am Soc Nephrol. 2016;27(8):2357–69.
78. Kaushal GP, Shah SV. Challenges and advances in the treatment of AKI. J Am Soc Nephrol. 2014;25(5):877–83.
79. Cao H, et al. Mesenchymal stem cells derived from human umbilical cord ameliorate ischemia/reperfusion-induced acute renal failure in rats. Biotechnol Lett. 2010;32(5):725–32.
80. Lu G, et al. Umbilical cord mesenchymal stem cell transplantation ameliorates burn-induced acute kidney injury in rats. Int J Low Extrem Wounds. 2013;12(3):205–11.
81. Fang TC, et al. Renoprotective effect of human umbilical cord-derived mesenchymal stem cells in immunodeficient mice suffering from acute kidney injury. PLoS One. 2012;7(9):e46504.
82. Tsuda H, et al. Transplantation of allogenic fetal membrane-derived mesenchymal stem cells protects against ischemia/reperfusion-induced acute kidney injury. Cell Transplant. 2014;23(7):889–99.
83. Qiu Z, Zhou D, Sun D. Effects of human umbilical cord mesenchymal stem cells on renal ischaemia-reperfusion injury in rats. Int Braz J Urol. 2014;40(4):553–61.
84. Rodrigues CE, et al. Human umbilical cord-derived mesenchymal stromal cells protect against premature renal senescence resulting from oxidative stress in rats with acute kidney injury. Stem Cell Res Ther. 2017;8(1):19.
85. Fahmy SR, et al. Therapeutic efficacy of human umbilical cord mesenchymal stem cells transplantation against renal ischemia/reperfusion injury in rats. Tissue Cell. 2017;49(3):369–75.
86. Cao Q, Harris DC, Wang Y. Macrophages in kidney injury, inflammation, and fibrosis. Physiology (Bethesda). 2015;30(3):183–94.
87. Li W, et al. Macrophages are involved in the protective role of human umbilical cord-derived stromal cells in renal ischemia-reperfusion injury. Stem Cell Res. 2013;10(3):405–16.

88. Morigi M, et al. Life-sparing effect of human cord blood-mesenchymal stem cells in experimental acute kidney injury. Stem Cells. 2010;28(3):513–22.
89. Park JH, et al. Early, but not late treatment with human umbilical cord blood-derived mesenchymal stem cells attenuates cisplatin nephrotoxicity through immunomodulation. Am J Physiol Renal Physiol. 2017;313(4):F984–96.
90. Perin L, et al. Protective effect of human amniotic fluid stem cells in an immunodeficient mouse model of acute tubular necrosis. PLoS One. 2010;5(2):e9357.
91. Al-Husseiny F, et al. Amniotic fluid-derived mesenchymal stem cells cut short the acuteness of cisplatin-induced nephrotoxicity in Sprague-Dawley rats. Int J Stem Cells. 2016;9(1):70–8.
92. Ashour RH, et al. Comparative study of allogenic and xenogeneic mesenchymal stem cells on cisplatin-induced acute kidney injury in Sprague-Dawley rats. Stem Cell Res Ther. 2016;7(1):126.
93. Condor JM, et al. Treatment with human Wharton's jelly-derived mesenchymal stem cells attenuates sepsis-induced kidney injury, liver injury, and endothelial dysfunction. Stem Cells Transl Med. 2016;5(8):1048–57.
94. Rabelink TJ, Little MH. Stromal cells in tissue homeostasis: balancing regeneration and fibrosis. Nat Rev Nephrol. 2013;9(12):747–53.
95. Choi SJ, Kim JK, Hwang SD. Mesenchymal stem cell therapy for chronic renal failure. Expert Opin Biol Ther. 2010;10(8):1217–26.
96. Sedrakyan S, et al. Injection of amniotic fluid stem cells delays progression of renal fibrosis. J Am Soc Nephrol. 2012;23(4):661–73.
97. Monteiro Carvalho Mori da Cunha MG, et al. Amniotic fluid derived stem cells with a renal progenitor phenotype inhibit interstitial fibrosis in renal ischemia and reperfusion injury in rats. PLoS One. 2015;10(8):e0136145.
98. Sun D, et al. Therapeutic effects of human amniotic fluid-derived stem cells on renal interstitial fibrosis in a murine model of unilateral ureteral obstruction. PLoS One. 2013;8(5):e65042.
99. Huang D, et al. Distribution of infused umbilical cord mesenchymal stem cells in a rat model of renal interstitial fibrosis. Ren Fail. 2013;35(8):1146–50.
100. Liu B, et al. Human umbilical cord-derived mesenchymal stem cells conditioned medium attenuate interstitial fibrosis and stimulate the repair of tubular epithelial cells in an irreversible model of unilateral ureteral obstruction. Nephrology (Carlton). 2017.
101. Peng X, et al. Human umbilical cord mesenchymal stem cells attenuate cisplatin-induced acute and chronic renal injury. Exp Biol Med (Maywood). 2013;238(8):960–70.
102. Piccoli GB, et al. Low-protein diets in diabetic chronic kidney disease (CKD) patients: are they feasible and worth the effort? Nutrients. 2016;8(10):E649.
103. Park JH, et al. Delayed treatment with human umbilical cord blood-derived stem cells attenuates diabetic renal injury. Transplant Proc. 2012;44(4):1123–6.
104. Park JH, et al. Human umbilical cord blood-derived mesenchymal stem cells prevent diabetic renal injury through paracrine action. Diabetes Res Clin Pract. 2012;98(3):465–73.
105. Malvezzi P, Rostaing L. The safety of calcineurin inhibitors for kidney-transplant patients. Expert Opin Drug Saf. 2015;14(10):1531–46.
106. Razonable RR. Not the usual viral suspects: parvovirus B19, West Nile virus, and human T-cell lymphotrophic virus infections after kidney transplantation. Semin Nephrol. 2016;36(5):428–34.
107. Serrano OK, et al. Post-transplant malignancy after pediatric kidney transplantation: retrospective analysis of incidence and risk factors in 884 patients receiving transplants between 1963 and 2015 at the University of Minnesota. J Am Coll Surg. 2017;225(2):181–93.
108. Dharnidharka VR, et al. Lack of significant improvements in long-term allograft survival in pediatric solid organ transplantation: a US national registry analysis. Pediatr Transplant. 2015;19(5):477–83.
109. Casiraghi F, et al. Mesenchymal stromal cells in renal transplantation: opportunities and challenges. Nat Rev Nephrol. 2016;12(4):241–53.
110. Perico N, et al. Autologous mesenchymal stromal cells and kidney transplantation: a pilot study of safety and clinical feasibility. Clin J Am Soc Nephrol. 2011;6(2):412–22.
111. Baulier E, et al. Amniotic fluid-derived mesenchymal stem cells prevent fibrosis and preserve renal function in a preclinical porcine model of kidney transplantation. Stem Cells Transl Med. 2014;3(7):809–20.
112. Feng J, et al. Immune tolerance of amniotic fluid stem cell-induced rat kidney graft and influences on oxidative stress. Transplant Proc. 2013;45(9):3394–401.
113. Parikh SV, Rovin BH. Current and emerging therapies for lupus nephritis. J Am Soc Nephrol. 2016;27(10):2929–39.
114. Yu F, et al. Redefining lupus nephritis: clinical implications of pathophysiologic subtypes. Nat Rev Nephrol. 2017;13(8):483–95.
115. Gu Z, et al. Transplantation of umbilical cord mesenchymal stem cells alleviates lupus nephritis in MRL/lpr mice. Lupus. 2010;19(13):1502–14.
116. Chang JW, et al. Therapeutic effects of umbilical cord blood-derived mesenchymal stem cell transplantation in experimental lupus nephritis. Cell Transplant. 2011;20(2):245–57.
117. Ruan GP, et al. Establishing a tree shrew model of systemic lupus erythematosus and cell transplantation treatment. Stem Cell Res Ther. 2016;7(1):121.
118. Chen C, et al. Mesenchymal stem cells upregulate Treg cells via sHLA-G in SLE patients. Int Immunopharmacol. 2017;44:234–41.
119. Deng D, et al. A randomised double-blind, placebo-controlled trial of allogeneic umbilical cord-derived

mesenchymal stem cell for lupus nephritis. Ann Rheum Dis. 2017;76(8):1436–9.
120. Peev V, Hahm E, Reiser J. Unwinding focal segmental glomerulosclerosis. F1000Res. 2017;6:466.
121. Malaga-Dieguez L, et al. Novel therapies for FSGS: preclinical and clinical studies. Adv Chronic Kidney Dis. 2015;22(2):e1–6.
122. Ma H, et al. Human umbilical mesenchymal stem cells attenuate the progression of focal segmental glomerulosclerosis. Am J Med Sci. 2013;346(6):486–93.
123. Zickri MB, Fattah MM, Metwally HG. Tissue regeneration and stem cell distribution in adriamycin induced glomerulopathy. Int J Stem Cells. 2012;5(2):115–24.
124. Zhang DW, et al. Repair effects of umbilical cord mesenchymal stem cells on podocyte damage of IgA nephropathy. J Biol Regul Homeost Agents. 2015;29(3):609–17.
125. Tsuda H, et al. Allogenic fetal membrane-derived mesenchymal stem cells contribute to renal repair in experimental glomerulonephritis. Am J Physiol Renal Physiol. 2010;299(5):F1004–13.
126. Guillot PV, et al. Transplantation of human fetal mesenchymal stem cells improves glomerulopathy in a collagen type I alpha 2-deficient mouse. J Pathol. 2008;214(5):627–36.
127. Moschidou D, et al. Human chorionic stem cells: podocyte differentiation and potential for the treatment of Alport syndrome. Stem Cells Dev. 2016;25(5):395–404.
128. Gheisari Y, et al. Stem cell-conditioned medium does not protect against kidney failure. Cell Biol Int. 2011;35(3):209–13.
129. Burger D, et al. Human cord blood CD133+ cells exacerbate ischemic acute kidney injury in mice. Nephrol Dial Transplant. 2012;27(10):3781–9.

Prenatal Mesenchymal Stem Cell Secretome and Its Clinical Implication

13

Lu Liang

Abstract

Mesenchymal stem cells (MSCs) possess paracrine characteristic involving in the main therapeutic effects. The soluble factors secreted by MSCs are responsible for the paracrine activity, which have attracted more and more attention. These protein factors named secretome could be related to the biological processes such as wound healing, inflammatory response, angiogenesis, cell proliferation, chemotaxis, and neurogenesis. And they have bypassed the confounding issues of MSCs themselves. The MSC secretome in MSC-conditioned media (CM) could be manufactured and used as protein drugs. The cost of manufacturing, transport, and storage of secretome is lower than that of MSCs. Prenatal MSC secretome, especially placenta MSC secretome, has higher concentration and stronger capacities than that of other origins, which is a promising alternative to stem cell therapy for various diseases treatment.

Mesenchymal stem cells (MSCs) are promising cell type for cell therapy, its therapeutic effects has been explored in many clinical implications. MSCs possess robust differentiation capacity, which was considered as the main mechanism to repair the tissue damages. However accumulating data showed that the therapeutic effect of MSCs was mediated by paracrine factors rather than their differentiation capacity, as the engrafted cell in vivo was rarely found 4 weeks after transplantation, especially in immune-competent animals [1, 2].

MSCs were able to express numerous cytokines, chemokines, growth factors, extracellular matrix proteins, and angiogenic molecules, collectively named secretome, that accumulated in MSC culture media. The MSC secretome could promote the tissue regeneration through regulating the tissue microenvironment niche, inhibiting inflammation, stimulating angiogenesis, and activating endogenous stem cells [3].

Although the tissue regeneration after MSC transplantation has been suggested mainly mediated by the paracrine effect, the paracrine factors produced by MSCs have not been systematically or comprehensively profiled. Here we summarized the major findings of secretome of MSCs.

13.1 Identification of MSC Secretome

MSCs have been shown to secrete a broad spectrum of cytokines, chemokines, growth factors, and extracellular matrix proteins. The secretome was released in MSC-conditioned media (CM), which was easily collected and preserved. Many proteins in MSC-CM were hardly detected by liquid

L. Liang (✉)
Tianjin Bion Bio-tech Co., Ltd.,
Tianjin, China

Z. C. Han et al. (eds.), *Perinatal Stem Cells*, https://doi.org/10.1007/978-981-13-2703-2_13

chromatography tandem mass spectrometry (LC-MS/MS) because the expression was too low or molecular weight was small. Only 19 proteins have been detected via LC-MS/MS in CM of murine MSCs [4]. Estrada R et al. have assessed the proteome profile of murine MSCs, and 54 secreted proteins have been identified by LC-MS/MS [5].

A cytokine antibody array was properly suitable for MSC-CM analysis which was highly sensitive and specific; the proteins present even in a low picogram per milliliter could be detected. Ionescu L et al. has reported 96 proteins in murine bone marrow-derived MSC-CM, which was composed of chemokines (31%), binding proteins (27%), cytokines (21%), growth factors (14%), cytokine antagonists (4%), and molecules involved in ECM repair (3%). The proteins with higher expression were cytokine-induced neutrophil chemoattractant (KC), monocyte chemoattractant protein (MCP)-1, macrophage inflammatory protein (MIP)-1 alpha, MIP-1-gamma, MIP-2, soluble tumor necrosis factor α receptor (sTNFαR) I, sTNFαRII, chemokine (C-X-C motif) ligand (CXCL) 16, insulin-like growth factor (IGF)-1, RANTES, vascular cell adhesion protein (VCAM)-1, TIMP-1, osteoprotegerin, monocyte chemotactic protein-1 (MCP)-5, VEGF, PF4, eotaxin, eotaxin-2, stromal cell-derived factor (SDF)-1 alpha, MIP-3-alpha, IL12, MDC, Fcg RIIB, IL-6, and LIX [6]. Parekkadan B et al. found 69 proteins in human bone marrow-derived MSC-CM using a high-density protein array. Cluster analysis of MSC-CM revealed the following distribution: growth factors (37%), chemokines (30%), cytokines (19%), extracellular matrix (ECM) repair (7%), binding protein (6%), and cytokines antagonist (1%). Among these proteins, IGFBP-1, leptin, CCL2, IL-8, BMP-4, FasL, FGF-6, TNF-b, CNTF, IL-9, IGFBP4, TGF-α, SCF, CCL22, TGF-b3, CXCL9, CCL27, IGFBP2, CXCL11, CXCL6, MCP-4, TNF-α, CCL20, GDNF, BNP-6, TGF-b2, BDNF, CXCL12, CXCL13, CXCL-16, IL-6, TIMP-2, bFGF, CCL15, GFG9, GCSF, IL-7, TGF-b1, CCL8, M-CSF, and angiogenin had higher expression [7].

Although the secretome profiles in murine and human were different, the large fractions in both were chemokines, and the functions were similar. These proteins are related to the biological processes such as wound healing, inflammatory response, angiogenesis, cell proliferation, chemotaxis, and neurogenesis.

13.2 Function of MSC Secretome

Accumulating references demonstrated that MSC-CM have therapeutic effect equal to MSC itself. Although only picogram level of each factor was contained in MSC-CM, cooperative reaction of multiple factors resulted in specific functions. These solute factors were categorized into functional groups: antiapoptotic, immunoregulatory, regeneration, angiogenic modulating, and endogenous stem cell recruitment.

13.2.1 Antiapoptotic Function

In vitro experiments have demonstrated the antiapoptotic activity of MSCs via soluble factors. MSC-CM suppressed cells apoptosis through the major cell cycle regulatory pathways. The Bcl-2/Bax ratio was increased by MSC-CM, and the release of cytochrome c from mitochondria was reduced into the cytosol [8]. Bone marrow-derived MSC (BM-MSC)-CM protected hepatocytes from apoptosis by overproducing IL-6 and FGL1 [9]. MSC-CM promote cell proliferation by activating PI 3-kinase and ERK1/2 pathway [10]. MSC-CM could protect against tendon damage [11] and myocardial infarction [12] through preventing cell apoptosis and accelerating functional recovery.

13.2.2 Immunoregulatory Function

The immunosuppressive properties of MSC were mediated by the soluble factors. TGF-β and HGF mediated the suppression of T-cell proliferation in a mixed lymphocyte reaction [13]. IL-10 played a key role in the inhibition effect of MSC on T-cell proliferation [14]. IDO [15] and PGE2 [16] induced T-cell anergy. IL-6 and M-CSF in MSC-CM contribute to the inhibitory effect of DC differentiation [17]. TGF-β and PGE2 secreted by MSC were important mediators for the inhibition of NK-cell proliferation [18]. MSC-CM contained anti-inflammatory factors such as IL-10 and TSG-6 to inhibit the inflammation in injured organ [19].

Macrophages play a crucial role in the recovery phases of tissue injury and promotion of tissue repair [20]. The immunoregulatory effect of MSCs on macrophage differentiation was mediated by soluble molecules such as IGF-1 produced by MSCs. MSC-CM promotes alternative macrophage activation to an anti-inflammatory M2 "healer" phenotype [6]. The differentiation of M2 macrophages was mediated by different soluble factors such as IL-1RA, IL-4, IL-6, IL-10, IL-13, M-CSF, VEGF, TGF-β1, LIF, and PGE2 while coculturing with MSCs [21–24]. MSC-CM modulate the cytokine expression profile in macrophages by enhancing extracellular signal-related kinase (ERK) signaling and suppressing nuclear factor-κB (NF-κB) signaling mediated by PGE2 [25].

13.2.3 Regenerative Function

MSC-CM has been used for tissue repair of different organs. Many cytokines have been proven to take effect in tissue regeneration and cell proliferation such as HGF, VEGF, SDF-1, KGF, FGF, PGF, MCP-1, and IGF-1 [26]. The migration of skin fibroblasts could be significantly stimulated by VEGF, bFGF, and PDGF-AA, and the proliferation of that could be significantly stimulated by bFGF and EGF in adipose MSC-CM [27].

Sphingosine 1-phosphate (S1P) secreted by MSCs stimulated myoblasts and satellite cell proliferation to promote skeletal muscle regeneration [28]. Park et al. suggested that tonsil-derived MSCs expressed and secreted cyclin 1 (CCN1) in response to endodermal differentiation. MSC-CM could induce the activation of JNK-p38 which was a positive regulator of proliferation and migration [29]. MSC-CM promoted cell proliferation through phosphorylation of Akt and ERK1/2 [10].

13.2.4 Angiogenic Modulating

Angiogenic factors secreted in MSC-CM such as VCAM-1, IL-8, IGF-1, and VEGF account for angiogenic activity of MSC [30]. IGFBP2 were also predominant proangiogenic factors in MSC-CM [31]. Besides, Ang1 and Tie2 in MSC-CM contributed to vascular remodeling [32].

13.2.5 Endogenous Stem Cell Recruitment

SDF-1 released in MSC-CM was critical for the recruitment of endogenous stem cell, involved in inflammatory diseases and bone healing [33]. MCP-1 and SDF-1 were essential to accelerate callus formation by regulating the recruitment and differentiation of MSCs and other progenitor cells [30]. C-X-C chemokine receptor 4 (CXCR4), C-C chemokine receptor 2 (CCR2), and c-met were determined in the MSC-CM, and SDF-1/CXCR4 and MCP-1/CCR2 axes were involved in the chemoattractive activity of UC-CM [30]. MSC-CM can protect against bone defect through mobilization of endogenous mesenchymal stem cells [34]. MCP-1 and MCP-3 mediated BM-MNC migration, and IL-3 or IL-6 significantly stimulated the HUVEC migration [35].

13.3 Exosomes in MSC Secretome

Exosomes were nano-sized vesicles of 30–200 nm that were derived from the fusion of multivesicular bodies to plasma membranes. Exosomes contained protein, RNA, DNA, and lipid cargo. Placenta MSC-derived exosomes were composed of proteins such as fibronectin, syncytin, Wnt/βcatenin-related molecules, galectin-3, and HLA-G, bioactive lipids such as the immunosuppressive PGE2, the PPARγ ligand 15d-PGJ2, and microRNAs such as miR-19a [36]. It was recently reported that exosomes were effective carriers of siRNA and miRNA, and functional RNAs can be transferred from one cell to another via the exosome [37]. Exosomes secreted by MSC possessed multiple biological characteristics.

Exosomes were positive for CD63 but not for calnexin; these were specific markers of exosome; besides CD9 and CD81 were characteristic markers and membranous components of exosomes [38, 39].

Exosomes derived from MSCs mediate regenerative functions. Mesenchymal stem cell-derived exosomes protected against organ injury such as acute tubular injury [40], liver injury [41], and myocardial ischemia [42]. MSC exosomes accelerate fracture healing owing to the effect of MCP-1, MCP-3, and SDF-1 [37]. MSC exosomes enhanced cartilage repair through supplying GAG and type II collagen [43]. Multiple miRs in MSC exosomes were responsible for reducing neonatal cardiomyocytes apoptosis and restoring cardiac contractile function [44]. MSC exosomes promoted the angiogenesis in ischemic diseases, which were partly due to the direct stimulation of endothelial cells [31].

Exosomes were contained in CM, and it was reported that biological activities of MSC-CM were partly due to thc direct contribution by exosomes such as angiogenic effects [31]. The function of MSC-CM and exosome was similar; however the mechanism of action was different. The multiple cytokines cooperated to enhance biological process in MSC-CM, while the level of cytokines in MSC exosome was low, and miRNAs in MSC exosome might contribute to the dynamics of tissue regeneration by influencing the microenvironment at the repair site [37].

13.4 Therapeutic Potential in Animal Models

13.4.1 Nervous System Disease

MSC-CM has been used to treat animals suffering nervous system disease such as stroke and Parkinson disease. Umbilical cord-derived MSC-CM has been used for treatment of animals suffering stroke. The ischemic brain was improved, and vascular was remodeled by increased expression of Ang1 and Tie2 and decreased expression of Ang2 [32].

MSC-CM protect neuro in model Parkinson disease (PD) with upregulation of the antioxidant enzyme sirtuin 3 (SIRT3) and dissertation of ROS generation [31]. BMSC-CM improved motor recovery after spinal cord contusion.

13.4.2 Organ Injury

MSC-CM have a therapeutic rescue function in treatment of kidney disease, livery injury, myocardial infraction, and lung disease. MSC-CM could support renal repair and hence renal function and thus reduce disease progression. MSC-CM increased the number of glomerular endothelial cells and reduced glomerulosclerosis and tubular damage. MSC-CM treatment rescued acute liver injury through reducing apoptosis of hepatocytes and promoting their regeneration [9]. Human MSC-CM treatment enhanced capillary density and reduced myocardial apoptosis and oxidative stress in a porcine model of myocardial infarction [45]. VEGF, HGF, FGF, IGF-1, and TB4 in MSC-CM protected myocardial cells from ischemia and promoted them regeneration [46]. MSC-CM rescued LPS-induced lung injury by attenuating lung inflammation and promoting M2 macrophages shifting to anti-inflammatory phenotype [6].

13.4.3 Wound Healing

Human amniotic fluid-derived MSC-CM enhanced wound healing, which was mediated by the activation of TGF-β/SMAD2 and PI3K/AKT [47]. MSC-CM stimulated skin wound healing by affecting fibroblast and keratinocyte migration and extracellular matrix deposition. Walter et al. identified that TGFb1, IL-6, and IL-8 played a key role to enhance wound healing [48].

13.4.4 Fracture Repair

MSC-CM promoted new bone callus formation in a mouse distraction osteogenesis model by accelerating the recruitment of endogenous mBMSCs and EC/EPCs. MCP-1 and MCP-3 in MSC-CM contributed to BMSC recruitment and bone formation [35]. These results were consistent with other studies; MCP-1 and SDF-1 were demonstrated to be essential in the initial phase of fracture healing by regulating the recruitment and differentiation of MSCs and other progenitor cells [49, 50].

13.5 Cell Processing-Affected Secretome

The secretomes varied from different cell types, different origins, and different culture duration, culture condition, or processing method. Prenatal MSCs were demonstrated to secrete higher concentrations of chemokines, pro-inflammatory proteins, and growth factors compared with adipose tissue-derived MSCs or bone marrow-derived MSCs. Prenatal MSCs had the strongest immunomodulatory and immunosuppressive potential [51]. The mechanism of action of pro-angiogenesis was different in CM from different origins. BM-MSC-CM promoted angiogenesis via VEGF, whereas placenta-derived MSC-CM enhanced angiogenesis via IGFBP2 [31]. Even in the same origin, MSCs from different fractions have different secretome. In the placenta, MCP-1 (CCL2) was expressed only in the amnion MSCs but not decidua-derived MSCs, and GROa and sICAM levels in amnion MSCs were higher than the expression in the decidua MSCs [52].

The biological process ontology "response to stress" of MSCs resulted in different secretion profile, such as exposing to hypoxic environment or different cytokines. The extent of environmental oxygenation affects the innate characteristics of MSCs, including their stemness and migration capacity; MSCs cultured in hypoxic environment secreted significantly higher amounts of bFGF, VEGF-A, IL-6, and IL-8 compared with normoxic conditions [53]. Factors released from MSCs under hypoxic condition could improve tendon repair [11] and myocardial infarction recovery through prevention of apoptosis and acceleration of functional recovery [12].

Biological processes of MSCs were specifically activated in the presence of bFGF. Many proteins were oversecreted by MSCs after bFGF stimulation, such as SDF-1, PF4, MCP-2, ceruloplasmin, small-inducible cytokine A6, ribonuclease 6, and Timp2. The immune response, the inflammatory response, and the response to wounding and chemotaxis were all upregulated in bFGF-stimulated MSC [54].

Sample processing might influence the phenotype of cells and thus the secretome; different secretomes were obtained from different culture durations, culture conditions, or processing methods. It is important to optimize the sample processing and to check that it does not induce modifications of the cellular phenotype to avoid changes in protein secretion profiles [55].

13.6 Prospect of MSC-CM

The studies to reveal the therapeutic effect of MSC-CM not only illuminated the paracrine activity of MSC but also opened novelty therapeutic perspectives of paracrine factor treatment. MSC-CM will be an alternative therapeutic approach to previous methods using MSCs. This will bypass the present confounding issues associated with cell transplantation, such as immune compatibility, tumorigenicity, xenozootic infections, and costs. And the manufacture of MSC-CM is less complex, and storage, transport, and end use of MSC-CM have low requirements. It can be stored and transported easily as drugs and no requirement for ultra-low temperature storage, such as that the stem cells need.

MSC-CM, especially prenatal MSC-CM, could be used in regenerative medicine to repair injured tissues by proproliferation, anti-apoptosis, proangiogenesis, and endogenous stem cell recruitment. Prenatal MSC-CM will be a better choice as the higher concentration and stronger abilities. Such an approach will have a greater potential for the development of "off-the-shelf" therapeutics with better quality control and consistency.

References

1. Liang L, Li Z, Ma T, et al. Transplantation of human placenta-derived mesenchymal stem cells alleviates critical limb ischemia in diabetic nude rats. Cell Transplant. 2017;26(1):45–61.
2. Creane M, Howard L, O'Brien T, et al. Biodistribution and retention of locally administered human mesenchymal stromal cells: quantitative polymerase chain reaction-based detection of human DNA in murine organs. Cytotherapy. 2017;19(3):384–94.
3. Gnecchi M, Danieli P, Malpasso G, et al. Paracrine mechanisms of mesenchymal stem cells in tissue repair. Methods Mol Biol. 2016;1416:123–46.
4. Sarojini H, Estrada R, Lu H, et al. PEDF from mouse mesenchymal stem cell secretome attracts fibroblasts. J Cell Biochem. 2008;104(5):1793–802.
5. Estrada R, Li N, Sarojini H, et al. Secretome from mesenchymal stem cells induces angiogenesis via Cyr61. J Cell Physiol. 2009;219:563–71.
6. Ionescu L, Byrne RN, van Haaften T, et al. Stem cell conditioned medium improves acute lung injury in mice: in vivo evidence for stem cell paracrine action. Am J Physiol Lung Cell Mol Physiol. 2012;303(11):L967–77.
7. Parekkadan B, van Poll D, Suganuma K, et al. Mesenchymal stem cell-derived molecules reverse fulminant hepatic failure. PLoS One. 2007;2(9):e941.
8. Xiang MX, He AN, Wang JA, et al. Protective paracrine effect of mesenchymal stem cells on cardiomyocytes. J Zhejiang Univ Sci B. 2009;10:619–24.
9. Xagorari A, Siotou E, Yiangou M, et al. Protective effect of mesenchymal stem cell-conditioned medium on hepatic cell apoptosis after acute liver injury. Int J Clin Exp Pathol. 2013;6(5):831–40.
10. Nakahara M, Okumura N, Kay EP, et al. Corneal endothelial expansion promoted by human bone marrow mesenchymal stem cell-derived conditioned medium. PLoS One. 2013;8(7):e69009.
11. Huang TF, Yew TL, Chiang ER, et al. Mesenchymal stem cells from a hypoxic culture improve and engraft Achilles tendon repair. Am J Sports Med. 2013;41:1117–25.
12. Hwang HJ, Chang W, Song BW, et al. Antiarrhythmic potential of mesenchymal stem cell is modulated by hypoxic environment. J Am Coll Cardiol. 2012;60:1698–706.
13. Di Nicola M, Carlo-Stella C, Magni M, et al. Human bone marrow stromal cells suppress T-lymphocyte proliferation induced by cellular or nonspecific mitogenic stimuli. Blood. 2002;99:3838–43.
14. Beyth S, Borovsky Z, Mevorach D, et al. Human mesenchymal stem cells alter antigen-presenting cell maturation and induce T-cell unresponsiveness. Blood. 2005;105:2214–9.
15. Meisel R, Zibert A, Laryea M, et al. Human bone marrow stromal cells inhibit allogeneic T-cell responses by indoleamine 2,3-dioxygenase-mediated tryptophan degradation. Blood. 2004;103:4619–21.
16. Aggarwal S, Pittenger MF. Human mesenchymal stem cells modulate allogeneic immune cell responses. Blood. 2005;105:1815–22.
17. Nauta AJ, Kruisselbrink AB, Lurvink E, et al. Mesenchymal stem cells inhibit generation and function of both CD34+−derived and monocyte-derived dendritic cells. J Immunol. 2006;177:2080–7.
18. Le Blanc K. Immunomodulatory effects of fetal and adult mesenchymal stem cells. Cytotherapy. 2003;5:485–9.
19. Khubutiya MS, Vagabov AV, Temnov AA, et al. Paracrine mechanisms of proliferative, anti-apoptotic and anti-inflammatory effects of mesenchymal stromal cells in models of acute organ injury. Cytotherapy. 2014;16:579–85.
20. Murray PJ, Wynn TA. Protective and pathogenic functions of macrophage subsets. Nat Rev Immunol. 2011;11:723–37.
21. Svensson J, Jenmalm M, Matussek A, et al. Macrophages at the fetal-maternal interface express markers of alternative activation and are induced by M-CSF and IL-10. J Immunol. 2011;7:3671–82.
22. Jeannin P, Duluc D, Delneste Y. IL-6 and leukemia-inhibitory factor are involved in the generation of tumor-associated macrophage: regulation by IFN-gamma. Immunotherapy. 2011;3:23–6.
23. Buechler C, Ritter M, Orso E, et al. Regulation of scavenger receptor CD163 expression in human monocytes and macrophages by pro- and antiinflammatory stimuli. J Leukoc Biol. 2000;67:97–103.
24. Heusinkveld M, de Vos van Steenwijk PJ, Goedemans R, et al. M2 macrophages induced by prostaglandin E2 and IL-6 from cervical carcinoma are switched to activated M1 macrophages by CD4+ Th1 cells. J Immunol. 2011;187:1157–65.
25. Asami T, Ishii M, Fujii H, et al. Modulation of murine macrophage TLR7/8-mediated cytokine expression by mesenchymal stem cell-conditioned medium. Mediators Inflamm. 2013;2013:264260.
26. Kuraitis D, Giordano C, Ruel M, et al. Exploiting extracellular matrix-stem cell interactions: a review of natural materials for therapeutic muscle regeneration. Biomaterials. 2012;33:428–43.
27. Pawitan JA. Prospect of stem cell conditioned medium in regenerative medicine. Biomed Res Int. 2014;2014:965849.
28. Sassoli C, Frati A, Tani A, et al. Mesenchymal stromal cell secreted sphingosine 1-phosphate (S1P) exerts a stimulatory effect on skeletal myoblast proliferation. PLoS One. 2014;9(9):e108662.
29. Park YS, Hwang S, Jin YM. CCN1 secreted by tonsil derived mesenchymal stem cells promotes endothelial cell angiogenesis via integrinαvβ3 and AMPK. J Cell Physiol. 2015;230:140–9.
30. Shen C, Lie P, Miao T, et al. Conditioned medium from umbilical cord mesenchymal stem cells induces migration and angiogenesis. Mol Med Rep. 2015;12(1):20–30.
31. Komaki M, Numata Y, Morioka C, et al. Exosomes of human placenta-derived mesenchymal stem

cells stimulate angiogenesis. Stem Cell Res Ther. 2017;8(1):219.

32. Zhao Q, Hu J, Xiang J, et al. Intranasal administration of human umbilical cord mesenchymal stem cells-conditioned medium enhances vascular remodeling after stroke. Brain Res. 2015;1624:489–96.
33. Kitaori T, Ito H, Schwarz EM, et al. Stromal cell-derived factor 1/CXCR4 signaling is critical for the recruitment of mesenchymal stem cells to the fracture site during skeletal repair in a mouse model. Arthritis Rheum. 2009;60(3):813–23.
34. An JH, Park H, Song JA, et al. Transplantation of human umbilical cord blood-derived mesenchymal stem cells or their conditioned medium prevents bone loss in ovariectomized nude mice. Tissue Eng Part A. 2013;19(5–6):685–96.
35. Ando Y, Matsubara K, Ishikawa J, et al. Stem cell-conditioned medium accelerates distraction osteogenesis through multiple regenerative mechanisms. Bone. 2014;61:82–90.
36. Record M. Intercellular communication by exosomes in placenta: a possible role in cell fusion? Placenta. 2014;35(5):297–302.
37. Furuta T, Miyaki S, Ishitobi H, et al. Mesenchymal stem cell-derived exosomes promote fracture healing in a mouse model. Stem Cells Transl Med. 2016;5(12):1620–30.
38. Théry C, Ostrowski M, Segura E. Membrane vesicles as conveyors of immune responses. Nat Rev Immunol. 2009;9:581–93.
39. Mathivanan S, Ji H, Simpson RJ. Exosomes: extracellular organelles important in intercellular communication. J Proteomics. 2010;73:1907–20.
40. Bruno S, Grange C, Deregibus MC, et al. Mesenchymal stem cell-derived microvesicles protect against acute tubular injury. J Am Soc Nephrol. 2009;20:1053–67.
41. Herrera MB, Fonsato V, Gatti S, et al. Human liver stem cell-derived microvesicles accelerate hepatic regeneration in hepatectomized rats. J Cell Mol Med. 2010;14(6b):1605–18.
42. Lai RC, Arslan F, Lee MM, et al. Exosome secreted by MSC reduces myocardial ischemia/reperfusion injury. Stem Cell Res (Amst). 2010;4:214–22.
43. Zhang S, Chu WC, Lai RC, et al. Exosomes derived from human embryonic mesenchymal stem cells promote osteochondral regeneration. Osteoarthr Cartil. 2016;24(12):2135–40.
44. Yu B, Kim HW, Gong M, et al. Exosomes secreted from GATA-4 overexpressing mesenchymal stem cells serve as a reservoir of anti-apoptotic microRNAs for cardioprotection. Int J Cardiol. 2015;182:349–60.
45. Timmers L, Lim SK, Hoefer IE, et al. Human mesenchymal stem cell-conditioned medium improves cardiac function following myocardial infarction. Stem Cell Res. 2011;6(3):206–14.
46. Gnecchi M, Zhang Z, Ni A, et al. Paracrine mechanisms in adult stem cell signaling and therapy. Circ Res. 2008;103:1204–19.
47. Jun EK, Zhang Q, Yoon BS, et al. Hypoxic conditioned medium from human amniotic fluid-derived mesenchymal stem cells accelerates skin wound healing through TGF-β/SMAD2 and PI3K/Akt pathways. Int J Mol Sci. 2014;15(1):605–28.
48. Walter MN, Wright KT, Fuller HR, et al. Mesenchymal stem cell-conditioned medium accelerates skin wound healing: an in vitro study of fibroblast and keratinocyte scratch assays. Exp Cell Res. 2010;316(7):1271–81.
49. Toupadakis CA, Wong A, Genetos DC, et al. Long-term administration of AMD3100, an antagonist of SDF-1/CXCR4 signaling, alters fracture repair. J Orthop Res. 2012;30:1853–9.
50. Ishikawa M, Ito H, Kitaori T, et al. MCP/CCR2 signaling is essential for recruitment of mesenchymal progenitor cells during the early phase of fracture healing. PLoS One. 2014;9:e104954.
51. Li X, Bai J, Ji X, et al. Comprehensive characterization of four different populations of human mesenchymal stem cells as regards their immune properties, proliferation and differentiation. Int J Mol Med. 2014;34(3):695–704.
52. Hwang JH, Lee MJ, Seok OS. Cytokine expression in placenta-derived mesenchymal stem cells in patients with pre-eclampsia and normal pregnancies. Cytokine. 2010;49(1):95–101.
53. Chen L, Xu Y, Zhao J, et al. Conditioned medium from hypoxic bone marrow-derived mesenchymal stem cells enhances wound healing in mice. PLoS One. 2014;9(4):e96161.
54. Tasso R, Gaetani M, Molino E, et al. The role of bFGF on the ability of MSC to activate endogenous regenerative mechanisms in an ectopic bone formation model. Biomaterials. 2012;33(7):2086–96.
55. Clabaut A, Grare C, Léger T, et al. Variations of secretome profiles according to conditioned medium preparation: the example of human mesenchymal stem cell-derived adipocytes. Electrophoresis. 2015;36(20):2587–93.

MIX
Papier aus verantwortungsvollen Quellen
Paper from responsible sources
FSC® C105338

If you have any concerns about our products,
you can contact us on
ProductSafety@springernature.com

In case Publisher is established outside the EU,
the EU authorized representative is:
Springer Nature Customer Service Center GmbH
Europaplatz 3, 69115 Heidelberg, Germany

Printed by Libri Plureos GmbH
in Hamburg, Germany